Uneasy Careers and Intimate Lives

THE DOUGLASS SERIES
ON WOMEN'S LIVES
AND THE MEANING OF GENDER

UNEASY CAREERS AND INTIMATE LIVES

Women in Science 1789–1979

edited by PNINA G. ABIR-AM
and DORINDA OUTRAM

with a Foreword by MARGARET W. ROSSITER

Rutgers
University
Press
New Brunswick,
and London

Second paperback printing, 1989

Manufactured in the United States of America

Library of Congress Cataloging-in-Publication Data

Uneasy careers and intimate lives.

(The Douglass series on women's lives and the
meaning of gender)
Bibliography: p.
Includes index.
1. Women in science. 2. Women scientists.
3. Women scientists—Biography. I. Abir-Am,
Pnina G., 1947– . II. Outram, Dorinda.
III. Series.
Q130.U525 1987 500 87-9660
ISBN 0-8135-1255-7
ISBN 0-8135-1256-5

British Cataloging-in-Publication Information available.

Dedicated to our children,

ESTEE and BEN.

May you have an equal opportunity
to pursue "easy careers" while
finding happiness in your intimate lives.

Contents

List of Illustrations

Foreword

AS ONE WHO well remembers the early days of work in the area of women in science, just a decade ago, when the only woman scientist anyone could name was Madame Curie and every name beyond hers was a triumphant discovery, I have been delighted to watch this volume develop. It is a great joy to see twelve such fine essays, solidly based on manuscript and other sources, on so many creative women in different fields and countries over nearly two centuries. The two editors are to be congratulated for envisioning, commissioning, and through their "gentle prodding," as Peggy Kidwell puts it, bringing this book to speedy completion. There is no other work like it both for breadth of coverage and for specific focus on individuals' career choices.

Although there have been for decades career guides addressed to ambitious young scientists (assumed to be men), advising them that the one true path to advancement in science is through long, uninterrupted hours of concentrated work in the field or the laboratory with a minimum of social or family life (which can only distract them), such models may not in fact fit everyone. We can even surmise that such demands or expectations may have brought misery to many and forced some otherwise worthy aspirants out of science. By contrast, these essays show not only that through ingenuity, careful planning, and a willingness to defy social convention, some women have devised ways to combine intensive work in science with time for family duties and pleasures, but also that "science" and "family," both broadly defined, have been closely connected for at least two centuries. Thus the authors here find science in such previously unstudied private realms as the salon and the breakfast room as well as in the more usual laboratories and observatories.

Upon inspection, the life-styles of the women described here (not a representative sample) were quite diverse and included communal living, quasi-marriages, and artificial or "fictitious families," a phrase that recurs in many of the essays. Similarly, the women's domestic roles, which varied both individually and over time, included a full range of dutiful daughters, single women, lovers, collaborative wives, divorcées, single parents, widows, and on occasion a hint of homosexuality. Here also are many unusual husbands, including, besides Pierre Curie, a "fictitious" one, a common-law one, and an outright tyrant, the villain of the piece. The children in these essays range from the neglected and resentful Charitas Dietrich to the studious Irène Curie, her mother's protégée and successor.

These essays demonstrate that the interactions of "science and gender," as the topic is commonly phrased today, have been so complex that they defy reduction to easy theories. They also remind us how important and even exciting it is to find new materials on all types of women in science, for who after reading these essays can feel anything but enriched, one's universe expanded for knowing of the struggles and experiences of Amalie Dietrich, Maria Mitchell, Clémence Royer, Marie Curie? Who would not be amused and instructed by the ornithologists' daily determination not to let their housework take precedence over their observing? These essays also remind one of historian Anne Firor Scott's recent justification for studying women of the past: not only does it teach one "much about the possibilities and mysteries of human existence," but it also shows one "how partial and incomplete is even the most careful reconstruction of lives, events, and social movements."[1] Thus, though the ultimate task remains difficult and elusive, this volume is a giant step toward the goal of a better, more inclusive, more humane and relevant history of science. When such an approach is applied to the whole range of working scientists, it will help reshape our understanding of what science has been and what it can become.

MARGARET W. ROSSITER

Acknowledgments

WE WISH to express our profound gratitude to the following colleagues for their help in promoting this collection through its numerous stages of development. Margaret W. Rossiter and Joy Harvey were most generous with advice on both potential authors and intermediate products. Donna J. Haraway and two anonymous referees offered perceptive comments and constructive criticism on earlier stages of our project. Karen M. Reeds recognized the potential of our collection from the very beginning and promoted it, through various stages, with great insight and confidence. And the collaboration and enthusiasm of our contributors surpassed the initial scope we envisioned for this collection.

Uneasy Careers and Intimate Lives

THE DOUGLASS SERIES
ON WOMEN'S LIVES
AND THE MEANING OF GENDER

Introduction

IT IS BY NOW a truism that women's careers in science face particular difficulties compared to those of men, and even special difficulties compared to those of women in other kinds of work.[1] Our aim in this collection is to examine just how this unique situation has arisen in the past two centuries. We are especially interested in analyzing how the interplay between career and personal life has affected the participation of women in science.

This project seems all the more important because the precise interrelationship between the personal lives and career patterns of women contributing to science, and the ways in which that interrelationship has affected their subjective experience of and substantive contribution to science, has attracted relatively little attention from historians of science or scholars working on women's history.

The strategy of this collection, however, is not simply to make relevant historical information available for a wide range of women's experiences in science—wide in geographical spread, scientific disciplines covered, and variety of family situations—but also to reconsider received wisdom in both history of science and women's studies. The original sources and their interpretations in these essays also compel us to raise several historiographical points.

Our case studies, both of pioneering and outstanding individual women and of groups of relatively obscure or supportive casts of women, most obviously challenge the view that a scientific community can be defined as wholly male, even for the earlier part of the period covered by our collection. Our case studies also challenge the assumption implicit in most work in the history of science, that the personal lives of scientific practitioners are of no explanatory value for the nature of their work. Finally, we also question here the idea that the development of modern science can be understood in terms of progressive "professionalization."[2]

Let us discuss these points in order. First, the case studies in this collection show how much women have contributed to science, whether measured in terms of conceptual innovation or the steady accumulation of small-scale discovery. The women included in this collection range from a scientist twice awarded the Nobel Prize to less well-known "amateurs," most notably botanists and ornithologists.

Second, all the case studies in this collection focus on the ways in which family situations—whether defined as marital relations, relations with parents and siblings, or relations with mentors when they are explicitly substituted for absent biological family—imposed problems and induced strategies and approaches to scientific work that were specific to women.

Third, these first two points, taken together, suggest that substantial modification needs to be made to the usual picture of the history of the structures of modern scientific organization as largely determined by a gradual process of professionalization. A minimal definition of this term is the appearance of paid posts for the full-time pursuit of science, within institutions rather than at the whim of patrons, as well as the certification of scientific competence through formal examination by recognized groups or institutional bodies. Clearly, it is difficult to fit the lives and achievements of most of the women discussed in our collection into this model. To what extent, then, and from what date, and in which scientific disciplines did the professionalization of modern science affect our view of the contributions of men and women to science?

In the early nineteenth century, the predicament of both men and women in relation to science was similar in many important ways. Approaches to science were enormously influenced by personal and family situations. Even in such countries as France, where the process of institutionalization was relatively advanced by the beginning of the nineteenth century, the actual production of science was still heavily dependent on the support of family members, both immediate and distant, often living in the same house.[3]

In other words, for the majority of scientific practitioners—especially in Britain and America, where the tradition of the gentleman/woman amateur remained intact well into the nineteenth century[4]—the experience of involvement in science was not as radically demarcated between the sexes as it became later in the century. Although actual scientific posts were wholly monopolized by men, such posts were few in number; not all men producing science held such posts; and most amateurs, male or female, worked from a domestic base. Often other family members greatly affected the actual resources of time, energy, and assistance available for scientific work of any kind.

It is one of the purposes of this collection to call attention, through carefully documented empirical studies, to the process by which science became lifted out of the domestic, amateur context and became transformed into an activity at odds with the domestic, intimate family lives of all of its practitioners, male or female, but in ways that affected a woman's position far more than a man's.

This collection also seeks to fill major lacunae in the field of women's studies, and especially in the area of women in science. Studies on women and science have often been conducted over enormous time spans.[5] Anxious to establish "founding mothers," or to demonstrate the continuities of a putative essential female experience, such studies tend to neglect the insights that may be gained from in-depth examinations of women's lives over a shorter and more carefully defined period of time. Yet, it is only such an approach that enables us to discover if, in fact, there are basic continuities in female experience, or whether such experience can be liberated from a mythical "eternal feminine" and seen to be as dependent as masculine experience on particular historical circumstances.

Yet another important perspective on women in science has focused on philosophical, especially epistemological, problems presented by cultural characterizations of nature as "feminine" and science as "masculine," while further adding the resonance of a long list of dichotomies underpinning scientific discourse (e.g., subjective versus objective, "soft" versus "hard," dependent versus independent). Such characterizations have made the women scientists involved in "conquering" nature culturally anomalous and thus easy to marginalize by the exclusion by other and oneself.[6] Although the philosophical perspective on scientific knowledge as gender-biased has provided an important contribution to an understanding of both scientific and feminist ideologies as constraining and enabling women's participation in science, it has been impaired by lack of empirical data on relations between women and science in precisely delineated historical and sociopolitical contexts. It has failed to account for historical diversity in women's scientific experience. More important, it has totally overlooked the possibility that the gender structure of modern science, with its massive underrepresentation of women, comes not so much from the exclusion of women from science, but rather from the exclusion of the *domestic* realm from science, and the incidental concomitant exclusion of women.

But to understand why the exclusion of the domestic realm from professionalizing science should have had such different consequences for the male and female elements in the scientific community since the early nineteenth century, we must situate the case studies in this collection in a far wider historical context. The historical period covered by this collection stretches from the French Revolution to the end of the post–World War II economic boom in Europe. This is a period of crucial developments not only in the social organization of science, but also of that of the family in Western Europe and North America. It was in this period, too, that a crucial dichotomy, the establishment of a cultural and institutional polarity between public and private worlds, came to affect science. History of science has ignored these changes, which have now become a recognized area of study in general history.[7]

How did science in this period change in ways germane to our theme? In 1800 science all over the Western world was weakly institutionalized, attracted relatively little state funding, and struggled for cultural parity with the classical literary and philological fields of study. In this situation, it remained normal for much science to be produced (even by the few who held positions as scientists) at least partly if not wholly in domestic surroundings, and to be dependent to some degree on the input of family members.

In the early nineteenth century many scientists of both sexes thus experienced as "normal" working conditions that the late nineteenth century and certainly the twentieth were to confine to women and to amateurs. In the early part of the period patronage, too, operated in science, as in all other areas, in a highly personalized way through patrons rather than through institutions. Almost all scientific practitioners were dependent on personal patronage, whether from another family member or from a patron who was in fact regarded as a substitute for the biological family, for the advancement of their careers (see Chapter 1).

By the twentieth century, or even by the late nineteenth, as the essays in this collection make clear, such a personalized patronage situation remained largely the province of women scientific practitioners, who became overwhelmingly reliant in their work on the support of other family members, whether parents, husbands, children, siblings, or a combination of these. In other words, women remained fixed in what had been in the nineteenth-century a situation typical for *both* sexes, while men in science moved on to a quite different one: in the twentieth century, small-scale, deeply personalized patronage situations came to be replaced by Big Science, which was dominated by public policy. The interwar years were, in particular, a transition period in which both patronage and policy operated at research foundations such as the Rockefeller Foundation, which both enabled and constrained transdisciplinary innovation, including such activity by women (see Chapter 12).[8]

The gradual transition from domestic to institutionalized production of science also explains why the only outstanding female patron in the early part of our period, the American astronomer Maria Mitchell (see Chapter 7), should have emerged at precisely that junction in history of science when a relative fluidity was still possible. It is no accident that Mitchell's difficulties increased with the growing institutionalization of science.

We must, however, consider a second set of general historical developments if we are to understand fully why the professionalization of science acted to exclude the personal and the domestic, and why that exclusion operated so much more to the detriment of female than of male practitioners. The answer is to be found in changes in the nature of the family, especially the middle-class family, which took place in Western Europe in our period, and especially dramatically toward its beginning.[9]

These changes in the family were an essential part of the constitution of public versus private worlds in the nineteenth century that introduced the very dichotomy between domestic life and profession that is the theme of this book. The broad picture that emerges is a change both in the ethos of marriage and in its actual demography at the end of the eighteenth century. More infants survived and the life expectancy of middle-class adults rose dramatically. In these new circumstances, it became possible to invest emotion in other family members in a way inconceivable before. A new conception of the family emerged at the same time. No longer was it seen primarily as an economic, financial, or dynastic arrangement that required no special emotional attachment among its members. Instead of stressing the property or juridical aspects of marriage, it became fashionable in Western Europe and North America to describe marriage primarily as a means of forming a private world of intense emotional bonds, cushioning its participants from an "outside" world that was increasingly seen as impersonal, confusing, and demanding.

At the heart of this new definition of the family lay the idea, also new in the eighteenth century, of childhood as a special stage of human life characterized by weakness and dependency and demanding special nurturing that could only come from women by virtue of such "natural" maternal functions as breast-feeding. Thus, the new definition of the family encouraged the identification of women, through their role in the nurturing of children, with the private, emotional role of the family. At the same time, it encouraged the identification of men with the external, public, "objective" world, a world increasingly defined, as the upheavals of the nineteenth century placed greater stress on the definition of the public realm, as being something the family was not. Of course, this new definition of the family, linked as it was also to such complex changes as industrialization and demographic shifts, did not take hold everywhere at the same rate and was completely in place in the West only by the mid-nineteenth century.

Dorinda Outram's essay (Chapter 1) discusses a transitional stage in this process, when science itself was not seen as a strictly objective activity unconcerned with the subjective, private natures of its practitioners and aspiring practitioners. For the scientific families in Napoleonic France that she describes, emotional bonds and public functions were inextricably linked.

But in the nineteenth century, most of the women portrayed in this volume experienced a sharp division between the domestic and professional world, a division that ascribed domestic roles to women rather than to men. Those roles trapped women into relying on domestic resources at precisely the time male scientists were leaving such restrictions behind to practice "objective" science in public institutions, and at a time when science itself struggled for validation and funds precisely in virtue of its impersonality and public consequence. Women did not fit into this situation. Their problems were worsened by their efficient internalization of these domestic roles. Despite the growth of organized feminism throughout the nineteenth century (see Chapter 8), few made their experience of female constraints in science into grounds for a radical revision of thinking on women's access to the public sphere. The strategies of Pierre and Marie Curie, a scientific and marital couple unusually aware of these problems (see Chapter 10), are interesting precisely because they devised a life-style to cope with the interaction of domestic and scientific dimensions far more consciously than did other couples (contrast the Curies with the couples discussed in Chapter 6). Surely part of the explanation of Marie Curie's extraordinary success, outranking that of any other woman scientist, must be this unusual, conscious management of her roles as wife, widow, mother, and prominent scientist.

It remains to be discussed how each individual essay in this book contributes to our overall strategy. Outram's opening essay acts as a control on almost all the others by portraying an early period of scientific patronage before women were excluded from scientific production because they were considered amateurs or because they were trapped in a family setting held to have nothing to do with a public profession. Family and professional objectives repeatedly overlapped in these Napoleonic scientific marriages.

Ann Shteir's essay on British botanists in the early nineteenth century (Chapter 2) describes women scientific producers at work within an overwhelmingly domestic context. Her essay shows that it is a mistake to see the family as, necessarily, a restrictive environment for intellectual production. Botany in particular, as the achievements of the Hookers and the Sowerbys show, could well be run as a family industry.

Regina Morantz-Sanchez's essay on American women physicians (Chapter 3) illustrates the ways they sought to reconcile their pioneering medical careers with a variety of choices for intimate bonds, ranging from marital partnerships to companionship with other women practitioners and the adoption of their children. Medicine itself incorporated the domestic ideology on the "appropriate" social roles of women by channeling their professional options to fields compatible with that ideology, such as social medicine and pediatrics. So did the women physicians for whom professional options were and still are influenced by their parallel quest for family and personal lives.

Marianne Ainley's essay on North American women ornithologists (Chapter 4) highlights the participation of women in a field not yet fully "disciplinized" and professionalized and contributes to a better understanding of scientific production in disciplines that accepted many "amateurs." Moreover, her close examination of the women ornithologists' self-perceptions, together with Morantz-Sanchez's comparable examination, enables us to confront the issue of women as autonomous agents in the making of their own historical predicament. As these and other essays in both sections show, women's position in science is as much the product of their own perception of that position as of the "objective" fact of domination by a male scientific establishment or by various cultural characteristics of "nature," "femininity," or "objectivity."

Nancy Slack's essay on American women botanists (Chapter 5) explores their botanical work in relation to their familial status as single persons, widows, or partners (to various degrees of subordination) in marriages to other botanists. While a variety of familial situations proved compatible with botanical production by women, Slack's essay also demonstrates how completely the scope and nature of that production depended on the women's familial circumstances.

Marilyn Ogilvie's essay on collaborative marriages (Chapter 6) explores the dual dimensions—constraining and enabling—of such marriages for the female spouse in German botany and in British astronomy and physics. In all these cases, marriage gave the female spouse an otherwise unlikely access to a lifelong pursuit of a scientific vocation, while benefiting from the resources and reputation the husbands had accumulated prior to their marriage. Yet the experience of each female spouse in science was different. In the case of the German botanist Amalie Dietrich, the wife, after separation, surpassed the accomplishments of her initially better educated husband. With the British astronomers Margaret and William Huggins, the wife's work was permanently incorporated into her husband's oeuvre; she obtained recognition only as her husband's helpmate. In the case of the British physicists Hertha and William Ayrton, the wife emerged as an independent and well-recognized scientist in her own right. This occurred as a result of legitimizing steps, including marital collaboration, the subsequent illness and death of her husband, and the capacity of Hertha Ayrton to continue the work alone.

The essays in Part I, in their different ways, suggest it was not the family itself that necessarily made it difficult for women to contribute to science: marriages, after all, could contain elements of negotiated support as well as of institutional restriction. Rather, it was changes in the ethos of science itself that impinged on women, as the biographical studies of outstanding women scientists in Part II make clear. These women, unlike their counterparts in Part I, all had independent careers ranging from lectureships and professorships, directorships of laboratories, or as self-managing grantees (often the first female recipients of a grant in their discipline). Their biographies reveal how science's increasing public image as a profession regulated by a self-contained disciplinary tradition and ethos, a variety of familial circumstances ranging from *union libre* to alternative phases of single and married status, specific sociocultural resources stemming from their transnational backgrounds, and last, their individual consciousness of gender all enabled and constrained the trajectories of their careers in time and space, but always with the effect that they lagged behind men of comparable potential and accomplishments in securing positions, recognition, and rewards, and above all, legitimate scientific authority.

The obstacles to women's exercise of power within science were only matched by the perhaps even greater obstacles of legitimizing such power, once it accrued to them through succession (see Chapter 10) or prolonged service (see Chapter 11). The profound difficulties faced by women as holders of scientific authority may also explain their minor showing in major scientific debates, theoretical advances, or large-scale empirical projects, all of which require potential and actual leadership of many men. The fate of those women who did venture into these high-stakes domains (see Chapter 12) is indicative of the intimate relationship between scientific authority and the cultural representation of gender. Their head-on collisions and bitter confrontations with the scientific consensus of their time, a consensus sustained by disciplinary traditions guarded by male empiricists, can be viewed as an alternative, albeit a risky and possibly desperate attempt to capture scientific authority. Since their gender constrained their integration into mainstream—that is, disciplinary and empiricist—science, those talented women felt compelled to resort to transdisciplinary and theoretical strategies of claiming scientific authority. Though such strategies better enabled them to express their repressed scientific creativity, they also triggered complex problems of reception, to the effect that scientific authority remained largely elusive for these women (see Chapters 8 and 12).

The scientific record of these outstanding women reflects the convergence of changes in the scientific, familial, sociocultural, and gender-ideological context of their careers. This occurred not because they possessed some special feminine or feminist "Archimedian lever" from which they could lift, shake, or modulate the male-dominated scientific consensus and its pretenses at objectivity, but rather because they suffered indirectly the results of complex, ongoing, and contingent processes of social and intellectual marginalization.

It is true that a great deal of this marginalization (rather than outright exclusion, a mechanism that became increasingly unpopular in the twentieth century, as evidenced by legislation against sexual discrimination in most countries) derived from societal obsession with gender-related cultural taboos as basic maps of social order. But it is equally important to grasp that gender alone does not account for the variability of women's trajectories in science documented by the detailed biographical studies in Part II of this collection. The very rarity of scientific careers for outstanding women for most of the period covered in Part II suggests a key role of gender ideology in access to and pursuit of scientific careers. Yet, at the same time, the case studies show that some women did become integrated into the mainstream of hierarchical and male-dominated science, usually as a result of a special relationship with a progressive male mentor who liberated himself from prevailing gender prejudice and transcended his structural position as potential oppressor. This progressive attitude often related to some form of emotional involvement and to the fact that the dominant group of men has never been structurally homogeneous. The effect of such mentor-related integration is that the scientific record of these women will tend to reflect the integrative collaboration rather than the precarious position in science they otherwise would have faced if they had tried to advance without such a protective cover. This is illustrated in Part II by the careers of mathematician Sofia Kovalevskaia (1850–91), described by Ann Koblitz in Chapter 9; the chemical physicist Marie Curie (1867–1934), described by Helena Pycior in Chapter 10; and the astronomer Cecilia Payne-Gaposchkin (1900–1979), described by Peggy Kidwell in Chapter 11.

Other women who lacked such opportunities and thus increasingly became objects of marginalization by contrast often displayed an excessively versatile scientific record characterized by "peculiar" contributions. These women's contributions were likely to be theoretical rather than experimental because their instruments and resources of scientific production could depend less on their position within a hierarchy of apprentices. Such a record reflects both their more erratic trajectories in science and their greater gender consciousness, often expressed in literary productions with a distinct utopian flavor, in which the repressed female experience becomes central through social innovation. For example, the French anthropologist Clémence Royer (1830–1902), described by Joy Harvey in Chapter 8, and the Anglo-American mathematical biologist Dorothy Wrinch (1894–1976), described by Pnina Abir-Am in Chapter 12, confronted head-on the scientific consensus of their time—a consensus constituted by male empiricist guardians of disciplinary traditions—with new theories of their own. Those theories were products not only of their scientific imagination and resources, but also of their structural position of intense marginalization.

The case studies in Part II also illustrate how alternatives to prevailing notions of traditional marriage were required to sustain active, ongoing, and prominent participation in science by women. Those alternative arrangements included abandonment of marital options or participation in informal marriage or *union libre*. Sally Gregory Kohlstedt's essay (Chapter 7) on the American astronomer Maria Mitchell (1819–99) illustrates the decision not to marry, whereas Harvey's essay on Clémence Royer and Koblitz's essay on Sofia Kovalevskaia illustrate the other possibilities. Royer was the lifelong companion of Pascal Duprat, a married colleague from the Parisian Société d'Anthropologie with whom she had a son and shared residence in both France and Italy. Kovalevskaia contracted a fictitious marriage so she could leave Russia for studies in Germany, where she benefited from unique educational opportunities as a part of the household of her unmarried mentor, Karl Weierstrass.

For both Royer and Kovalevskaia, the freedom from legal and social dependence on a formal marriage enabled them to express their scientific creativity. Yet they did not escape the ideology of marriage, in whose shadow such alternative relationships had always been conducted, or its implications for wifely subordination. Thus, Royer often assumed supportive functions for her companion, such as serving as hostess, and Kovalevskaia reverted to the position of socialite and scientifically inactive wife upon her return to Russia, the consummation of her marriage, and the birth of her daughter. The impact of such an alternative relationship on the career of the woman varied: Royer's companion served her as a social-scientific resource, whereas Kovalevskaia's husband remained a liability until his death.

In addition, both women shared the experience of single parenthood or primary responsibility, including financial, for child rearing. Royer often had to assume responsibility for their son from her own resources. Similarly, Kovalevskaia's daughter remained her primary responsibility as a result of her husband's erratic preoccupations and early death. A different set of personal ties came to her aid: she was greatly assisted in rearing her child by women friends with whom she had formed a commune in her early days.

The other three women scientists discussed in this section who pursued their careers largely in the twentieth century—Marie Curie, Cecilia Payne-Gaposchkin, and Dorothy Wrinch—provide examples of a combination of both single and unconventional marital status. Their marriages involved some degree of scientific collaboration, ranging from the pooling of complementary skills (the Curies) to occasional coauthorship of papers or books (Payne-Gaposchkin, Wrinch). In each case, the wife succeeded in maintaining an independent scientific credit rather than being assimilated into the husband's reputation, either by publishing most of her papers as sole author, by collaborating with others (Curie, Payne-Gaposchkin, Wrinch), or by maintaining her continuous creativity beyond the duration of the marriage (Curie, Wrinch). The independent reputation of the wife was further reinforced by the relatively short duration of her marriage as a result of her husband's premature death (Curie), illness and death (Wrinch), or by the relative lateness of the marriage (Payne-Gaposchkin). The uninterrupted scientific activity of the female spouse before, during, and after her marriage required the scientific community to relate to her directly rather than to rely on the sociocultural convenience of husband mediation. This was especially true when the topical focus of the female spouse's work overlapped only partially with that of her husband (Curie, Payne-Gaposchkin, and Wrinch).

In all these cases, the consolidation of the women's independent status as researchers derived both from their great talent and determination and from the husbands' progressive, liberal, or unconventional beliefs and attitudes. The husbands not only accepted their wives' postmarital scientific activity, but actively promoted their wives' professional opportunities (Curie and Wrinch; Payne-Gaposchkin and her husband enjoyed an interesting reversal of the usual situation: she was able to increase his professional opportunities in America, which had been impaired by his status as a recent immigrant). The husbands' motivations in stepping out of the conventional marriage patterns was rooted in their relative security and their liberal beliefs, as well as their interest in their wives' sexual assets as attractive women.

For example, Kovalevskaia, Curie, and Wrinch were not only intelligent and ambitious, but also attractive women who led colorful, romantic lives. Thus in addition to supporting the career of an able scientific companion, the husbands were also creating their own socially sanctioned monopoly, through marriage, of the emotional and sexual assets of their attractive wives. Nevertheless, the negotiating of personal relationships among men and women, scientists or otherwise, and its implications for career, remains complex in the case of women (and men) who seek a dual, intellectual and emotional partnership, because it is conducted in the shadow of traditional exclusive associations of one gender with intellect and the other with emotion. Women's scientific careers have been adversely affected by this culturally pervasive dichotomy.

Access to a husband's scientific resources was an advantage of spouse collaboration for these scientists, but such collaborations often diverted the wife from conceivably more relevant collaborations with other scientists, a phenomenon especially pronounced in the case of Wrinch and Payne-Gaposchkin. Yet spouse collaboration was important to these women for social reasons: to signal the compatibility of marriage and scientific vocation. This was of great importance in view of the adverse social perception of the combination of career and marriage and its effect on women scientists.[10]

The number of children each woman had depended in part on the duration of the marriage and its timing. Curie married at age twenty-eight and gave birth to two daughters during her thirteen years of marriage; Wrinch, also married at twenty-eight, had one daughter in the eight years of her marriage. Payne-Gaposchkin, whose marriage at age thirty-four lasted forty-five years, had two sons and one daughter, all born within the first six years of marriage. Once again, educational and financial responsibility for the children fell disproportionately on the female spouse, either because of the relatively short duration of the marriage (Curie and Wrinch) or because of her superior earning capacity (Payne-Gaposchkin). Not surprisingly, this strain upset the women's scientific work to various degrees, ranging from delays in completing work (e.g., Curie's invocation of younger daughter

Eve's illness in 1919 as the reason for delays in responding to Rutherford's request for data) to social embarrassments (e.g., Payne-Gaposchkin being told not to bring the children to the observatory's library) or anxiety over the way moves for the sake of a career could affect the child (e.g., Wrinch's concerns for her daughter upon moving to America at the outbreak of World War II).

A category closely related to marital status is the degree of gender consciousness displayed by the pioneering women in our case studies, particularly in relation to their aspirations and opportunities in science. As these women were all the "first" woman in a particular position, field, degree, or scientific accomplishment, and met few other female colleagues in their scientific career, they inevitably confronted the problem of women's compatibility with scientific pursuits.

Some responded to the problem of their access to science with a lifelong activism on behalf of women's education (e.g., Maria Mitchell's activities as a teacher at Vassar College and an activist in women's organizations and scientific societies, and Clémence Royer's lectures on behalf of women's education in science), while others proposed utopian solutions to their gender-related personal-cum-social problems in novels (Kovalevskaia and Wrinch). Still others used their successful combination of family life and career as a public argument for the cause of women in science, although they did so only when approached by journalists or others interested in the public dimension of women's anomalous presence in science (e.g., Curie, Wrinch, and Payne-Gaposchkin). Only Royer incorporated her acute sense of gender discrimination into her professional writings and forced her male colleagues to face the challenge of women's experience as they formulated their theories of social science in general and anthropology in particular.

On the whole, the nineteenth-century women scientists (Mitchell, Royer, and Kovalevskaia) were more socially active on behalf of women's education and access to science than their twentieth-century counterparts (Curie, Payne-Gaposchkin, and Wrinch); the latter were more absorbed with consolidating their position as scientific stars and subscribed to extreme versions of scientism, regarding science as the supreme outlet for their exceptional talents. They were also shielded to some extent from gender discrimination as a result of their husbands' supportive attitudes. They practiced rather than preached feminism through a lifelong insistence and personal demonstration that devotion to science is not incompatible with a woman's gender or her familial opportunities and choices.

The marginal status associated with gender and familial circumstances was compounded for most of these pioneering women scientists by the experience of transculturation and geographical displacement as they sought educational opportunities and jobs. Five of the six left their native countries to seek greater access to scientific careers. Thus, the French-born Royer received part of her education and pursued some of her scientific activities in Italy and Switzerland, though she ultimately returned for the greater part of her career to France. Russian-born Kovalevskaia studied for her Ph.D. in Germany and eventually obtained a professorship in mathematics in Sweden. When she was twenty-four, Polish-born Marie Sklodowska Curie came to France, where she studied and pursued a scientific career. Payne-Gaposchkin and Wrinch, Britons educated at Cambridge women's colleges, pursued their careers to varying extents in America. Only Mitchell stayed close to home.

As members of a gender barely represented in science and as foreigners, these women confronted a minority status twice over in the scientific institutions that accommodated them, where local traditions were shaped by native male scientists. In two cases, marriage to a scientist who was a native of the woman's adoptive country (Pierre Curie and Wrinch's second husband) facilitated entry into the new scientific community.

Conceivably, the experience of transculturation endowed these women with additional cultural and social resources, derivative of their ongoing contacts with family or friends from the country of origin. Thus, Curie used Polish relatives to help raise her two daughters, Kovalevskaia used Russian friends to educate her daughter in Russia while she pursued a career in Sweden, and Payne-Gaposchkin and Wrinch used their British friends for emotional and scientific support throughout their American careers. But in general, strong native patronage was required to offset the disadvantages of being a foreign-born woman in science, as Kovalevskaia in Germany and Sweden and Curie in France discovered.

Opportunities for women to study science or do research varied markedly from country to country, and these differences entered into women's decisions to leave their native countries. Opportunities related to research in universities, both state and privately endowed, were especially important.[11] It is significant that all these pioneering women scientists first made their reputation in research; only later were some of them able to obtain university positions, and then always through the intervention of husbands (Curie, Wrinch) or mentors (Kovalevskaia, Payne-Gaposchkin). Access to status as university faculty members for these women depended on legislative and national differences. England granted such a status to women in the 1920s, thus enabling Wrinch to be (in principle, if not in practice) a university (rather than college) lecturer after 1926. Kovalevskaia's and Curie's professorships in Sweden and France, respectively, were exceptional actions, testimony not to progressive university policy, but rather to the cultural prominence of their patron friends. In the United States, which had no national university, local traditions and political cultures shaped women's options. Mitchell remained at a women's college despite her achievements in astronomy; Payne-Gaposchkin eventually became one of Harvard's first female professors, but not until the 1950s. Wrinch, too, was confined for the American phase of her career (the 1940s and 1950s) to a women's college, where she had relatively limited contact with doctoral students.

Women's difficulties in converting their scientific accomplishments into paid professional positions stem to a large extent from science's association with the university system, which had remained a bastion of male-dominated, exclusive loci of knowledge reproduction (or transmission across generations). While scientific institutions did incorporate women, it was difficult for women to obtain university positions commensurate with their capabilities.

The subcultures and technologies of different scientific disciplines have also affected the opportunities for women. The six pioneering women scientists in this section span a number of disciplines: astronomy (Mitchell and Payne-Gaposchkin); mathematics (Kovalevskaia and Wrinch); anthropology and social science (Royer); chemical physics (Curie); and mathematical biology (Wrinch). They are further distributed among observational sciences (Mitchell, Payne-Gaposchkin), theoretical ones (Royer, Kovalevskaia, and Wrinch) and experimental ones (Curie). Apparently it was easier for women to gain entry and make a reputation in nonexperimental sciences, where the resources of production were less controllable or manipulable, than in experimental science, where the interaction with nature was mediated by man-made complex instrumentation whose manipulation required a prolonged and dependence-bound apprenticeship. These points, however, must be explored in a quantitative way before we can conclude that the construction of scientific facts in the laboratory had become more of a male-controlled activity than their "mere" observation or theoretical invention. Still, the ultimate power of science continues to lie in its practitioners' claims that the "experimental method" is the source of facts whose authority or truth is grounded in suprasocial or natural reality.[12]

Finally, we ought to mention some other variables that could shape the interaction between family life and career of women scientists. The relationship between father and daughter is crucial; all of these women except Payne-Gaposchkin, who lost her father at an early age, had a father who displayed strong interest in his daughter's education and achievements. Another common feature is a middle-class social background that became unstable through a variety of political, economic, social, or natural events. Hence, a family's decreased capacity for social conformity often allowed more educational freedom for its daughters.

All of these factors—as well as additional ones that cannot be discussed here for lack of sufficient data; for example, patterns of national and international recognition—need to be explored in a quantitative way before we can ascertain statistically significant patterns in shaping the interaction of career and familial context for women scientists.[13] Until more studies, especially of complementary coverage of disciplines and countries, become available, this collection remains a unique source of in-depth archival documentation of the experiences of both exceptional and ordinary women in nineteenth- and twentieth-century science. Whether viewed collectively (as in Part I) or biographically (as in Part II), the lives of these women scientists highlight the complexity of the interaction between science and gender in time and space, while underscoring the urgent need for additional studies of a historical and anthropological orientation.

Philosophical simplifications are necessary guidelines to urgent political action on behalf of women's liberation. Yet, as scholars in the history and sociopolitics of science, we also wish to explore, illuminate, and contextualize the vast diversity of historically contingent human action underlying the unity of structures of gender discrimination in both science and society at large.[14] For only such a sensitivity to a well-documented and interpreted historical record can lead us to a better understanding of and policy toward the relationship between the evolution of objective knowledge, epitomized by science, and the persistence of gender bias at the core of the prevailing sociopolitical orders. In that context of global processes and outcomes, women and men alike have always been transient actors, forever aspiring to transcend their personal sets of past legacies, present constraints, and future utopias.

PNINA G. ABIR-AM
DORINDA OUTRAM

ONE
Social-Historical Studies

1

DORINDA OUTRAM

Before Objectivity: Wives, Patronage, and Cultural Reproduction in Early Nineteenth-Century French Science

"THEY ARE NEVER BORED," wrote the composer A.E.M. Grétry of his female contemporaries, "because they are ceaselessly innovating."[1] He, like many others in early nineteenth-century France, was not short of ways to acknowledge and emphasize the debt cultural innovation in general owed women. Traditionally, history of science has been the least capable of all the histories of acknowledging the contribution of women either to its substantive or to its social development. In this essay I will assess that contribution for a period of history when there were so few women (just as there were relatively few men) gainfully engaged in science that women's contribution to science can only be examined through their interaction with the social organization of science, and with the value systems they might have been able to inject into it.

As I have argued elsewhere, this period is one in which science was as much organized through highly personalized patronage systems as it was through institutions.[2] Accordingly, it makes sense to examine women's role in that system, and in doing so extend our understanding of its operation from the traditional male dyad of patron and protégé to a new view of a triad including patron, patron's wife, and protégé.[3] Such a dyad also performed far more complex functions than the simple transmission of power traditionally portrayed in accounts of the patron-client relationship. While most essays in this book implicitly or explicitly view marriage and career for women as reconcilable only with difficulty, this essay twists the perspective and explores marriage among scientific elites, and the female contribution to those in marriage patterns, as actually constitutive of scientific ethos and the practice of scientific patronage.

Late eighteenth-century thinking on marriage and on the scientific vocation crisscrossed in many ways, some of which demonstrate the intimate dependence of scientific ideologies on the field of concepts surrounding them. Powerful currents of thought in this period rejected the world (*le monde*) and vaunted the simple, the natural, and the good as a means to the making of authentic, uncorrupted human beings.[4] A true sensibility came from such a rejection of the world and was also seen as an integral part of the image of the savant and the philosophe. Scientists themselves were continually enjoined to a direct contact with nature, and rejection of the corrupt world of intrigue and advantage, as the price of their ability to see nature as it was.[5] Both the clichés and the technical philosophy of the time thus tended to resolve the perceiver into that which was perceived. Scientific objectivity hardly made sense as a value against this ideology. The pure in heart were also the pure in eye. Against this background marriage fit in at best uneasily. Some could agree, like the naturalist L. Ramond, that marriage was the social institution that lay nearest to "nature."[6] For others, the inescapable companion of marriage, however neutralized by Rousseauist idealization, was *le monde:* the world of social connection and social leverage, and above all, as romantic ideas of marriage as companionship gained ground, of lifelong negotiation with another human being, with all of its implied threat to the blessed objectivity of the man of science. There were many who saw the marriage of the savant as nothing but that "perilous leap" against which Jean d'Alembert warned Joseph Lagrange.[7] And just as parental dismay traditionally attended the awakening of a scientific vocation, sure prelude to poverty, trial, and social marginalization, so parental alarm was easily triggered at the thought of marriage with a man of science. The well-known nervousness of the Marquis de Condorcet's future parents-in-law must have been paralleled in many other cases.[8]

Marriage, in short, with its aura of compromise and negotiation, with its inevitable links with all the world of social production, enjoyed an ambivalent relationship with the official ethos of science even when, as was true in this early period, there was little threat of professional competition from the side of the wife. While family was linked, as was the ethos of science itself, to ideals of retreat, purity, and sensibility, marriage in the real world disrupted that ideal person whom the man of science was supposed to be. In an age much given to the insistent construction of figures of representative virtue as guides to the business of living, this was a serious drawback, and one that, as we will see, was fully reflected in the marriages of individual men of science.

But precisely because of its ambiguous ideological setting, marriage could also perform a valuable mediating role. The wife typically took on roles essential to the real-world maintenance of scientific organization, such as the handling of protégés, thus preserving intact the claims of her husband to exert his "scientificness" entirely outside the tainted world of career making, patronage, and advantage.[9] At an even deeper level, it was one of the functions of the woman to reconcile yet another profound conflict in scientific life: the demands of the individual vocation and of its social implantation. As noted earlier, the discovery of a scientific vocation by a young man was hardly ever a cause of parental rejoicing in eighteenth-century France. Such conflicts were so frequent that they were almost an identifying mark of the true savant. Georges Cuvier said publicly: "It is with such similar conflicts over the choice of a career that the stories of many of our colleagues begin."[10] Such a rupture with the family forced the fledgling savant to confront an enormous paradox: the freedom to pursue innocent knowledge, and thus the freedom to become who that person authentically was, could occur only as the result of an act of archetypal guilt—the rejection of parental authority. The eighteenth century insisted strongly on submission to parental, and in particular paternal, authority as a moral duty analogous to submission to the will of the heavenly Father. Rejection of parental authority was thus in a true sense an act ushering in chaos, for it upset a chain of power that stretched from man to God. This was an ideology that the conservative state of Napoleon was no less concerned to preserve than had been the "divine right" monarchy of Louis XVI.

The only way out was to reenter a family grouping, for only this could cancel the original act of betrayal of the young scholar's biological family.[11] The finding of a second family was always fused in contemporary accounts with adoption by an effective patron, the sine qua non of the effective implementation of an intellectual vocation, as it was, of course, of entry into any other form of public role. The young savant at this point is allowed to be the person he authentically is by the man who becomes at once his public spokesman and supporter, as well as his "good" father. Contemporaries agree on ascribing a transforming power to the moment of adoption by the patron.[12] Accounts telescope the grueling process of the search for an effective patron into a single miraculous moment of acceptance, a moment that wipes out former inadequacy and doubt at the same time it dissolves all the claims of the biological father over the young protégé. Eleuthère Du

Pont de Nemours, for example, scarred by bitter quarrels with his father after the early death of his adored mother, credited François Quesnay with transforming him from an infant into a man; Louis Daubenton released the twenty-year-old Etienne Geoffroy St. Hilaire from parental authority by the sudden words "I have a father's authority over you."[13] The self-image of the protégé damaged by conflict with parents, and by the struggle for survival in a world that saw his vocation as of marginal importance to its great concerns, is reaffirmed in these encounters with an embodiment of intellectual authority in the precise lineaments of another human being.

Yet other contemporary accounts also strongly acknowledge the importance of fictive "mothers," whether single women or the wives of male patrons. These women are credited by contemporaries with a crucial role in the discovery and enforcement of vocation, which to my knowledge has never been examined in a more than anecdotal way by any historian of intellectual groups of this period. Here I can only begin to suggest what such a study should entail. But it does seem clear that we need an approach to the problem of the formation of patronage networks that is not exclusively tied to the two-person relationship of patron-protégé; we need to acknowledge the numbers of men who had no single obvious male patron, but who, like Georges Cabanis or André Morellet, remained linked with a female patron on a lifelong basis.[14] We need also to recognize that other male patrons were married to women to whom they often delegated some of the most vital parts of the operation of patronage, and that the protégés often established relationships with these women that were far more important to their development and advancement than were those with the husband; the circle around Mme. de Prony, wife of the director of the Ecole Nationale des Ponts-et-Chausées (a prestigious college for civil engineers), is a case in point.[15]

In expanding our analysis in this way we also acknowledge contemporary ideas about the importance of women in the cultural field. Social and legal historians have recently emphasized the declining position of women in general under the Directory and Empire; in the early years of the Revolution legislation on marriage and the family was systematically abrogated in favor of a return to a patriarchal family system.[16] These arguments are important and valid, but they do not accord with other evidence from this period on the importance contemporaries attached to women's impact on the cultural field and on the shaping of individual character.[17] We should not mistake this emphasis for what it became in the later nineteenth century, when according women supreme importance in the realm of personal relations, home and family, and the consumption of cultural goods was strongly linked with efforts to restrict them to a purely privatized existence.[18] Culture was vitally important because it produced the symbols through which the public realm conducted the great debates of this period on the nature of power and the relationship between individual and state. Thus, to say that women had an important role in this field of culture is not to describe a trivialized group excluded from real participation in the public realm; it is to describe one of the substantive, innovative sectors of that realm at work.

Contemporaries were clear about the ways in which a relationship with a female patron, with its exposure to specifically female ways of exploring individual personality, was a powerful force in encouraging the young savant. The writer J. B. Suard, for example, stressed in his panegyric on his "fictive mother," Mme. d'Houdetot, how her particular way of getting to know people emphasized and displayed their strengths: "It seemed as if nature had given her an insight all her own; she found out quickly and surely the best that there was in any person she talked to, just as she did in any book that she read."[19] Or as Benjamin Franklin had earlier remarked to Mme. Helvétius: "In your company we are not only pleased with you but better pleased with one another and with ourselves."[20] This sentence neatly reveals how the enabling and reinforcing role of such women was not simply confined to individuals, but also affected the groups of their habitués, giving increased confidence and cohesion to this newest section of the French elite. Typically, groups of individuals linked to such women tended to form the social institution known as a salon. Unfortunately, in spite of the importance of the salon in promoting the cohesion of the intellectual elite, and in channeling individuals into the network of patronage relations, we possess no general account of the salon as a dynamic social institution for the period of the Revolution and the Empire. The fine works of Deborah Herz on Germany, Carolyn Lougée on seventeenth-century France, and Daniel Kors on the "d'Holbach coterie" have either concentrated on the history of one salon or have tried to write the salon into the history of women's intellectual emancipation.[21]

My aim is not to contradict these insights, but to suggest that salons need to be integrated into a full appreciation of their use as power bases both for the patron and his wife.[22] The salon was where the patron, or "father," first encountered his "sons," or protégés, through the medium of his wife, the fictive mother. At any one time it would contain individuals who were known and accepted members of the "family" and others, less familiar, who were on trial, waiting to be evaluated before their adoption was certain. We possess a detailed account of this process as it was carried out by

Mme. de Prony, wife of engineer and physicist Gaspard de Prony. The account was written by her favorite fictive son, the economist and technologist Charles Dupin:

> Mme. de Prony, who could have chosen her friends from the highest ranks of society, yet showed no eagerness to receive the homage of men who had little to recommend them except birth and title. It was personal, intellectual merit, and, even more, merit of the heart, which she looked for in her friends above all else. . . . She had a sort of tact all too rare today, whereby she could assign to each person, from the first to the last of her intimates, the precise degree of esteem and consideration they deserved, without in the least seeming to deprive anyone of the courteous regard that was their due. This equality of treatment is the only way to make the world of society tolerable. Mme. de Prony could raise anyone to the precise rank in her esteem to which their age, merit, and social position entitled them, while never denigrating anyone. She loved above all to show her goodwill to young men who, with some hope of success, were just starting in their careers.[23]

The salon provides one of the strongest examples of how an informal social institution, dominated by women in their actual management, even if ultimately financed by the husband, and using the most subtle interpersonal discriminations as a guide to access to the public power of the patron himself, was in fact an integral part of the making of a career and the gaining of reputation. Henri Beyle, the contemporary expert on the interconnection between public and private politics, and himself a dedicated habitué of salons, made the point unequivocally in his autobiography:

> What a difference it would have made if M. Daru or Mme. Cambon had said to me in January 1800: Dear cousin, if you want to have an assured place in society, you must have twenty people who have a good reason to recommend you. So, choose a salon, go there every Tuesday (or whatever its day is) without fail, make absolutely sure that you are winning, or at the very least polite, to everyone who goes to that salon. You will be something in the world, you can even hope to win the favor of an amiable woman, when you are backed by two or three salons; and at the end of ten such years, these salons will open every door to you.[24]

The institution of the salon, crucial to the creation of the fictive family of protégés gathered around the patron and his wife, was also heavily dependent on the demography of the intellectual elite. The salon could only have functioned as it did within a given sort of family structure. There have, however, been no previous studies of patterns of family and marriage among the scientific elite.[25] To this end, I have collected demographic information for a distinct sample defined by membership of the Institut de France between 1795 and 1815, in order to compare, with some crudeness, the resulting group of some individuals with information already available on the demographic behavior of other sections of the Napoleonic elite. Interesting differences emerge.

Possibly the largest single study of the Napoleonic elite is Louis Bergeron and Guy Chaussinand-Nogaret's exhaustive examination of the lists of provincial notables. From this emerges a picture of an elite characterized, as were the members of the Institut, by late marriage—around age forty-three for men, compared to twenty-eight in the general population.[26] In both the case of the notables and of the Institut members, differences in age between husband and wife tended to be large, often of more than ten years, as was to be expected if the group was to reproduce itself. Among the notables, many remained unmarried at age fifty. The Institut sample tends to reinforce these trends. The number of lifelong bachelors is even higher than among the notables, partly due to the many priests and former priests among the intellectual elite.[27] We need to modify this picture by recognizing the sizable numbers from the Institut group whose chose to remain outside formal marriage, but produced children within more or less stable informal liaisons.[28] For many members of the Institut, having children within marriage was clearly low on their list of benefits to be gained from personal relations. This picture is reinforced when we consider the family size of those members who were legally married. The Institut scientists produced far fewer than the 2.83 children per marriage noted by Bergeron and Chaussinand-Nogaret among the notables, which is itself around half the average in the general population. The Institut sample is also remarkable for the high number of marriages within its ranks that produced no children at all.[29] If normal fertility in this group is assumed, deliberate choice against the production of children is clearly at work. All these factors mean that for the Institut members marriage was not the reproductive machine it appeared to be for many of the French elite in the later nineteenth century.[30]

Typical family structure in the Institut, on the other hand, facilitated, if not necessitated, the formation of the fictive family of protégés so strongly encouraged by the treatment, in the vocational ideology of the savant, of the patron as equivalent to the rejected and rejecting biological parent. The low numbers of children produced in the average Institut family meant that the fictive family had space to accumulate. Since the wife was not preoccupied with the care of many children, she had the time and emotional space to devote to her husband's protégés and, in doing so, to influence their careers.

The fictive family was also, it can be argued, a necessary social device for this specific elite, which is characterized by the wide disparities in income between its members and a lower average income than any other section of the elite.[31] Relatively few in this group seem to have married to remedy a deficiency of capital. In these circumstances, to have produced higher numbers of children would not only have meant higher investment in their early years, but also high single-time expenses, such as the provision of dowries for daughters. It was clearly far less expensive to maintain the fictive family. Without the fictive family, the demography of the Institut group would also be ineffective in ensuring transmission of cultural styles and research programs into the next generation.[32]

We have now entered an area in which the demographic evaluation of these marriages has to be reinforced by a qualitative study. It is clear that in the main we are dealing with marriages not primarily regarded by those who enter into them as reproductive machines or devices whereby the husband may accumulate capital from a rich bride. If we wish to examine these functions, we have to turn to the fictive family of protégés, that multiplier of the patron's credit and earning capacity, not to the biological family.[33] How, then, are we to view these marriages? The usual historical approach toward marriage as a strategy of advantage is inappropriate in many cases.[34] Instead we have to look more carefully at attitudes toward marriage and at the kind of women who married professional savants, members of the Institut.

In collecting information on wives and marriage in this group, we encounter many different sorts of women and many different types of marriage. We need to examine these wives as a group to find out what sort of women they were, what social classes they originated in, what sort of bond they constituted between Institut members, and what other sections of the elite they linked them to.

No attempt has ever been made to produce a collective portrait of the wives of any sector of the Napoleonic elite, let alone of the scientific elite.[35] Partly this is a result of historians' tendency to produce a history in which only men appear as the protagonists and thus cannot offer a picture of their personal relations as in any way constitutive of their historical role. This failing has been taken to the opposite extreme by those who have written about women; they also tell their story without taking into account the marriages in which their protagonists passed the greater part of their lives. The problem in constituting a collective portrait of the wives is thus not primarily that of lack of information, for many, in particular nineteenth-century historians, contributed much to them; rather, it is one of establishing an approach to the evaluation of the importance of marriage in an elite whose marriage patterns cannot be subsumed under the analysis of strategies of advantage of the more obvious kind. The first positive point to make, however, is that a sizable proportion of Institut wives wrote novels, plays, pictures, translations, and political writings. And these were not merely ornamental exercises. Mme. Daubenton's novel, *Zélie dans le désert,* went through twenty-one editions after its first appearance in 1787. Mme. Biot published a translation of Ernst Fischer's standard work as *Physique mécanique;* it went into four editions. Mme. Fleurieu was the author of two comedies and a novel. From a preliminary list of similar achievements,[36] there emerges a group of women who to a large extent inhabit the same intellectual world as their husbands; they do not relegate themselves to a mysterious background world of children and household that has no impact on the world of the man. Marriage with such women meant a marriage of companionship rather than a marriage of biological reproduction.

From learning more about the Institut wives, another striking point emerges. One cannot fail to be struck by the extraordinarily disparate social origins involved, which range from Desfontaines's assistant gardener and Joseph Gay-Lussac's shop assistant, to Gaspar de Prony's, Pierre Charles Lemonnier's, and Pierre Leroy's marriages into the nobility, Jean-Baptiste Delambre's and Nicolas Deyeux's into the financial elite.[37] Historians of the Napoleonic elite have pointed out that elite status and income seemed to have little to do with each other in imperial France.[38] But the Institut, even disregarding disparities of income, does not show a stable class structure formed by men uniformally marrying "up" in a way that strengthens their claims to belonging to an elite. Rather, the Institut demonstrates an insouciance about class in the choice of a marriage partner that tended to dissociate prestige, wealth, and, as far as we can use this terminology today, social class.

Historians of ancien régime elites have frequently remarked on their tendency not to intermarry, despite financial interests in common or common membership of institutions.[39] When we turn to the Napoleonic Institut, however, we find that although not all members are related to each other by marriage, an option in any case ruled out by the high numbers of at least nominal bachelors, a substantial amount of intermarriage does occur. Men who married a former colleague's widow, or who were the brother-in-law of a colleague, or who married their daughter to a colleague—a tenable option when all elite groups were accustomed to marriage partners widely different in age—all are relatively common in our group[40] Such intermarriage cuts across subject boundaries embodied in the four classes of the Institut. This fits in with what we know about the lack of definition of discipline boundaries in this period.[41] Members of the Institut thus had little incentive to use marriage in a widespread way in order to increase their credentials within a well-defined scholarly group.

More interesting questions are raised when we consider possible reasons for the relative frequency of intermarriage among the scholarly elite of Napoleonic France compared to what we know of its ancien régime counterpart. We may speculate on the effects of increased pressure from the Napoleonic state on the Institut to create a greater feeling of identification among its members, which could be accomplished through increasing intermarriage. Or this could be one of the first stages in the growth of a self-conscious intelligentsia like that of post-1830 France.[42] More probably, however, increased intermarriage should be viewed as one of the first indications that the relatively undifferentiated elites of the Napoleonic state were gradually changing into the vertical elites of modern France.

This brings us naturally to a consideration of the extent to which marriage linked the members of the Institut to other sectors of the elite. This is not simply a measure of elite fluidity in the Napoleonic state as a whole, but also of the relative status of the Institut *in* that elite: rarely does a man willingly marry his daughter to someone he considers his social inferior, preferring rather to marry her to his equal or, even better, his superior. The question of where the intellectual elite as such stood in the pecking order of Napoleonic France, its perceived importance by other sectors of the elite, has rarely if ever been considered, even in social studies of such groups. Here, however, we encounter a difficult problem. Many members of the Institut performed multiple roles within the state; a few had wealth deriving from successful business enterprises.[43] How are we to establish whether such men were viewed by their prospective father-in-law, let alone by their prospective bride, as members first and foremost of the Institut, or as important bureaucrats or wealthy men who happened to have the right to wear an academician's uniform?

Put like this, the question is almost impossible to answer. We can approach it in two other ways, however: by pointing out that membership in the Institut, at the very least, does not seem to have deterred other sectors of the elite; and by using those members of the Institut who had no other occupation as a control group. There are no recorded examples of any member solely dependent on his vocation for income marrying "up" into nobility or higher bureaucratic families.[44] Institut membership by itself therefore conferred little chance of acceptance into the wider elite. But we must also note that membership in the Institut was usually connected with bureaucratic involvement of some kind. As one of the major new pools of talent in the Napoleonic state, a state in continuous expansion, bureaucratic employment was the normal fate of the intellectual elite.

I have argued that patronage in science cannot be fully understood without insisting on the role played by women. They not only provided the social setting in which future protégés could collect within the ambit of their patron, but also played a crucial role, one fully acknowledged by contemporaries, in the sifting and sorting of the patron's clientele and in the unfolding of individual consciousness of a scientific vocation. In an era before the full institutionalization of science, women, just as much as the male patron, provided by conscious exertion of social art and psychological insight a medium through which the aspiring young savant could locate his authentic self, the self his biological parents had refused to allow him.[45] That authentic selfhood was seen as the sine qua non of the ability to view the natural world correctly.

Women, dominant in the social world but largely excluded from formal decision-making structures, were capable of reinforcing authentic vocation precisely because their input was remote from the world of alliances, intrigue, and advantage in which the male patron, however unwillingly and ambivalently, was forced to make his way, a world seen as the enemy of personal authenticity. If patronage systems are means of cultural reproduction as much as of transmission of authority, then the system reproduced all the ambiguities of the scientific ideal. On the one hand, the otherworldly qualities of the true savant, located in authenticity, were enhanced by the women involved in the patronage network; on the other, the male patron provided entrée into *le monde,* the world that threatened personal corruption and the loss of the true self, and yet supplied the indispensable real world, means for the actual implementation of vocation.

Women in the scientific elite were able to perform these functions because of the specific demography of the group. They also performed them by virtue of a certain style of marriage, one whose function was seen as companionate activity rather than biological reproduction. Given the trend of much current research to identify marriage and career as hostile entities for aspiring professional women scientists from the mid-nineteenth century on, it is worthwhile to remember that in an earlier epoch, when no "professional" careers in science were given women and very few were open to men, the values and functions of marriage were able to mesh together relatively easily. Presented by many eighteenth-century thinkers as the most "natural" of the social institutions, wives and husbands, in their different ways, were able to heal the many breakings of bonds that characterized, in contemporary estimation, the typical career of the budding savant, and they were able to overcome the many contradictions in the scientific ethos.

Why was the position after mid-century so different? It cannot be wholly ascribed to the new problem of professional careers for women, or some idea of "necessary" conflict between career and family. It was much more that the nature of marriage itself changed after the impact of industrialization not only on economic life, but on value systems as well. Historians have often described these changes as an increasing tendency among the middle class to allocate the domestic, "intimate" interior to the wife, and at the same time to see the cultural role of "the home" (i.e., the wife's role in marriage) as a conservative rather than an innovative one. In setting up an automatic opposition between career and marriage we are also forgetting that the full professionalization of science was a long drawn out process, productive of insecurity at every level. It was not, after mid-century, that "all men" were preoccupied to deny "all women" their rightful place in the scientist's profession (although some clearly were); at a deeper, structural level, we can see that they were encouraged to make that exclusion more by changes in the nature and ideology of marriage, as well as by the continuing insecurity in science itself, both as ethos and as career.

Another consequence of industrialization was an increasing emphasis on ideals of objectivity, rarely posed as such in the science of the late eighteenth and early nineteenth centuries. However one defines objectivity, such an ideal has two clear consequences: to separate subject from object, observer from that which is observed; and to insist on the ejection of emotion from the processes of cognition. In doing so, it removed the ideological base from the important female input into the development of scientific vocation. If the inner world of the observer had no impact on the quality of his observing, the drawing out of individual self-awareness was of comparatively little importance, and science became far more the *doing* of a certain set of activities than the *being* of a certain sort of person, who could be helped to full authenticity by the emotional insights of women.

It was in this very different set of expectations that women were to struggle for careers in science after mid-century. A vision of women as complementary to science collapsed at the same time as did the culture that linked personal authenticity to the struggles of the savant.

2

ANN B. SHTEIR

Botany in the Breakfast Room: Women and Early Nineteenth-Century British Plant Study

> I was quietly pasting down botanical specimens, in the breakfast-room upstairs, when . . .
>
> MARY KIRBY,
> *"Leaflets from My Life": A Narrative Autobiography* (1887)

IN EVELYN FOX KELLER'S biography of Barbara McClintock, belatedly named Nobel laureate for her work in plant genetics, we meet a woman who stood for a long time on the periphery of her field.[1] Keller's book brings McClintock from the periphery into the center and tells the story of modern biology from McClintock's vantage point, locating her within carefully detailed local and intellectual contexts. Although McClintock herself sees science as transcending gender, and although she herself would seem to disallow any historical account that explains her ideas and her career in terms of her sex, one must wonder how much McClintock's story is that of the woman scientist in the early twentieth century, or how much that of women in the history of botanical science. McClintock, born in 1902, came of intellectual age at a time when women in America had a fair amount of access to higher education, science training, and scientific occupations. Her British counterparts in plant study were similarly finding their places within the world of science, studying botany, teaching botany, and doing laboratory work. But what might have been the biography of

Barbara McClintock had she been born in 1802 rather than in 1902? This essay concerns her British foremothers, women who were botanically active in the early nineteenth century, but were peripheral to mainstream botanical science as it then was being shaped and codified, and as we find the codifications in all histories of botany to date.

My discussion makes the peripheral central and looks at nineteenth-century British botany from the vantage point of women who were active participants in botanical culture at that time. When we place women at the center of the history of botanical culture in nineteenth-century England, when we enable women to speak as much as possible for themselves, then we can better trace a picture of the place of one science in relation to everyday life. The most startling change in the picture that results is that the locus of science shifts from the public sphere to the domestic sphere. Home becomes the geographical locus for learning and for botanical activity. We then see the family nature of preprofessional botany, the presence of women in botanical culture, and the centrality of mothers in botanical education.

Seeking to trace pathways for women into botanical work, I use biography to address both individual women who worked botanically and the collectivity of women. I also direct us to autobiography, to voices of women who report on the place of botany in their lives. I approach the topic of women's work in British botany from a literary angle, through the narrative format of introductory botany books, a corpus of popularization in which women were prominent as authors and as characters. Through these paths we can establish something of the place of serious botanical work in the lives of some women in England, in the period when this science was still amateur and when hardly anyone, female or male, thought of botany as a career.

Eighteenth-century botany was largely a taxonomic project, concerned with collecting and categorizing. It was tied closely to a natural history tradition of observation and empirical detail. The taxonomic orientation dominated botanical debate into the nineteenth century, when it was supplanted by interest in plant physiology. In turn, morphology and plant geography became central areas of nineteenth-century disciplinary debate. Although the natural history tradition far from disappeared, it was nudged aside in many mainstream quarters by an experimentalist tradition.

A lively botanical culture grew up in England in the second half of the eighteenth century and continued to develop well into the next century.[2] In part this was due to traditional British enthusiasm for natural history, in part to the developing cachet of science. The taxonomic work of Linnaeus, with its easy sexual system of plant classification and nomenclatural scheme, helped to facilitate plant study at various educational and social levels. The Linnean Society was a forum for botanical contributions and discussion from its formal establishment in 1788, as were provincial literary and philosophical societies and the British Association for the Advancement of Science later on. In general, botany stood at the cultural nexus of popular science and fashion from the late eighteenth into the nineteenth century. It benefited both from Enlightenment teaching about the importance of self-improvement and from Romantic and Victorian sensibilities about nature.

While women took an active part in public botanical culture, their contributions were invisible in histories of botany until recently. The reasons are like those for the historiographical invisibility of women in other areas of intellectual and cultural endeavor. Conventionally, histories of botany have been written from the perspective of present concerns, charting the emergence of botanical theory, or have highlighted "great men" and "great achievements."[3] Women do not appear in these canonical accounts, for, so far as has been recorded, women were not among the conceptual shapers of botany, made no noteworthy taxonomic contributions, gave no groundbreaking papers, and presided over no formative scientific institutions. Yet women were all over the map of nineteenth-century botanical culture as cultivators of science, to adopt a widely used typology. They corresponded with such leading botanical figures as Sir James E. Smith and Sir William Hooker and sent reports to the *Journal of Botany* regarding observations. They contributed to the compilations of local observations that made up county flora.[4] They collected and drew plants, and did illustrative plates for botanical publications.[5] But these were not great achievements, as conventionally understood. (If conventional histories offer no window onto women and botany, one also finds none in recent new-style French internalist work on botany.)[6]

An approach via social history goes some distance toward restoring women to visibility in the botanical tableau. David E. Allen has been the pathbreaker in writings on the social history of nineteenth-century natural history.[7] His recent study of the small membership pool of notable women in the Botanical Society of London at mid-nineteenth century is important for its methods and for its many details about the backgrounds of the women members and the circumstances of their affiliation.[8]

During the years before women had routine access to university or professional training, families were central to the history of women's work in science. As in the history of other fields, women who were active or who made public contributions to natural history or science often were part of families that did the same.[9] Examples from eighteenth-century British botany include: Elizabeth Linné, daughter of Linnaeus himself, who made botanical observations that were published; Jane Colden, daughter of a British official in New York State (himself a prominent botanical enthusiast), who entered into correspondence with Linnaeus and collected, dried, and drew plants; Anna Blackburne, well known in botanical circles of the late eighteenth century, who worked with her botanist father on projects of mutual interest.[10]

The most multifaceted example over a few generations, drawn from the following century, is that of the wives and daughters of the extended family of J. S. Henslow, Dawson Turner, J. D. Hooker, and W. J. Hooker, four men at the heart of nineteenth-century botany. Mary Turner (1774–1850) did drawings and engravings for her husband's monograph on British seaweeds, *Fuci* (London, 1808–1819); her daughters, Maria and Elizabeth, did drawings and engravings for W. J. Hooker, and Maria Turner married him. Their son's first wife was Frances Henslow (1825–74), daughter of J. S. Henslow, the professor of botany at Cambridge, who was mentor to the young Charles Darwin; Frances Henslow Hooker translated from the French an influential French textbook on botany that appeared under the editorship of J. D. Hooker. In these cases, their work was, in effect, for the family firm, and family projects offered them a means of developing skills and going public with contributions to botany. Daughters-in-law were part of the family firm too; Margaret Roscoe (fl. 1820s–30s) did most of the illustrations for William Roscoe's major work, *Monandrian Plants of the Order Scitaminae* (1824–29), after which she went on to publish her own book, *Floral Illustrations of the Seasons* (1829–31).[11]

In all these cases, father (or father-in-law) seems to have been the central shaper of interests (although the biographical record is too scanty for us to be fully certain). For example, Elizabeth Linné's mother is said not to have approved of too much education for girls, and we may surmise that Carl Linnaeus himself established the tone for his daughter's botanical reporting. It may be that in cases where mothers were alive, they represented more conventional views on appropriate activities, and that it was fathers who encouraged daughters in some degree of botanical work. It has become something of a truism in the current psychological literature that a supportive father is centrally important in the life patterns of women who achieve suc-

Opposite: An illustration from Emily Ayton's *The Children and the Flowers* (1855). Courtesy of the Osborne Collection of Early Children's Books, Toronto Public Library.

"Well, little Fanny, and how many varieties of leaves and stems and roots have you to shew me?" *Page* 85.

cess.[12] Certainly, paternal encouragement gave an opportunity for the women I have mentioned to make their scientific contribution.

It is not, to be sure, only daughters who derived an impetus to botanical study from their fathers. A botanical counterpart to other famous Victorian fathers and sons is the example of the elder and younger Hooker.[13] Nor is it only male relations who set the stage and the tone for daughters in botany. As we shall see later in this essay, mothers were formative in cultivating interests on the part of daughters and sons. Scratch a male botanist in the early nineteenth century and one can expect to find, for example, botanical relatives such as Eliza and Marianne Boswell, aunts of J.T.I. Boswell, curator of the Edinburgh Botanical Society in 1850 and editor of the 1863–72 edition of Sowerby and Smith's *English Botany*. The Boswell aunts collected plants near their home in Scotland, put together a plant list and herbarium, and assisted other botanists in a work on the plants of Edinburgh.[14]

For both daughters and sons, nieces and nephews, scientific interests could develop within a family context, but the public expression of these interests shows more among sons than daughters. Sons had more chance for interests that could be developed further, at school or in forms of apprenticeships appropriate to their class. Sons had more opportunities for enlargement in universities, or in active public engagements. Brothers sought growth through travel, but most sisters traveled only through books. Boys went off to school, but girls were educated at home. Margaret Gatty (1809–73), for example, well-known algologist and author of *British Seaweeds* (1863), sent her boys to public schools and educated her daughters at home herself, with help from occasional tutors and with the older girls teaching the younger.[15]

Within family networks male mentors made a difference to women in learning about botany and pursuing botanical activities. Prior involvement of male relatives or family friends brought many women to botany. One example is Anne Pratt (1806–93), author of the five-volume *Flowering Plants and Ferns of Great Britain* (1855) and of other popular botanical books in the 1840s and 1850s, including the much reprinted *Wild Flowers* (1852) for children. She was introduced to botany in her younger years by a doctor who supervised her plant studies.[16] A family friend also introduced Margaret Gatty to botany—although she was not young and delicate, but was, rather, middle-aged and suffering from the effects of too many pregnancies. For health reasons, Gatty moved from Liverpool to the seaside and there began her years of absorption in collecting and studying marine plants.

An uncle was important to the developing botanical interests of Lydia Becker (1827–90), a prominent nineteenth-century feminist who is best known to history for her work as national secretary of the Society for Women's Suffrage and as editor of the *Women's Suffrage Journal.* Lydia Becker was the author of *Botany for Novices* (1864), a short exposition of the natural system of plant classification, and she also gave a botanical paper at the 1869 meeting of the British Association for the Advancement of Science on the effect of a fungus parasite on the sexual development of a particular plant.[17] The eldest child of a large and prosperous manufacturing family, she grew up in the country and was educated at home. She became active in botanical work in the early 1850s and was encouraged in this by an uncle who was enthusiastic about botany. He corresponded with her about plants; tutored her on collecting, drying, and exchanging plants; sent her specimens; and gave advice about botany books for her to read. He also guided her toward studying plant structure rather than just classifying her specimens.[18] Lydia Becker's father was not himself interested in botany before this time, and there is no evidence of her mother's shaping hand. Her uncle, by comparison, seems to have set her on a serious botanical course.

Networks of male family friends also were central to how Mary Kirby (1817–93) came to botany and came to do the *Flora of Leicestershire* (1848), the work for which she is known and the only nineteenth-century county flora compiled by a woman.[19] Mary Kirby came from a prominent merchant family with connections sufficiently well placed to secure them a privately guided tour of the British Museum. She grew up in the general climate of science and natural history enthusiasms of the mid-nineteenth century. In her family, impetus to learning came from her father and from male family friends, who taught her languages and brought her books. Men are the central figures in her botanical story, which began when she was a young woman. Active in botanical work in the early 1840s and living in a country village for reasons of health, she was assisted by a male friend who collected plants and mosses for her, which she then examined with his microscope. Her county flora arose from correspondence with a clergyman and naturalist who had coauthored several local plant lists, encouraged her undertaking, and contributed one section to it. Another clergyman-naturalist also contributed material and did editorial work.

Botanical brothers, too, belong in the picture, in an idealized relationship that also finds a place in Victorian mythology. Consider Anne Elizabeth Baker (1786–1861), younger sister of George Baker, both unmarried. He was author of the multivolume *History of Antiquities of Northampton* (1822–41), a county history that had an arduous gestation over many decades. In a contemporary review of this work, she was described as "the companion of his journeys, his amanuensis, his fellow-labourer, especially in the natural history."[20] In the service of her brother's project, she developed expertise in philology, geology, and botany. Of particular interest are plant lists and botanical notes she made along with drawings and engravings. Through her older brother she developed scientific interests and talents. She became his botanical helpmate.

Botanists often had wives as helpmates. We saw this earlier with the Hooker clan. Jane Loudon (1807–58), another well-known example, served as secretary and amanuensis to her husband, John Claudius Loudon. She had neither botanical knowledge nor botanical interests before they married, but she developed both under his tutelage and went on to help him in his prolific horticultural journalism, in his books, and in his periodical publications.[21] Thereafter all her writings were on plants, and she wrote various books after his early death.

Of course not all wives of botanists became involved in botanical work themselves. Pleasance Smith (1773–1877), wife of Sir James E. Smith, founder and president of the Linnean Society, was at the heart of British botanical culture during the first quarter of the nineteenth century. She supported her husband's efforts to spread public interest in botany and edited his memoirs and correspondence after he died. Her own intellectual interests, however, were literary, historical, and religious. In her unpublished commonplace books, plants figure as objects of religious and moral contemplation rather than of taxonomic consideration.[22] (As she put it once in a letter to her husband, "You have your delights and I have mine—I am now in raptures with Gibbon's works.")[23]

But the formative influence of a father, husband, or brother with botanical interests is not enough to account for the wide degree of involvement by women in botany in the early nineteenth century. Botanical culture was wide-ranging, and natural history hobbies were abundant for everyone. In addition, however, botany had broad social acceptability as a good area of study for girls and women. It became part of the social construction of femininity.

By the 1790s botany was among the subjects girls learned in middle-class families, and this continued through the Victorian period. An activity that promoted self-improvement and accomplishment, botanical instruction was considered appropriate to a feminine ideal and thus acceptable for young women, wives, and mothers. It had its place in the schoolroom as a way to teach moral lessons.[24] It was a sensible way to prevent idleness. It offered a path to religious contemplation. It was among the activities thought safe for young women to pursue. Victorian advice manuals and other prescriptive material saw self-education as part of how young women should prepare themselves for their future vocations as wives and mothers,[25] and botany was a topic on that agenda. While class-specific, to be sure, botany was in fact welcomed into everyday family life and integrated into the daily round of girls and women from the late eighteenth century well into the Victorian period.[26]

Autobiographical and anecdotal materials from this time show that botany was a versatile resource for some women and was among the central activities of their lives. Emily Shore (1819–39), for example, was already a botanophile by age twelve, able to describe and classify plants according to the Linnaean categories.[27] Her journals are a window onto the thoughts and daily routines of a serious girl in the 1830s whose circumstances of family and health enabled her to pursue intellectual interests. Educated at home, her father a private tutor, she was an assiduous student of Greek and Latin and of natural history. However she came to learn her botany—whether from her father, her more shadowy mother, or a girl cousin—she became a keen observer and collector of plants. By 1836, when poor health began to curtail her activities, she had amassed hundreds of ferns, which she sewed and pasted into portfolios. Journal entries show how much she enjoyed recording botanical names, citing descriptive details, and consulting botanical handbooks.[28]

Lydia Becker's deep involvement with botany dates not from her girlhood, but rather from her middle years. During the 1860s she won a national prize for a collection of dried plants (having devised a system for drying the plants that retained colors well), and she made other public contributions, cited earlier in this essay. Once she embarked on her suffrage work, botany lost its centrality, but remained an important part of her private life. Her biography reports that Lydia Becker's "best refreshment" when political work became too fraught "was a run down to the gardens and conservatories at Kew."[29] In a letter from 1887 containing many botanical particulars, Lydia Becker wrote: "I hope I do not tire you with these details, but plants are my hobby, and when I am mounted I never know when to stop."[30] Her fervor about the place of botany in life was political as well as personal. She was an eloquent advocate of science education for women during the decades when institutionalized schooling was replacing home-based education for the middle classes, and when opportunities for female science education were opening up in some schools and universities.[31] In 1867 she was elected to the new Manchester School Board, the first woman elected to that position in England. In 1869 she argued against sex segregation in teaching women science at the university level.[32]

Botany came to Jane Loudon's life through marriage, and plant work became central to daily life in all that she wrote when she was a wife and a widow. She had been a published writer of fiction before her marriage in 1830 to the most energetic and prolific horticultural writer of the day, but she was neither interested nor proficient in botany. In fact, as she reports in the autobiographical preface to her *Botany for Ladies* (1842), she was antipathetic to Linnaeus's botany as a child. Embarrassed, however, that John Loudon's wife did not know about plants, she embarked on a course of self-instruction, studying the natural system and proceeding from observation of particular beautiful flowers to an understanding of the plant orders. She also attended botanical lectures at the Horticultural Society in late 1831.[33] Her own difficulties of access to knowledge about plants became the basis for the many books she wrote on botany and on gardening.[34] She was adept at finding the right level of language and tone for fledgling students like herself.[35] Jane Loudon's books aim to show women how to develop plant knowledge and how to integrate it into their daily lives. In her case, plant study was part of both her personal and professional life at home. It was family life that guided Jane Loudon into being a successful professional writer of popular botanical and horticultural books in the late 1840s and 1850s, and family finances were often a spur to the projects she undertook.

By the later eighteenth century, botanical writing had become a way for women to earn money. Publishers cultivated the natural history market for young readers and female readers, and women authors produced introductory books for both these audiences. Prominent among the introductory botany books were Priscilla Wakefield's *Introduction to Botany* (1796), Maria Jacson's *Botanical Dialogues* (1797), and Sarah and Elizabeth Fitton's *Conversations on Botany* (1817). Priscilla Wakefield's much reprinted book of 1796 (eleventh edition, 1841) came about in part because of the author's conviction about the benefits of botany for female education and moral improvement, but mostly because of strained family finances in her Quaker household during the mid-1790s. Priscilla Wakefield (1751–1832) saw writing as part of her family responsibilities, and many of her books date from times of specific worry about the financial well-being of her husband, her grown children, and their families.[36]

Mary Kirby also harnessed botanical interests to economic concerns. Her autobiography shows her to have been, like many a young woman of her day, a fervent student of botany, collecting seaweed during a seaside holiday and being much affected by the "botanical mania" in the 1830s and 1840s.[37] She began writing about plants after the death of her father in 1848, partly, it seems, for financial reasons. She compiled the *Flora of Leicestershire* with the help of various male mentors and helpers, but publication of her flora did not lead her into a career as a writer of mainstream botany books. It is unclear whether she found empirical or systematic botany beyond her, whether she was uneasy in a public botanical forum, or whether she neither found nor sought institutional ways of being active in botanical work.[38] During the 1850s she embarked on a career as a professional juvenile writer, collaborating with her sister. Mary and Elizabeth Kirby coauthored many books between the 1850s and the 1870s, ranging from natural history books to moral tales and adaptations of the classics.[39] Her autobiography reflects the personal and professional rewards writing brought her, including her pleasure in earning money. Botany was part of Mary Kirby's private avocation, and she made it part of her public career as a writer as well.

Mary Kirby worked at home (note the quotation from her autobiography at the beginning of this essay), as did Priscilla Wakefield, Jane Loudon, and many other nineteenth-century writers. Home was where natural history enthusiasts of both sexes and various social classes pursued their hobbies, studying plants, consulting handbooks, and corresponding with the like-minded. But home was particularly the workplace for women, and the domestic sphere was the geographical locus for much of women's botany in early nineteenth-century England.

The family was the institution through which women could learn botany during the period under discussion, taught by family networks, and brought into family projects. By mid-century, children increasingly were learning their natural history in school, and adult beginners were benefiting from the spread of institutionalized popular education. J. S. Henslow, for example, taught botany to village boys and girls in a parish school in the 1850s, the same decade in which Elizabeth Twining was lecturing on botany to young women attending women's classes at a working men's college.[40] But from about 1780 through the 1830s science education took place principally at home, and women were central to the educational project.

In the domestic sphere women were both learners and teachers. It was mothers, or mother substitutes, who taught children their early botany at the beginning of the nineteenth century. Botany was part of home schooling, and responsibility for teaching it at the entry level was the mother's. Many introductory botany books were addressed specifically to mothers, to assist them in teaching botany to their children.[41] Introductory books usually taught young readers by using a narrative format of conversations at home between mothers and children, or of letters between sisters, and they set learning into the context of daily life.[42] Much of the biographical and autobiographical evidence amassed in this essay reflects the importance of fathers and male figures in the lives of women who were botanically active. In some cases, mothers are dead or ill; in some cases, mothers are disturbingly elusive. But the literary evidence of the introductory botany books suggests a different story, in which botany becomes a woman-centered activity, with maternal instruction on botany a fundamental part of family life. According to these depictions, botany is part of the family routine, and mother's interest in botany is fully in keeping with her other responsibilities. Botany in fact becomes a part of good mothering. While the narrative pattern of these early books likely reflected historical practice in some families, the prescriptive dimension is particularly noteworthy, for it mirrors a world of social concern well beyond botany alone. We find a call for maternal attention and a restoration of family life, for example, in such late eighteenth-century writers, otherwise ideologically opposed, as Mary Wollstonecraft and Hannah More.[43] The emphasis on maternal instruction in early nineteenth-century England was intellectually enabling. It gave women latitude to study botany, develop expertise in it, and act as cultural intermediaries, passing their knowledge along to others. Botanical motherhood became its own career.

If we are to gain insight into how women's lives and work intersect with the history of British botany, feminist approaches to the history of science will be particularly helpful.[44] How, for example, might a gender-based analysis illumine the life and career of Eliza Brightwen (1830–1906), who wrote natural history books for children and whose autobiography chronicles years as a Victorian neurasthenic?[45] Closer to our own day, how did morphologist Agnes Arber (1879–1960) come to botany, and did she have the feeling for the organism that Barbara McClintock had? Along with research on women whose botanical contributions have been acknowledged, we need archival work to search out women whose contributions may have been poorly reviewed or excluded from canonical accounts of the history of botany.[46] There are prosopographical projects of many kinds to be undertaken on, for example, the first group of women admitted to the Linnean Society in 1904. Local contexts should be traced for women's relationship to botanical study and botanical culture.[47] Women's contributions to science in the nineteenth century come into sharper focus when we enlarge our ideas about scientific work to include activities at the conventional margins of professional and protoprofessional involvement. We should draw, for example, on wider textual resources, such as on popularizations of botany and on botanical writing for children and women.[48] In this way, the history of botany would reflect how many women lived and worked in relation to the science culture of early nineteenth-century England.

As a cultural activity, botany is linked to women's education, work, and lives—in both theory and practice. We need a much broader interpretive purview to understand how women worked *within* conventional social constraints in any period. There is only a limited applicability to the argument that women with botanical interests were mainly passive elements in the exercise of social control and ideological manipulation. Rather, botany can be understood more as a resource for women in ways that have not been sufficiently identified and explored. The student and her teachers may have construed the practice and purpose of botany in her life in very different ways. Feminist biography in science is one way of getting at the place of gender in the definition of science and in the pursuit of scientific projects. Biography from a feminist perspective is partly interested in achievement as defined by a given scientific community; one does want to know about the enabling circumstances or about barriers successfully surmounted. But we also should seek to locate scientific interests within life as a whole—not only in the public realm, but also within the sphere of everyday and domestic life.

This essay has sketched familial dimensions of botanical culture into the mid-nineteenth century. Through family links, some women came to contribute to the public face of botany. But early nineteenth-century social ideals and practices also brought botany into the family domain. In the preprofessional years botanical culture had a strong family locus, and women had pride of place there. In later years, the family locus of botany fades. Narrative mothers disappear from introductory books, and women-centered texts teaching lessons about life are replaced by impersonal, noncontextual expository textbooks. Botanists, their teachers, their fathers, and even their mothers came to prefer the high road toward laboratory work and toward science on a male model. Although some women had enhanced access to mainstream botany in that public sphere, more women probably participated in botanical work when it was in the domestic sphere than when it moved away from home. During the early nineteenth century in England, it was in some ways easier for women to combine family and botany than it became later on, for science was part of family routines, and the interest of girls and women in scientific work was not disjunctive with family life. Seen from the vantage point of the breakfast room, the picture of the place of women and family in nineteenth-century botanical culture is much more thickly textured than we earlier thought.

3

REGINA M. MORANTZ-SANCHEZ

The Many Faces of Intimacy: Professional Options and Personal Choices among Nineteenth- and Early Twentieth-Century Women Physicians

NINETEENTH-CENTURY prescriptive literature, responding to and perhaps even helping to shape alterations in family life resulting from industrialization, gave voice to an elaborate domestic ideology that made woman a powerful symbol for stability in a world that was changing all too rapidly. Domesticity emphasized the importance of the woman-as-mother, urging all women to shape their lives according to the canons of an emerging middle-class ideology. That ideology stressed small, emotionally intense families with specifically delineated roles for men and women. While the public world was reserved for men, the private one, linked to the public by the central importance of female child nurture in creating responsible citizens in a democratic republic, became middle-class women's domain. Yet at the very historical moment when public ideology insisted that woman's place was in the home, aspiring female physicians left the home to study medicine and establish professional careers.

In the 1850s the medical profession was still immature and lacked most of the characteristics it would display seventy years later. Sharing no common intellectual base, it was plagued by competing sects with contending theories of disease causation and treatment. Educational standards were haphazard, sharing only the characteristic that they were uniformly low. Generally unable to control entry into the field, physicians stood by helplessly as "undesirables"—including women—earned a license to practice. Weak and ineffectual, professional societies had yet to develop either a strict professional ethos or a common base of shared ethical assumptions.

In addition, the role of the physician was shaped by a traditional system of belief and behavior understood by both doctor and patient. Sickness was viewed as a total condition of the entire organism—the body in disequilibrium with its environment. The "art" of medicine consisted of the doctor selecting the proper drug in the proper dose to bring on the proper physiological changes. The best physicians knew their patients well, drawing not only on each's history and unique physical identity to make a decision, but also on the family's constitutional idiosyncrasies. Environmental, climatic, and developmental conditions were also important variables to be considered. Thus, the locus of practice was usually the patient's own home, with doctors often treating entire families.[1]

Though this rationalistic framework was labeled "scientific," no honest physician in the nineteenth century would have denied the importance of intuitive factors in successful diagnosis and treatment. As Professor Henry Hartshorne put it to the graduating class of the Woman's Medical College of Pennsylvania in 1872: "It is not always the most logical, but often the most discerning physician who succeeds best at the bedside. Medicine is indeed a science, but its practice is an art. Those who bring the quick eye, the receptive ear, the delicate touch, intensified, all of them, by a warm sympathetic temperament . . . may use the learning of laborious accumulators, often, better than they themselves could do."[2]

At mid-century in particular, when advances in European laboratory science increasingly began to discredit traditional heavy dosing of medicine, the intuitive aspects of medical practice became even more important. Many physicians, believing in the self-limiting quality of most diseases and skeptical of the efficacy of traditional intervention, sought merely to minimize pain and anxiety and "wait on nature." In a society that increasingly emphasized women's superior intuitive and empathic abilities, such a situation made the profession especially vulnerable to the entrance of females.

Never themselves entirely free from the powerful assumptions of nineteenth-century domestic ideology, women physicians created their own version of it. Wedding the belief in special female qualities to traditional assumptions about the role of the physician, they argued that they were naturally suited for the work. And while not directly challenging the importance of women's role in the family, they argued that mothers and wives needed expert advice to perform their roles properly. Women doctors could provide this aid by linking the advances of science to the everyday lives of women. Thus armed with a philosophy that accepted women's central role in family life while rejecting the interpretation of domesticity that barred women from the public and professional realm, thousands of young women studied and practiced medicine in the second half of the nineteenth century.

Despite their career aspirations, however, most women physicians had their own sentimental attachment to Victorian family life. Though they had entered medicine with the idealistic notion that their medical work would make an important social contribution, few could be satisfied with lives devoted totally to that work. Naturally they sought intimacy and connection in the private sphere. And just as they defied social convention and created new public roles for women, so they also exhibited much creativity and energy in the pursuit of more personal goals. This meant struggling to find ways to be creditable professionals without surrendering some of their most fundamental female desires: for some it was for a husband, home, and family; for others, intimate and long-term relationships with other women gave them satisfaction; still others became single parents by adopting children. All of them coped for the first time with dilemmas that have become central to the lives of modern women, the most significant of which was balancing work and family.[3]

The historian can gain much insight into how different women approached this problem by examining a unique collection of stories published in 1897 by the students and alumnae of the Woman's Medical College of Pennsylvania. In *Daughters of Aesculapius*—a volume of fiction and nonfiction—four short stories in particular suggest how various women physicians viewed their personal and professional options.[4]

Probably the most amusing of the four is the piece "One Short Hour." This tale of a medical student forced to choose between her career and the man she supposedly loves was written, not surprisingly, by a graduating student, Rosalie Slaughter. Indeed, the modern reader's amused response was clearly not intended by the author, whose melodrama was deadly serious.

The scene is set in the sitting room of Rachel, the hapless medical student heroine, as she impatiently awaits her fiancé. Obliged to cut a class in order to satisfy his request for a visit, she tries to study her anatomy to assuage her guilt. Soon her young Howard, a professor of English, arrives. After the appropriate words and gestures of love, he haltingly reveals the nature and gravity of the meeting: he wishes her to give up medicine and marry him at once. When she protests that he had promised to support her professional choices, agreeing that "marriage and medicine need not conflict," he demurs, admitting that he can no longer "feel just as I felt three years ago." Indeed, he confesses to her, his best friend's marriage has done much to modify his views: "His home is so cozy, his bride so charming; whenever I call they give me so cheery a welcome that I've gone home to think of how different it would all be if our comfortable chat were interrupted by a clang of the bell and the query, 'Is the doctor in?' . . . I have come to believe that a wife's place is always with her husband—sheltered."

Crushed for a moment, the courageous young woman finds both her voice and her anger: "Men talk about their sweethearts' happiness as the one object in life and all that, but when it comes to testing this nobly generous spirit it becomes quite another story." Refusing to relinquish the fruits of years of hard work or her future professional goals, she suddenly sees her fiancé with new eyes:

> "The choice seems to lie between marriage and medicine."
>
> His eyes answered. The solemn chimes rang out, the clock in the gray college tower pealed forth six. She crushed the rose in her hand, its petals fell among the cups.
>
> "I have chosen, Howard—farewell!"[5]

Providing a proper balance to Slaughter's story of disappointment is another by an alumna of the class of 1883, Dr. Hester A. Hewlings. In "Dr. Honora," Hewlings reassures her readers that there are other men besides the Howards of the world. Dr. Honora, a woman physician who settled in a small town and has had discouraging results since opening her practice three years earlier, is one day contemplating quitting the profession as she walks and muses by the side of the road. Suddenly she becomes witness to a fateful accident—runaway horses that throw both a grandfather and a granddaughter. Rising to the occasion, she immediately and successfully mends the man's dislocated shoulder and is eventually called in as well by the family's physician, the handsome Dr. Bragg, to care for the ailing girl. Only by her vigilance and careful treatment day and night does the little girl live. Having secured her professional reputation with one of the first families of the town, Dr. Honora is rewarded in love as well when Dr. Bragg asks her to be his wife. Agreeing to practice together, her gallant husband admits: "I had a contempt for professional women until I met you, but the calm way in which you handled Jessie's case, and kept yourself through all a sweet, true, modest woman, conquered me. I think now women should enter all the professions and if the ranks become overcrowded let the men step out. It might be a good thing for our profession." Pleased with his sentiments, of course, Dr. Honora requests that he send a written version of them to a leading medical journal "to learn how many of his brothers agree with him."[6]

A third short story in *Daughters of Aesculapius* suggests that there were other options open to women besides being successful or unsuccessful in heterosexual marriage. Indeed, even in "One Short Hour" Rosalie Slaughter's medical student chides her lover because he cannot understand the deep emotional satisfaction she gets from the knowledge that she has made a difference in the lives of her patients: "Ah, Howard, if only once you could hear the fervent 'Thank God, you have come!' that springs to the lips of the pain-wracked one, and could but see the gratitude in her eyes, you would realize that the tear on the hard-lined face and the hope in the trusting gaze are dearer—yes, forgive me—than caresses!"

Dr. Anna F. Fullerton's story, "Mater Dolorosa—Mater Felix, a Sketch from Hospital Life," plays out this theme more fully. In it, the woman doctor is a hospital resident physician, routinely making the rounds of the maternity ward one evening, when she comes upon a beautiful but delicate and tearful patient with a three-day-old baby. Confessing that she had falsely claimed to be married in order to be admitted to the hospital, the girl pours out her life story to the kindly doctor. Orphaned, alone, and "hungering for love and the sweetness of home-life," she had allowed herself to be befriended and ultimately compromised by a young man who was her social superior. Betrayed and forsaken, she knew of nowhere to turn. Refusing out of loyalty to divulge the name of her lover, the defenseless girl throws herself on the mercy of the doctor. Later, the young man comes to the hospital to inquire surreptitiously after the girl's welfare, but the doctor is not fooled. She confronts him with his crime and manages to convince him of his obligation to marry her. Her arguments emphasize the contrasts between female helplessness and masculine strength and the possibility of the girl being "molded" into a happy wife and mother. The story ends several years later with a visit from the girl—now a happy, refined, and elegant matron—to ask the doctor to see her through her second pregnancy. When the young lady thanks the doctor for changing her fate from a tragic to a happy one, the doctor responds by assuring her former patient that it was Providence who chose to do good through the "humble instrumentality of a woman doctor."[7]

Dr. Gertrude Walker's "The Greatest of These Is Love" provides still another personal option for the woman physician who chooses to remain single. In her story, Dr. Helen Brockway has given up a refined life of ease to study and practice medicine. While working at a dispensary for women and children, she is visited by a sick little toddler named Marion and her depraved mother, Mrs. Simmons. Warming immediately to the little girl, she admits the child to the hospital and in the course of the next two months of treatment comes to love her. During this interval, the mother, who berates the child constantly, neglects to visit her offspring even once, and the doctor, intending to "save" the toddler, determines to adopt her. (In her writing Walker expresses all the sentiments of paternalism and noblesse oblige that the middle and upper classes harbored toward the poor.) But before she can settle matters with the parents, the mother pays a visit and in a goodbye kiss on the child's mouth gives the poor little girl pneumonia. Though the doctor struggles desperately, no amount of solicitude can save her, and the child dies. Crushed, the doctor sends word to the mother, only to be told that the woman has moved and left no forwarding address.

Two weeks pass, and the doctor, having finished her residency at the hospital, is preparing to move to Pennsylvania to begin practice. Late one snowy night as she packs, she notices a thinly clad woman laboring up the hospital steps. A few minutes later Dr. Brockway is called to see the visitor and the woman turns out to be Marion's mother. Drawn to the woman because of their mutual connection through the dead child, the doctor invites her to sit by the fire. Slowly Marion's mother explains that as her baby recuperated in the hospital, she herself began attempts to take control of her life. Though from a good family, she had married a shiftless drunkard, who had slowly dragged her down. Six weeks before, he had been killed in an accident, and she had resolved to raise herself up and make a proper home for her daughter. She had moved to the other side of the city and "was tryin' to get a nice home ready for my little girl" when news of the child's death had reached her. Filled with emotion, Dr. Brockway determines to regenerate the mother in place of the child:

> "Mrs. Simmons, will you let me help you? I meant to ask you to let me take Marion to my own home. If her mother will come instead, I feel sure she will find life brighter and happier than the past has been. . . . Will you come and keep house for me?"
>
> "Doctor, I will go to the ends of the earth with you, and I will be your faithful servant."

In this highly emotional and sentimental final scene, the reader is led to understand that the bonds between the two women will last a lifetime.

Though overly dramatic and romanticized, these stories represent in an exaggerated fashion some of the real-life choices made by women physicians as they struggled to balance their personal needs for intimacy with the demands of their profession. For example, let us take the question of marriage. The decision to marry was not an easy one to make. Certainly when a woman decided to study medicine, she was well aware that she challenged conventional definitions of woman's role, even if she believed, as many did, that medicine was naturally suited to female talents. Conventional Victorian marriage neither promoted nor condoned a woman's freedom to pursue personal goals. Thus feminists often pictured marriage as a dangerous impediment to underdeveloped women who ought instead to seek to live and think independently.[8] Dr. Gertrude Baillie, examining the question of marriage for women physicians on the pages of the *Woman's Medical Journal* in 1894, also voiced suspicions that marriage and medicine did not mix. Though she rejected arguments of the alarmists who claimed that educated professional women were ruined for wifehood and motherhood, she believed that the conflict between family and professional roles would eventually prove too grueling.[9]

Some women physicians found Baillie's prediction to be correct. There were examples of women in both the nineteenth and early twentieth centuries who gave up medicine voluntarily when they married, or were forced to do so by their husbands.[10] Yet married women who ceased to practice medicine were distinctly in the minority, and in general the leading female medical educators cautiously endorsed marriage for their students. After completing a survey of the graduates of the Woman's Medical College of Pennsylvania in 1880, Dean Rachel Bodley concluded that it demonstrated that "womanhood of the noblest type can rise to the full possession of all its powers, and yet lose nothing in sweet grace or womanly dignity, lose nothing in love of husband or of children."[11]

Mary Putnam Jacobi, married and the mother of two children, gave the most balanced and sensible assessment of the problem. Ultimately, she argued, it was a matter of individual struggle and adjustment. Marriage "complicated" the life of a woman, even when she did not choose to become a professional, so it was bound to call for creative solutions if she decided to become a doctor. She observed further:

> Many married women will lose all interest in medicine as soon as they have children, as many now fail to develop the full needed interest precisely because they have no other, and are dispirited by isolation from family ties. Many will interrupt their practice during the first few years after marriage to resume it later. Whatever is done, either with or without marriage, can evidently be well done only in proportion as more complete intellectual development and more perfect training enables the woman to cope with the peculiar difficulties in her destiny.[12]

There were enough women physicians in this period who agreed with Jacobi to make their marriage rate disproportionately higher than that of other professionals, though they usually married at a noticeably older age than did their female counterparts in the general population. Between one-fourth and one-third of them married, and in 1900 that rate was double that of all employed women and four times that of women in other professions. Indeed, only in 1940 did female professionals catch up. Though Rosalie Slaughter, for example, despaired of finding a mate willing to support her interest in her career, she married just such a man a few years after having set up private practice in Washington, D.C. In fact, her real-life experience came closer to imitating that of Dr. Hester A. Hewlings's "Dr. Honora" than her own fictionalized heroine in "One Short Hour." In her autobiography Slaughter describes being romantically courted by a young lawyer who had attended two years of medical school. Her husband was delighted to see her continue in medical practice, though they both expected her to keep house, and Slaughter planned to "lay my work aside for a few years" should they have children. After moving from the nation's capital to New York City shortly after the turn of the century, he helped her study for the New York State Board examination and enthusiastically aided her in re-establishing her practice. Although the couple enjoyed several of what Slaughter termed "happy . . . uneventful, but never monotonous" years, their time together was tragically cut short by her husband's sudden death from an aneurism. Slaughter never remarried.[13]

Other examples of comparatively egalitarian marriages abound from impressionistic sources. Mary Putnam Jacobi, married to the well-known New York physician Abraham Jacobi for over thirty years, shared with her husband a devotion to their profession and prominent distinction in the New York medical community, while bearing two children. Similarly, Emily Dunning Barringer, a graduate of Cornell Medical College when it merged with the Woman's Medical College of the New York Infirmary in 1900, married a classmate. Her husband, a surgeon, gave her the kind of support she deemed essential to her success. Recalling their initial encounter, she mused in her autobiography:

> Fate was certainly very kind to me. I have often wondered what would have happened if I had met an average man . . . [instead of] the extraordinary one I had found. . . . I found him the aggressor in interest in my career. He was the one who crossed the boundaries, discussed, evaluated, and encouraged. And he did it, I found, not to please and flatter me, but because he was genuinely thrilled that the woman he wanted to be his wife was capable of that type of mental de-

velopment. My life both medical and personal took on a perspective and depth, color and meaning from that day. . . . Ben's love, pride and enthusiasm was the delicate adjustment needed to bring the machine into perfect timing and I settled down with a deep sense of power that brooked no opposition.[14]

Although a significant number of women physicians married men who were themselves doctors, by no means all of the successful marriages recounted in their diaries and letters were unions of two doctors. What these husbands did have in common, however, was a willingness to deviate from the classic patriarchal Victorian ideal. Interested not only in their wives' work, but in the full development of their talents, such husbands took pride in their spouse's achievements and showed a willingness to aid both materially and practically. Thomas Longshore, a teacher and philosopher of religion whose zeal for social reform, abolition, and women's rights was well known, was one such man. When he married Hannah E. Myers in 1841 she had already expressed a desire to study medicine. Although financial considerations forced her to postpone her plans for several years, even the birth of two children did not deter her enrollment in 1850 in the first class of the Woman's Medical College of Pennsylvania at the age of thirty-one. Longshore encouraged his wife throughout her long medical career. Hannah attended classes with her sister-in-law Anna, who came to live with the family and helped with household chores. Their daughter remembered that "Aunt Anna studied medicine at night and father helped." He, in fact, was "very instrumental in urging them on. He hunted all the notes for Dr. Longshore's lectures and wrote them for her." Later on, "when Dr. Longshore got busy," Mr. Longshore kept the books and compounded her medicines. Similarly, the son of Sarah Cohen, a graduate of the same school in 1879, remembered: "[My father] was quite proud of my mother. When she was teaching in medical school, she would dictate her lectures and he would copy them down in longhand for her."[15]

Though many women thus found the challenge of combining marriage and professional life exhilarating, others confessed to more difficulty. Dr. Pauline Stitt, a 1933 graduate of Michigan Medical School, never had the chance to try: her fiancé, also a medical student, threw her over for a nurse in a cruel and heartless manner. Stitt, much like the plucky Rachel in Rosalie Slaughter's short story, responded only with more determination and zeal to pursue her profession. Though she realized that her fiancé had been threatened by the possibility of competition with his wife and forgave him, she never married. In contrast, Mabel S. Glover, a promising Wellesley graduate and one of the three women to enter the first class at Johns Hopkins, fell in love with the school's new young anatomy professor, Dr. Franklin P. Mall, and gave up medicine to become his wife. Years later she wrote: "Dr. Mall always insisted that he made up his mind that first day that he was going to marry me as soon as possible." A year or two later Edith Houghton became engaged to Donald Hooker, a fellow classmate and later professor of physiology at the school. She, too, abandoned medicine. In 1916 Ernestine Howard, another Hopkins student, wrote to her parents in some distress about a favorite fellow classmate, Irma Goldman, who had recently become engaged to a man she had known for a very short time. The psychological stress had become unbearable for Howard's friend and she feared that "Irma . . . is on the way to flunk out." Goldman began to cut classes and became so worked up over Adolph Meyer's clinics in psychiatry that she began to believe that "she has a psychoneurosis." Howard continued: "All that in addition to getting engaged . . . is too much for her. If she'd attend to business, I think she'd be all right, but the little idiot hasn't got sense enough to do that."[16]

Dorothy Reed, of the Johns Hopkins class of 1900, was determined to have marriage as well as a medical career, but like many other medical women, the cost to her was dear. Drawn into an unhappy but passionate love affair during her last year at school, the experience brought her enormous emotional strain and led to her decision to leave Baltimore and take residency training in New York. A few years later she married a childhood friend, Charles Mendenhall, a professor of physics at the University of Wisconsin. Her diary suggests, however, that her decision to marry came less out of profound romantic attachment than out of a wish to share her life with a congenial mate she could respect and with whom she could raise a family. Although she never abandoned her medical career, she did subordinate it to those goals. Nevertheless, she accomplished a great deal working for the Children's Bureau in the area of public health and preventive medicine. But her brilliant promise as a student in William Welch's laboratory at Johns Hopkins, when she identified and isolated the "Reed cell," used in diagnosing Hodgkin's disease, seemed to her teachers to have been betrayed by her subsequent choices, and she was viewed by them as an example of "an able woman who had married and failed to use her expensive medical education." Mendenhall responded to their disappointment with bitterness.[17]

Marriage, sometimes successful, sometimes not, was not the only alternative open to women physicians as they sought intimacy in conjunction with an active career. Indeed, in spite of the comparatively high numbers of those who did marry, the majority of them did not in the late nineteenth and early twentieth centuries. How close were the life choices of this group to the patterns and possibilities held out by the fiction in *Daughters of Aesculapius?* Where did single women find relationship and connection?

For some women who chose to remain single, the decision exacted its price in loneliness. Anna Wessels Williams, a successful bacteriologist who graduated from the Woman's Medical College of the New York Infirmary in 1891, wrote in her diary of working hard to develop a "detachment from all disturbing longings" that she believed essential to being a good physician. She rejected marriage for herself, but not without some sadness. Unsure of herself in all relationships, intimate friendships of any sort came hard to Williams. In 1908 she recorded in her diary: "I was told today (by A.) that it was quite pathetic to think that I had no one particular friend. It's too true and tho it's probably largely my own fault, yet I do not know that I wholly regret it—considering the life I must lead."[18]

Similarly, Clelia Mosher, a Hopkins graduate who became resident physician for women at Stanford University, was plagued by loneliness for most of her professional life. Her notes and unpublished fiction suggest that she felt intensely the conflict between the needs and wants of the independent-minded professional woman, society's prescriptions, and her own romantic longings. In the notes for one story about a beautiful and accomplished woman, she wrote that she must bring out the "struggle in the woman's own soul" between her right to intellectual development, her "overwhelming and passionate love for a man who is her ideal, . . . the claims of her motherhood," and the fact that in "yielding to her love [she] loses some of her fineness."[19]

The pursuit of self-development not only produced loneliness, but guilt as well. Aware that the conventional role for women was to live for others, Harriet Belcher, an 1879 graduate of the Woman's Medical College of Pennsylvania, self-consciously caught herself after a lengthy and enthusiastic letter to a friend describing her life at medical school. "This letter is 'ego' from first to last," she apologized. "Well, I can't help it, in these days I am wrapped up in myself to a most ignoble extent, but you who are living so *in and for others* write soon to tell me all the news." A year later she commented revealingly to the same friend: "What a family you and Mary have on your hands, my dear! And yet you write as if you do not consider that you are doing much. Why, it seems to me a very heavy charge. . . . Thus far my professional life . . . has been far less onerous to me than my old housekeeping days."[20]

Guilt did not stop Harriet Belcher from relishing her work and her life as a single person, however. Like many women physicians, including the central character in Anna Fullerton's "Mater Dolorosa—Mater Felix," she viewed her relationship to her work as a kind of marriage. She eagerly wrote to her intimate friend Eliza Johnson how much she wished the latter could be present at her graduation, which she termed "my 'wedding day.'" Similarly, Harriet Hunt, who often spoke of herself as being wedded to medicine, celebrated her silver anniversary in the summer of 1860, after twenty-five years of practice. Finally, Emily Pope, a resident physician at the New England Hospital for Women and Children, insightfully labeled the institution the object of its founder, Marie Zakrezewska's, "most intense affection, the child of her prime and of her old age."[21] Indeed, Fullerton herself chose a form of marriage to her profession—missionary work in India. For much of her career she shared her personal life with other single female missionaries in a community of women that in some sense provided rewarding substitutes for the joys and sorrows of the conventional family.[22]

Other single women devoted more time to their private lives. Like Gertrude A. Walker's heroine, Dr. Helen Brockway, indeed like Walker herself, who remained single but extremely active socially and professionally, they satisfied their desire for intimacy by establishing relationships with other women, adopting children, or both. Both the Blackwell sisters and countless others adopted one or more youngsters and raised them to productive adulthood. Elizabeth Blackwell's Kitty, an Irish orphan, was a cross between a daughter, a servant, a companion, and a wife, playing different roles for Blackwell in different stages of their lives together. Cordelia Greene, a member of the first group of women to graduate from Cleveland Medical College in 1856 and the proprietor of her own "water cure" (the health spas of the nineteenth century), adopted six "offspring," all of whom called her "mother." When her children married she continued to involve herself in their lives and in the lives of her grandchildren. Similarly, the University of Michigan's first dean of women, Dr. Eliza Mosher, adopted a daughter from among the prisoners when she was the superintendent of the Massachusetts Reformatory for Women at Sherbon in 1877. Although little is known about her relationship with the girl, we do know from her letters that she took enormous pride in her daughter's accomplishments. Regarding the adoption, Mosher once wrote to her sister that were she never to achieve anything more, "I shall feel as if my life has not been in vain." Thus Dr. Brockway's decision to adopt the little Marion in Walker's story was a commonplace occurrence in the real lives of women physicians.[23]

Equally satisfying were strong ties with other women. Helen Morton, for many years an attending physician at the New England Hospital for Women and Children, retained a close relationship with Mary Elizabeth Watson, even after Watson became Mrs. John Prentiss Hopkinson. Morton's letters to Mary Elizabeth Hopkinson are full of expressions of love and emotion, for example a typical note ends: "Dear heart Good night. Sleep. hang, heavy on my eyes & tomorrow will be busy and it is now near 12. Your H.M." It is clear that Morton felt herself able to share in some significant "female" events through Hopkinson's agency. She often wrote to Hopkinson of those of her patients who were babies: "I've got a beauty of a baby on my list. . . . The daintiest piece of perfection I ever saw. She's etherial [*sic*] but she won't fly away. . . . I wouldn't miss seeing her for anything." Morton delivered both of Mary Hopkinson's daughters and sent her friend a long poem on the birth of her son. "I wonder" she wrote, "if you ever could know how I envy you your beautiful children. . . . You know I'm glad to hear all you tell me about your babies."[24]

Again imitating the fiction in *Daughters of Aesculapius,* women physicians, like Dr. Helen Brockway, frequently formed lifelong relationships with other single women. Often two women doctors lived together, practiced together, and shared work, leisure, and various degrees of emotional commitment, although the relationship was not always between two physicians. For example, Anna Fullerton lived with a fellow missionary, Alice Hamilton, in a vibrant and supportive female community at Jane Addam's Hull House, and Harriet Belcher lived with a friend who kept house for her. Some of these bonds resembled marriages in the degree of closeness and mutual sense of obligation. Lilian Welsh met Dr. Mary Sherwood soon after Welsh graduated from the Woman's Medical College of Pennsylvania. In 1889 the two spent a year together studying pathology in Zurich, returning to Baltimore in 1890 to live, practice, and work together for the next thirty years. The relationship ended only with Dr. Sherwood's death in 1935. Similarly, Emily Blackwell, Elizabeth's sister and dean of the Woman's Medical College of the New York Infirmary, enjoyed an equally lengthy and rich relationship with Dr. Elizabeth Cushier, a surgeon who also taught at the school. In a letter to Elizabeth Blackwell in 1888, Dr. Mary Putnam Jacobi commented about Cushier: "She is in fact a remarkably lovely woman, spirited, unselfish, generous and intelligent. I do not know what Dr. Emily would do without her. She absolutely basks in her presence; and seems as if she had been waiting for her for a lifetime."[25]

Many of these relationships were possibly homosexual. Others, like that of Eliza Mosher and her partner Lucy Hall, were less intense and might best be categorized as one of mentor and novitiate. Nevertheless, whatever their configuration, such creatively diverse solutions to the problem of loneliness and the hunger for human connection illustrate the prevalence of a wide spectrum of options for women.

The correspondence between Elizabeth Clark, an intern at the Woman's Hospital of the Woman's Medical College of Pennsylvania in 1910, and her friend Ada Pierce reveals the rich intimacy achieved by many women doctors who lived together. Clark shared the home of an older physician on the faculty, Dr. Emma Musson, and letters to their mutual friend Pierce, written between 1910 and 1913, are strikingly descriptive of a world in which men were generally absent but not much missed. Clark's newsy missives evoke a vivid picture of her medical work even as they recount everyday experiences, revealing a life happily balanced in work, love, play, and occasional disappointment. These women clearly knew how to enjoy themselves: the letters bristle with self-mocking irony, gentle humor, good-natured loyalty and good times. Musson, for example, who was nicknamed "Saint Juliana" by her friends, wrote to Pierce in 1910: "E is a joy to one's soul and a constant source of delight." Dr. Musson, in turn, called Clark "Izzie." Several other women physicians completed this lively circle of friends, but the primary relationship remained that between Musson and Clark. When Musson died of pneumonia in 1913, Clark wrote to her friend Ada: "My old heart is clean gone out of me forever & forever."[26]

THROUGHOUT THE 1850S the *Boston Medical and Surgical Journal* published several articles opposing the entrance of women into the medical profession. "The proposition that women, as a sex, cannot practice medicine—that their weak physical organization renders them unfit for such duties and exposures—that their *physiological condition, during a portion of every month,* disqualifies them for such grave responsibilities," argued its editors in 1856, "is too nearly self-evident to require argument." Beside being physically weaker, "nervous," and "excitable," the *Journal* believed that the duties of marriage were incompatible with those of the physician, while a single woman choosing not to marry was not only "unnatural," but a social and moral wrong.[27] Women physicians contended with such antagonistic social attitudes throughout the nineteenth and into the twentieth century. But the real-life stories—with or without marriage—belie both their own fears and those of their detractors that those first generations of professional women were maladjusted. Indeed, the evidence suggests the reverse. In spite of enormous obstacles to their success, women doctors in this period pioneered not only in proving that women could become creditable physicians, but also in finding creative solutions to balancing work and family —solutions that younger generations of women are still struggling to follow.

4

MARIANNE GOSZTONYI AINLEY

Field Work and Family: North American Women Ornithologists, 1900–1950

ORNITHOLOGY HAS an unusual status among the twentieth-century sciences. In ornithology, as opposed to most other disciplines, the nineteenth-century British amateur tradition is still alive, and because of its incomplete professionalization, paid positions for scientists of either sex remained scarce until the 1970s.[1] The measure of scientific activity and excellence, therefore, has not been confined to those with paid employment. Rather, recognition for contribution to science came with publication in prestigious ornithological journals.[2] Thus it is very difficult to arrive at a precise figure of how many women were involved in various ornithological activities. Many turn-of-the-century zoology teachers taught ornithology as part of a general zoology course. Their numbers are hard to ascertain, because most of them did not publish in ornithology journals. Women also assisted family members in ornithological research, and hundreds took part in large-scale bird-banding studies. Most of them never published papers under their own name, however.[3] The slow growth of ornithology before 1950 meant that only a few women ornithologists found paid employment; they worked at the Biological Survey, natural history museums, and colleges.[4] As in other sciences, marital status was a hindrance for women, although not for men; and women ornithologists, even those with a doctorate, were expected to resign their post when they married.[5]

Despite the lack of employment opportunities, women and many men who by their social or financial situation were forced into other occupational niches have been contributing to ornithology. In contrast to other sciences, such as astronomy, however, there has been no separation of tasks, resulting in what could be considered "women's work."[6] On the contrary, in spite of their lack of institutional position, women ornithologists conducted research that placed them among the most respected ornithologists of the twentieth century. They were among the pioneers of such diverse fields as taxidermy, life history, and population biology, and their research has been published in prestigious journals in North America and Europe. The contribution of women ornithologists has been acknowledged by their male colleagues in both scientific and historical articles. *A Centennial History of the American Ornithologists' Union* (1987) discusses their role in that scientific organization.[7] Since no comprehensive history of American ornithology exists, however, a short introduction to issues and research fields within the discipline is necessary to situate the experiences of women ornithologists in the general context of ornithological research.

Until the end of the nineteenth century ornithology was a branch of natural history concerned with the study of classification, distribution, and migration of birds through the examination of museum specimens. For most of this period, North American birds were collected by transient naturalists (explorers, army and navy personnel) and studied by European scientists. By the second half of the nineteenth century, increased leasure time for scientific activity and the economic importance of birds as food and as destroyers of insect pests in agricultural areas prompted new fields of inquiry. The establishment of several natural history museums—the California Academy of Sciences (1853), the U.S. National Museum and the Museum of Comparative Zoology at Harvard (1859), and the American Museum of Natural History (1869)—paved the way for large-scale collection of birds, which provided study material for research on continent-wide patterns of bird migration, distribution, food habits, and evolution.

Until the early twentieth century most ornithologists were self-educated specialists who acquired expertise both in the field and as apprentices to other ornithologists in museums.[8] There were so few natural history museums, however, that employment opportunities were rare and most ornithologists in North America and Europe were forced to contribute to the science as nonprofessionals. An analysis of the occupational status of the founders of the American Ornithologists' Union (AOU) in 1883 shows that in this discipline scientific excellence was not a measure of professional status.[9]

In the late nineteenth century ornithology began a transformation. An increased awareness for the need of conservation, the development of better field equipment (prismatic binoculars and relatively portable photographic cameras, for example), together with improved road networks to previously inaccessible areas, facilitated the study of living birds. Instead of measuring and comparing museum skins, ornithologists began to observe, note, and photograph aspects of the life history (including behavior) of birds. Since such data was practically nonexistent, long-term studies of the birds were needed and this provided a new wide-open research area for both male and female nonprofessionals, since the few scientists working at government institutions or in academia rarely had the time to carry out extensive field work, particularly in early spring at the beginning of the breeding season.

In this essay I will discuss the career and home life of three North American women ornithologists who took advantage of the opportunities provided by an incompletely professionalized science to pursue research as independent scientists. Although Althea R. Sherman, Margaret M. Nice, and Amelia Laskey were among the best women ornithologists of the twentieth century, their lives and careers still await full-length biographies. They followed in the footsteps of pioneer naturalist-ornithologists Grace Anna Lewis (1821–1912), Martha Ann Maxwell (1831–1881), and Florence Merriam Bailey (1863–1948), whose lives and scientific contributions have recently been documented.[10] Sherman and Nice are of British background, and Laskey was the daughter of German immigrants. All three were well educated, disliked housework, and resented interruptions of their scientific work. Sherman, born in 1853, is a good example of late nineteenth-century professional women who did not marry. We know nothing of her personal life in the early period, but in midlife she lived with a female relative. Nice was married and had five daughters; Laskey was married, but had no children. Her marriage retained vestiges of old attitudes concerning sexual divisions of labor. In contrast, Nice's marriage gradually became an example of what late nineteenth-century feminist leaders advocated: a union in which "marriage and practical life work [could be] reconciled by cooperation and organization."[11] Regardless of their marital status, these women ornithologists relied much on feminine friendship for intellectual, emotional, and moral support.[12] Although they received recognition from male colleagues, not all of them felt that they fully realized their scientific potential. The emotional price paid by women ornithologists to maintain independence as persons while combining scientific activity with family life remained high well into the twentieth century.

THE OPPORTUNITY to study birds came to Althea Rosina Sherman (1853–1943) in midlife. Born in Iowa of pioneer stock, both Althea and her older sister, Amelia, were college educated, Amelia in medicine and Althea as an art teacher. Nature provided many subjects for her art, and she was planning on becoming a professional artist. The serious illness of her parents changed her future; in 1895 she returned to National, Iowa, to care for them.[13]

In the early years of the twentieth century Sherman's interest in nature became scientific. Freed from the care of elderly parents, this fifty-year-old woman put her considerable energy into an investigation of the nesting biology of birds. Now paper and pencil were used to record detailed field observations, as the art specialist's trained eye observed, compared, and evaluated details of the life history of many species of birds nesting at what she called her "Iowa dooryard." She was particularly interested in the life history of cavity-nesting birds, such as woodpeckers, of which very little information was available at the time. Sherman's investigations dovetailed with other studies on the life history of birds then being conducted in North America by a growing number of ornithologists. Most early publications, however, were the result of only one or two seasons of observation. Long-term investigations of migration and breeding biology of birds, including courtship, nesting, and parental care (such as feeding and nest sanitation), required patient, careful, repeated observations through several seasons in a certain environment—and different geographic races even within the same species were known to exhibit differences in nesting biology.

Sherman was well situated for such studies: her Iowa dooryard, once prairie land, contained a variety of habitats, where many species of birds nested year after year. Her eye was trained for observations of minute differences and discrepancies. This, combined with her keen intellect, energy, and determination, provided the needed ingredients for successful scientific research. One of her friends wrote: "[In] her bird laboratory at National, Iowa, . . . she worked as regularly as a businessman goes to his office, but her hours were not fixed and vacations were not on the schedule. Her work was independent and thorough in a laboratory that had no duplicate."[14] In order to keep abreast of current ornithological research, Sherman joined the AOU in 1907 as an associate; five years later she was elected a member, testimony to her outstanding work at a time when there were still very few women members in that predominantly male organization.[15] Realizing the importance of exchanging ideas with other ornithologists, she attended scientific meetings, where she presented papers and exhibited her bird art. Her studies on cavity-nesting species (e.g., "At the Sign of the

Northern Flicker," *Wilson Bulletin* 22 (1910):135–171, and "Nest Life of the Screech Owl," *Auk* 28 (1911):155–168), were rapidly accepted for publication by the editors of the *Auk* and the *Wilson Bulletin*.

Despite the acclaim she received from the scientific community and though few paid researchers would have had the time to do as much research as she had done in thirty-five years, Sherman was not always satisfied with her own accomplishments. She had to overcome many distractions, such as visitors, household duties, and her domineering, exceedingly frugal sister, all of which, she felt, hampered her research. Like other women scientists, she had clearly defined priorities. She wrote to Margaret M. Nice: "I work all the time and get almost nothing done, always was *slow* and old age has not quickened my motions. As you are aware I waste no time on the styles and follies of life."[16] To keep up with the work of other scientists, she read at odd times, "when eating, while combing my hair, and when resting."[17] Careful management of her time was necessary for many reasons, such as the old-fashioned nature of the homestead, with outdoor plumbing and a windlass to draw water, since Amelia Sherman did not want to pay her share to modernize the home. Domestics were hard to find in rural Iowa, and Althea Sherman did much of the household drudgery herself, but she couldn't or wouldn't pay for all repairs on her own. "I live in the country which is a matter of my own choice," she told Canadian ornithologist William Rowan (1891–1957), "but the squalor all about me is not my choosing."[18] In another letter she wrote: "My slavery to work is because I own the house with a lazy . . . domineering miser. I should have left twenty years ago. . . . Should have done so if it had not been for the birds and the good chances they gave me to study."[19]

For independent women scientists field research had to be integrated with the demands of home life and gardening, particularly at the best possible times for migration and nesting studies. Winters were spent on writing up notes, preparing articles for publications, attending conferences and catching up with correspondence. Sherman had other difficulties, such as frozen pipes and icy roads. Living in the country, where there was little of the ready-made clothing found in most American cities, she had to do much of her own sewing, otherwise she found that "my clothing, both outside and underwear is becoming so threadbare that I should be soon des-

---

Opposite: Althea Sherman's chimney swift tower. She has her hand up to her head and her sister is looking up. Early 1920s. Courtesy of North American Women Ornithologists, 1900–1950.

THE CHIMNEY SWIFT TOWER

titute of covering if I did not sew." All the while, she rebelled "against spending time on menial tasks" that she felt hired help could do.[20]

One of Sherman's best-known studies was on chimney swift (*Chaetura pelagica*) nesting biology. In 1915, hoping to attract this species to her "laboratory," she had a 28-foot-tall tower built on the property. There was an artificial chimney inside, with windows and observation holes permitting unobstructed viewing of different parts of the chimney. From 1918, when the first swifts remained to nest, to 1936 she spent long hours every night of the breeding season observing their activities. She filled more than four hundred pages with careful documentation, which included spring arrival time, egg-laying schedule, incubation period, and care of the young. She even kept notes on the number of visitors, and although she deplored mere curiosity seekers, many of her friends were privileged to see her swifts and the work in progress.[21] Children were always well received, since she hoped to foster their interest in nature, but she dreaded visitors, such as Professor Lynds Jones (1865–1951) of Oberlin College, who for many years conducted field trips for his ornithology and ecology classes at the Sherman homestead. Raised to be hospitable, Althea felt that she was expected to cook for Jones and his students. Her sister may have argued with her over the expense of feeding so many people. After one such visit she complained:

> I several times told Dr. Lynds Jones that I was not fit to cook and wait on a crowd of strangers, and I have everything to do unaided, yet if he wanted to spend a day here with his class he might do so, but I could promise nothing. Of course if one can crawl at all one would have food ready for such tramps. The upshot was that Jones in a party of TWENTY ate three meals here. . . . I should have had some time for conversation instead of working like a common drudge.[22]

Althea Sherman's productivity (she studied each of forty species for many years) was achieved at considerable personal cost. By the mid-1920s she had to give up going to scientific conferences and restrict her personal correspondence. Constant friction with her sister left her bitter and exhausted. Unfortunately, no records remain to indicate to what extent these two women shared household duties. Neither do we know whether either ever contemplated marriage. Attractive, educated, strong-minded, brought up at a time when careers and marriage were considered incompatible, they may not have wanted to spend their productive years bearing and caring for children. Yet some of Sherman's letters to Margaret Nice, written when she was an old woman, speak wistfully of having daughters, as Nice did.

Sherman, wrote Margaret Nice, was a woman of "extraordinary vitality and ability to work" and was well into her eighties before she gave up ornithological research.[23]

"OUR HIGHLY educated gifted women have to be cooks, cleaning women, nurse maids. We who cherish things of the mind should . . . strive earnestly to give such . . . women a chance to make the highest contribution to society of which they are capable."[24] Margaret Morse Nice (1883–1974) wrote these words about Althea Sherman, but the sentiments reflect her attitude toward her own life and work. Nice was a trained zoologist, and this, together with lack of parental support in her original quest for a professional career, led to stresses and frustrations not encountered by other independent women ornithologists. Nice's early socialization forced her to exchange a promising scientific career for marriage, because parental pressure had been against her continuing her education beyond a certain level.[25]

She was born Margaret Morse, in Amherst, Massachusetts. Her mother was a graduate of Mount Holyoke Seminary and her father taught history at Amherst College. Margaret's happy childhood, spent among wildflowers and birds, was followed by frequent bouts of depression in her teens as she increasingly resented her parents' old-fashioned, overprotective attitudes. At a time when household help in New England was "plentiful and inexpensive, while labor-saving devices were few," the Morse sisters did very little work at home.[26] The parents did not believe that "their daughters should prepare themselves for professions," and they held up the ideal of "perfect housekeeper and homemaker" before them.[27] It seemed a dreary prospect, and Margaret and her sisters wished they were boys, able to choose exciting careers.

Training for the future, to become a better wife and mother, included getting a basic college education, and in 1901 Margaret Morse enrolled in Mount Holyoke College. Although the college years were an inspiration to her, affirming the importance of "things of the mind," she almost lost her interest in nature because the science courses bored her. Ornithology, as in most American colleges, was taught as part of the zoology course of studies and consisted almost entirely of the identification of dead specimens. Sixty years later Margaret wrote: "I could see very little connection between the courses in college and the wild things I loved. . . . I did not like to cut up animals."[28]

Morse graduated with no plans for earning a living; this was not required of her. She knew what she did *not* want to do: teach or conduct research in a zoology laboratory. In 1906 she became a "dutiful daughter-at-home as my parents wished."[29] After several attempts to escape her aimless life, she convinced her father to allow her to enroll in Clark University in Worcester, Massachusetts, in the fall of 1907. There she studied the food habits of the bobwhite quail (*Colinus virginianus*), a topic of great interest to both economic ornithology and entomology.

The research orientation of Clark University impressed Morse and she found that "the world was full of problems crying to be solved."[30] She also enjoyed the friendly and stimulating atmosphere and the easy companionship of men and women. Joining the AOU, she attended its annual conference in Cambridge, where, as she later described, "the gentlemen of the Union" were invited to a reception held in the private museum of William Brewster, while the ladies were separately entertained. "Evidently," she observed, "the role of the ladies was expected to be largely ornamental."[31]

In 1909 Margaret married physiology instructor L. Blaine Nice (d. 1974). Although she wanted to continue her research on the bobwhite for a Ph.D., the lack of encouragement from her family changed her plans. Her parents were happy to see her married, because they thought she would give up her career plans. What Blaine Nice thought is not recorded. In her old age Margaret Morse Nice recalled: "Sometimes I rather regretted that I had not gone ahead and obtained this degree."[32] At the time, however, the full impact of her changed life-style had not dawned on her. She enjoyed her independence, having her own home and time to write up her research results. Housekeeping was not yet a chore; her method consisted of "efficient preparation of good, simple meals, scalding water instead of dishtowels, sending out the washing and ironing, and dispatch in the matter of cleaning."[33] This way she had time for her intellectual life.

In 1913 the Nices moved to Norman, Oklahoma. A college student helped with dishwashing and baby-sitting their two daughters, which gave the young couple free time to explore the Oklahoma countryside. Recalling the psychology and language courses she had taken in college, Nice began to study the speech development of her daughters and their friends. By 1918 she had four daughters, aged six months to eight years; lived in what seemed to her "cramped quarters," without transportation, household help, or free time; and at thirty-five felt truly frustrated.[34] Whatever scientific work she was able to accomplish did not satisfy her. Moreover, the lack of a doctorate was a hindrance to publication and recognition in psychology, although as she found out later, not in ornithology. She resented the implication that her family had brains but she had none: "He taught; they studied, I did housework."[35] Her spirit was "smothered" for lack of continuous intellectual activity.

In 1919 her life changed dramatically. A secondhand car gave her increased mobility and she had the opportunity to do some field work. This was prompted by a newspaper article on the nesting season of the mourning dove (*Zenaida macroura*) which, she thought, contained inaccuracies. Nice decided to verify the facts and rediscovered her fascination with nature and the scientific problems it provided. To help her study the birds of Oklahoma she bought F. M. Bailey's *Handbook of the Birds of the Western United States.*[36] She rejoined the AOU, initiated correspondence with several ornithologists, and started feeding tests on captive birds. The discovery of the Oklahoma avifauna as a research area gave her life new purpose and filled her with excitement and joy. Her husband participated in some of the work, and retaining her interest in child development, she involved her daughters in caring for injured birds and climbing trees after birds' nests.

In her autobiography, *Research Is a Passion with Me,* Nice dealt separately with her scientific life and her personal life. Although her changes of study locations were dictated by her husband's professional moves, she hardly discusses her relationship with him.

In 1927 the family moved to Ohio, where she had access to a new outdoor laboratory, Interpont, an abandoned tract of land that became the study site of her long-term investigations of the song sparrow (*Melospiza melodia*). By 1930, working on 30 acres of land, she was banding individual sparrows and studying their nesting territories, song repertoire, song learning, and courtship season after season. Her painstaking research showed the meticulous scientific approach she had demonstrated in her graduate research and her studies on speech development. This, together with lucid evaluation and great care in verifying data published by other ornithologists, became the hallmark of her work.

Nice's excellent research was soon recognized by the larger scientific community, and European and American ornithologists praised her work. In 1931 she was elected a member of the AOU; in 1937 she became the first woman president of the Wilson Club. In spite of this she was excluded from a local organization, the all-male Wheaton Ornithological Club of Columbus, Ohio. Some male ornithologists, possibly unaware of her graduate training, or noting only her lack of professional position, referred to her work as that of an untrained housewife. This never ceased to annoy her, and any reference to it elicited an instantaneous and spirited response: "I am not a housewife. I am a trained zoologist."[37]

Lack of immediate scientific colleagues was a problem Nice shared with other independent researchers, male and female. As book review editor of *Bird Banding,* she had access to much of the ornithological literature published in Europe and in North America; and attending conferences, visits, and correspondence with peers assured the exchange of ideas. Nice was at the center of a network of women ornithologists whose scientific correspondence also served as a support system. She encouraged their research and publications, discussed their family problems, and exchanged information about other ornithologists.[38]

From her autobiography and her correspondence it is obvious that once Nice found an outlet for her research passion, she combined her scientific investigations and family life very well indeed. She did not believe in wasting time on menial tasks. Instead, she spent time on research and on associating with people she loved. Family reunions were considered "fun," although possibly these were arranged to suit her schedule. She had close relationships with her sisters and other members of her extended family. One of her daughters recalls that Nice always had time to talk to her children.[39] Blaine Nice remained a close companion, but after their initial study of the Oklahoma avifauna, he did not participate in her research. Moreover, Margaret Nice did not find satisfaction in being a nameless assistant or a participant in a two-person single career.[40] She and her husband followed their own scientific interests, while sharing equally the responsibilities of family and household.

Margaret Nice's ability to set her priorities and concentrate on what was essential for her research contributed to her success as a scientist. Forceful in writing, she was quiet in person, but could be fiery when defending her position. This characteristic was doubtless behind her failure to be elected to the post of editor of the *Auk* in 1942. Although she was nominated and had many supporters, the final decision was: "Mrs. Nice would make a good Associate Editor [but] we can scarcely pick a woman editor for the Auk."[41] Although the conservative AOU did not give her this post, in 1937 she was elected a fellow of that organization, and in 1942 received the Brewster Award.[42] A subspecies of the song sparrow was named *Melospiza melodia niceae* after her. All these accolades pleased her, as did the receipt of two honorary doctorates (Mount Holyoke in 1955 and Elmira College in 1962). She was very proud when in 1952 she became the first woman ornithologist to have a club named after her.[43]

NOT ALL women ornithologists were as successful as Nice in combining scientific activity with their "home duties." Amelia Rudolph Laskey

(1885–1973) could never completely suppress her early socialization, the object of which was to make her into the perfect wife and homemaker. The daughter of German immigrants, she was born in Bloomington, Indiana, and brought up in Chicago. Her father was a businessman; her mother kept house and spent much of her time gardening. After high school Amelia worked as a stenographer, but there is no evidence that she continued working after her marriage to F. C. Laskey in 1911. When the couple moved to Nashville, Tennessee, in 1921, Amelia began to develop her four-acre garden into a "natural home for wild things," including birds.[44] She spent her first few years in Nashville as a member of garden and bridge clubs and the local literary society, a not unusual life-style for white middle-class women with no children who had domestics to handle the housework.

Laskey joined the Tennessee Ornithological Club in 1928. Soon she began her bird observations in an outdoor "laboratory" consisting of the open fields, meadows, and wooded parks around her property. According to co-worker and friend Katherine Goodpasture, "Her study of birds matured and became an obsessive mental pursuit, an all consuming force that pervaded her life in later years."[45] During the next forty years she banded thousands of birds (using bird traps in her garden), contributing information to the continent-wide migration investigations of the U.S. Fish and Wildlife Service; she also studied the detailed life history and comparative behavior of captive and wild birds. Such investigations were begun in Europe in the early twentieth century by Oskar and Magdalena Heinroth. Few other ornithologists pursued this new line of research. By keeping injured birds, along with others she reared from nestlings to adulthood, Laskey was able to observe in detail their territorial and other behavior and compare this with similar behavior observed in wild birds. Captive birds were housed in large flying cages in the garden and in smaller ones in her study; a few could fly around parts of the house. Looking after these, while keeping her house clean, a necessity for her, resulted in extra work and entailed considerable "watchfulness"; it could only be done by restricting birds to "the service part of the house where there are no carpets or upholstered furniture."[46]

Laskey was known to local children as the "lady that fixes birds," and each spring and summer numerous injured and orphaned birds were brought to her home; some of these she tried to return to their nests, others she looked after herself. One year she had three chimney swifts in a makeshift chimney in her kitchen, and for a week she fed them by forcing their beaks open and inserting food. Since swifts are insect eaters, part of each day was occupied by catching insects for them. Other young birds were fed with pablum, yeast, egg yolk, fruit, and chopped nuts.

Laskey often rose at 4:00 A.M. to be able to do her gardening, housekeeping, and field work (which included long hours of observations of nesting birds). This tiring schedule, combined with the hot, humid Tennessee weather and Laskey's sinus problems, left her listless and exhausted. When illness made field work impossible, she worked indoors on her notes, publications, and correspondence. Visitors meant not only less time for research, but also a housecleaning blitz, since she could never get rid of her early conditioning to keep a clean and tidy home. Since her husband did not share home responsibilities with her, and by the time of World War II household help was rarer than "hen's teeth," Laskey had to work long hours to accomplish her self-perceived and self-appointed tasks. In 1944 she wrote to Canadian ornithologist Roy Ivor (1880–1979):

> So I have been working 12 to 15 hours in this heat doing all the manual labor of washing woodwork, cleaning, waxing, washing windows from a stepladder outside, etc. . . . I had to slip away in the evening or very early morning to look after the Bluebirds in Warner Park [where she had a series of nesting boxes under observation]. However, I am glad I was able to do all the strenuous work and have the house a great deal more presentable and livable.[47]

In that same year, much to her satisfaction, she learned that one of her banded chimney swifts was found in Peru. This first solid evidence of the winter home (long a topic of speculation) of this North American nester was an exciting, rewarding experience, since "the banding of many thousands of swifts has been dirty, messy work and has consumed hours upon hours of day and night, besides all the expense of [banding] operation and equipment."[48]

Laskey enjoyed her research, her correspondence with other ornithologists, and meeting peers at scientific conferences. An associate of the AOU since 1933, she was elected a member in 1951 and a fellow in 1966. Her work was widely quoted by other researchers, and the many papers she published in ornithological journals—12 each in the *Auk* and the *Wilson Bulletin,* 19 in *Bird Banding,* 104 in the *Migrant,* and 2 in the *Journal of the Tennessee Academy of Sciences*—are a testimony to her industry and determination.

Laskey had many field companions during the four decades of her career: scouts, college students, other ornithologists. Her husband was not among them. Although he tolerated her activities he helped her hardly at all until after he retired. Then, since he was a "lost man" without outside interests, he would drive her to areas where she conducted field work. This became increasingly important, because Laskey developed eye trouble, eventually loosing one eye, and could not drive in poor light.

There are no indications that Amelia Laskey as a young girl would have strived for a scientific career. Like most of the women discussed in this essay, she developed her ornithological interest in midlife. Then she exchanged a comfortable life for one of hard work and satisfying accomplishments. Although she made no concessions in collecting scientific data and in looking after her home, she often saved time and effort by eating in restaurants. Katherine Goodpasture said of her: "Amelia did not 'play' with ornithological pursuits. Whenever she deemed an activity to yield unsound or questionable results, or whenever she thought material might be unwisely used, she withheld her participation . . . when she saw opportunity to make a well-founded contribution, she applied her energy without stint."[49]

Laskey is another excellent example of a serious, committed, independent researcher. No doubt her old-fashioned marriage to Fred Laskey contributed to her maintaining vestiges of her early socialization, with its emphasis on being the perfect homemaker. In spite of this, with determination and hard work she was able to produce outstanding, original research in ornithology.

WOMEN ornithologists have participated for well over a century in a variety of scientific activities beyond the low-status "women's work" found in other sciences. The development of many scientific fields and the increased organization of scientific activity was rapid during the 1850–1950 period. The attitude of American society toward women's rights to education and careers other than motherhood was slower to change. Nevertheless, women managed to contribute to many sciences, often as independent researchers rather than as professionals, and they were particularly active in ornithology, a field, unlike botany, not regarded as suitable for women.[50]

The women ornithologists mentioned here lived during a period of transformation in the study of birds. In the early 1860s, when Grace Anna Lewis began serious study,[51] ornithology was still practically synonymous with classification. By the time she gave up research twenty years later, the shift had started toward field investigations of living birds. At first field observations were sporadic, short-term affairs (Martha Maxwell's and Florence Bailey's work were conducted along these lines). Later, life-history research developed into detailed, long-term investigations of many species; by the early 1920s studies had gradually changed from being merely descriptive

to being more analytical and controlled. Sherman, Nice, and Laskey were pioneers in this type of research.

The scientists discussed in this essay all possessed single-minded determination to carry out their studies despite all manner of difficulties and interruptions, but their background, education, family, and scientific experiences varied. Sherman, like Maxwell, was from a midwestern pioneer family and attended Oberlin College. Resourceful, hardworking, and intelligent, she had to make great sacrifices for family reasons.[52] Starting a second career at midlife involved giving up her independence, but despite constant disagreements with her sister, Sherman became a successful scientist. Nice, like Bailey, came from a comfortable, middle-class family and attended a New England women's college. Although as young women both suffered from the prevailing attitude that ladies should not pursue professions, both eventually made their bid for independence and became famous in the ornithological community. By the time she married, Bailey was a mature woman with well-defined scientific interests. Her husband, a fellow scientist, provided her with support and companionship. Nice gave up graduate school to marry and have a family. It was only many years later, when she resumed her ornithological work, that she received the cooperation of her husband and children. Laskey, the only one without some college education, was the least likely to rid herself of traces of her early socialization, with its emphasis on homemaking. Her husband accepted but did not encourage her interest and involvement.

Relationships with other women, whether family members or friends, provided an important source of emotional support for these women ornithologists. Throughout this period correspondence networks substituted for immediate colleagues; attendance at scientific conferences led to personal friendships, intellectual exchange, and visibility in the scientific community. There is no evidence that any of these women encountered more difficulty in publishing papers than did male ornithologists, even at a time when editorial policy and lack of finances greatly restricted the publications of lengthy articles in the *Auk*.

Women ornithologists *did* have certain problems that were sex-specific, arising from the nature of intensive field research. Twelve- and fifteen-hour days were common, and nesting birds do not distinguish between week-

---

Opposite: Martha Maxwell, the "Colorado huntress." Photo taken for her own exhibit of birds and mammals in the Kansas-Colorado Pavilion at the U.S. Centennial Exposition. Courtesy of Colorado Historical Society.

days and weekends. To accomplish such concentrated field work and carry on the routine of everyday life caused considerable stress. Strong organizational skills, stamina, flexibility on housekeeping matters, and a clear sense of priorities were needed if one wanted to combine science and family life successfully. On the other hand, independent male researchers rarely had to worry about children's schedules, regular meals for family members, or clean clothes when engaging in field work.

There are many other independent women ornithologists whose lives and careers deserve detailed study: for example, Annie Alexander (1867–1950), field explorer and museum builder; Amelia S. Allen (1894–1945), pioneer bird bander and first woman president of the Cooper Ornithological Club; life history experts Doris Huestis Speirs (1894–) and Louise de Kiriline Lawrence (1894–); museum ornithologists Elsie Binger Naumburg (1880–1953), Mary Ellen Terry (1888–1967), and Margaret H. Mitchell (1901–). My sample does, however, illustrate clearly the experiences, difficulties, and accomplishments of outstanding North American women ornithologists whose independent research would place them in any ornithological hall of fame. No history of North American ornithology should ignore their work. To students of the history of women in science their position in a field-oriented discipline is of particular interest. As the psychologist Martha S. White remarked, during the past century attempts to include women in the rigid professional structures of most sciences have met with failure. Because of this she advocates alternative career patterns.[53] The success of women ornithologists, who without funds or institutional position produced first-class research, proves that alternative modes of scientific research are possible.

5

NANCY G. SLACK

Nineteenth-Century American Women Botanists:
Wives, Widows, and Work

BY ONE ACCOUNT there were 1,185 women "actively interested in botany" in the nineteenth century.[1] This figure only includes those for whom records have been found. Almira Lincoln Phelps wrote in 1833 that her text *Familiar Lectures on Botany*[2] had sold 10,000 copies and was in use in female seminaries from New England to the Arkansas Territory.[3] It had sold 375,000 copies before the end of the century, mostly to women. Botany had been considered, nearly since the time of Linnaeus, a suitable and elevating subject for women. It had indeed become so feminized by 1887 that J.F.A. Adams could write an article in *Science* titled "Is Botany a Suitable Study for Young Men?"[4]

Yet in C. Earle Smith's lengthy discussion of nineteenth-century American botany, not a single woman is mentioned, although he included men who were alive until 1940. Harry Baker Humphrey, in his *Makers of North American Botany*, which covered 250 years, included only two women, Jane Colden Farquahar (1724–66) and Elizabeth Gertrude Knight Britton (1858–1934).[5] The 122 entries include many men who were not professional in the modern sense. They had only to be dead by 1961 and to have made "outstanding contributions to the field" or to have pioneered in some phase of botany. This is, of course, one man's selection, made before there was much material available on women's contribution to American science. Even Margaret Rossiter, however, in her comprehensive and perceptive study of American women scientists, listed only six botanists in her index, although others are mentioned in the book.[6] Her botanists, moreover, were not required to be dead.

Why are so many botanical women nearly invisible? Is it that they failed to make "outstanding contributions"? If so, what was it about the lives of these women in the nineteenth and early twentieth centuries that prevented them from doing so? Or was it that their contributions were simply not seen as outstanding? If so, what gave rise to our culture's collective blind spot? What of those who were not only "actively interested" but were active contributors, at least in terms of publications? Did they marry or remain single? Who were their husbands, families, mentors, and how did such people affect the women's professional work?

The women included here, with a few early exceptions, all lived and worked in the nineteenth and twentieth centuries; those born after 1900 are excluded. Owing to the unusual and unexplained longevity of women botanists, a few of those discussed here are still alive. Gertrude Douglas, a botanist with a Ph.D. from Cornell University and a long teaching career at Albany State died in 1986 at the age of 102; many of the women discussed here lived into their eighties or nineties.[7]

The word *career* in the modern sense does not apply to many of these women. To the extent that a career implies paid employment, this was not possible in American botany for women or men until the mid-nineteenth century. Before that time there were no fully professional botanists outside of Europe; eminent American botanists such as Torrey and Sullivant had other means of support. By the end of the century American botany had become largely professionalized; that is, positions were available at universities and in government agencies, botanical societies had been formed (and were increasingly excluding amateurs), and the Ph.D. was becoming a standard qualification.[8] In addition, several botanical journals were available for publication of research. All the women botanists included here published either botanical books or articles in botanical journals, meeting at least one requirement of the twentieth-century definition of a professional. Many of them, however, had neither advanced degrees nor paid employment, Britton, the most eminent woman botanist of this period, had neither.

It is interesting to note which groups of women do *not* appear in this account. There are no women of color. There are almost no immigrant women; although important in American science, they were in other fields, such as chemistry, or they appear later. The exception in this group is Katherine Esau, born in 1898 in the Ukraine. Working-class, Jewish, and Catholic women are largely absent, even in the early decades of the twentieth century, for reasons that become apparent in the accounts of the group that *is* present—white, middle- and upper- middle-class Protestants. There was also a lack of publishing women botanists south of Washington, D.C., and west of Chicago, or at least Nebraska, though they did appear in California near the end of the nineteenth century. Not a single woman was included in a lengthy enumeration of southern botanists, many quite obscure, written in 1893.[9]

In the eighteenth century and throughout most of the nineteenth century in the United States there were many effective barriers preventing a woman with an interest in any branch of science from actually becoming a scientist. Science was a male bastion, and we must look for chinks that did indeed allow some women to break through and take part in the ongoing work of science during that period. The views of society, which were also the views that most women assimilated in the process of socialization, made marriage and children the raison d'être of a woman's existence. When women's education was approved at all it was often thought of in terms of its value in the education of future sons. Even such highly successful women as Almira Phelps supported the "separate spheres" doctrine in her writings. Until the last quarter of the century at least it was not possible for a woman to support herself as a botanist, except perhaps by writing textbooks. By 1870 a male botanist could be a university professor, a professional collector, or a member of a government exploring expedition, all with the approbation of society.

Spinsterhood was a misfortune in the view of both the woman and society. In *Pride and Prejudice* Elizabeth is dismayed that her friend Charlotte Lucas has agreed to marry someone she doesn't even like, but for Charlotte in eighteenth-century England, marriage was the only way to make a life for herself. In America, even in the nineteenth century, this was still largely true. The Bible, as well as the churches to which most of these women belonged, supported the doctrine of separate spheres and the duty to procreate. Increased technology and the movement of men's work out of the home during the nineteenth century, as Ruth Cowan has shown,[10] left the wife in full charge of household work. This was true whether she did most of it herself or, in the case of upper-class women, managed the work others did. The church also provided such women with useful work within their sphere: charitable services to the poor. All of these aspects can be seen in the two marriages of Almira Phelps, discussed below.

Such responsibilities, which often included the care of many children, made original work in sciences, in this case botanical research, nearly impossible. Women might collect and examine flowers and perhaps make herbaria of their dried, pressed specimens. A few married women managed pedagogy and even the writing of botanical books. Some were able to work with their husband at home or in the husband's laboratory. This became a common pattern for women with Ph.D.'s in botany from the end of the nineteenth century until very recently.

Until the late nineteenth century a single woman rarely had "a room of her own" or the financial security to do her own work. A father or other male relative sometimes provided the security and mentorship—but only until marriage. Not until the founding of women's colleges, largely in the 1870s and 1880s, was it possible for a single woman to support herself in botany. Limited work opportunities for women in botany also became available at agricultural experiment stations by the 1890s[11] Many of the notable women in botany born after 1850 remained single. Most of the botanical women of the generation born in the decade after 1900, professional botanists with Ph.D.'s and academic positions, largely at women's colleges, chose to remain single. Was there any other choice?

The total number of American women who were born before 1900 and who did significant work in botany is difficult to assess. Although there were several hundred who published papers or books, the books were largely popular ones, often for children, and some of the botanical papers were published in such magazines as *St. Nicholas* and *Old and New*. Even in botanical journals, the *Bulletin of the Torrey Botanical Club* for instance, the type of botanical paper published in 1871 differed considerably from that published thirty years later, although even then, as now, not all were significant. Of Rudolph's 1,185 American women actively engaged in botany in the nineteenth century, most belonged to botanical clubs and/or made plant collections, but only nine were listed as having made scientific contributions "on a par with those made by men." Of these, all but one are also included in *Women in the Scientific Search.* This book includes nineteen additional women botanists, including all of those in the *Notable American Women (NAW)* volumes used as my basic sample. I have included a few additional women, two of whom are still alive and therefore excluded from the other works. We thus have a pool of about thirty five American women born before 1900 who are thought by at least one source, usually more, to have made significant contributions to American botany. Five of these were deemed worthy of stars in *American Men of Science* as being among the top one thousand scientists, but two fields included by some of these authors under botany—plant pathology and bacteriology—were purposely not awarded stars. The "significant" women are not all listed under botany in all sources, which makes counting difficult. The great majority of these women are at least mentioned in Rossiter's book, which covers approximately the same period as this essay, to 1940, though some of her characters and mine went on publishing long after that date.

Among these significant American women botanists there are single women who never married; women who worked in botany when they were single but not after their marriages; married women with children who managed important botanical activity, but in teaching or the writing of books, not in research; and women who, married to professional botanists, achieved various types of botanical work as part of a couple. In addition, there are widows; widowhood, ironically, often led to significant botanical activity.

Consider the example of Wanda Kirkbride Farr (1895–1967), the only botanist in Yost's biographical collection of American women scientists.[12] As a married woman, Farr worked in her husband's laboratory both at Washington University and at the University of Iowa, doing research on plant cells. At his early death, she was, like Marie Curie, given her husband's academic post; she went on to do important research on plant fibers, working in academia (as associate professor at the University of Maine), the government (USDA), and industry (the Boyce-Thompson Institute). A full life in science would not have been possible for her as a wife of a professor in Iowa, but as a widow with a child it was. She had at least some financial means left to her after her husband's death, as did several of the widows discussed in this essay. Perhaps more important, she had the approbation of society because she had married and become a mother. Society also approved her finding employment to support herself and her child, at least if she could not find another husband. Widowhood, moreover, seemed to provide, for a number of these women, a spur to ambition, along with the lessening of household and societal obligations.

In this essay I examine the relationship between marriage and botanical work for the women listed under "Botany and Horticulture" in the two editions of *Notable American Women.* Then I discuss in more detail the life of Almira Phelps, one of the two most important women in botany in the nineteenth century. Finally, three botanical couples and the differing relationships of the women to their work and their husband are examined. One of the couples is Nathaniel and Elizabeth Britton; Elizabeth Britton was the other best-known woman botanist in the era covered here, the nineteenth and early twentieth centuries.

How many of the American women botanists born before 1900 were single, how many married, and how many widowed? Seeking an unbiased sample, I have used the twenty one chosen for their contributions to botany and horticulture in the two editions of *Notable American Women.*[13] Apart from three eighteenth-century women, all were born between 1822 and 1889. Curiously, no "notable" women in botany appear to have been born during the nearly one hundred years between the births of Elizabeth Pinckney (1722) and Jane Colden (1724) and that of Lydia Shattock (1822). This gap probably represents the lack of opportunity for women to make public contributions to botany before the middle of the nineteenth century. Colden and Pinkney were in the tradition discussed by Ann Shteir in "Linnaeus's Daughters" for British women in botany;[14] they were both taught and in a sense sponsored by their fathers, though the work was their own. Two women in the NAW sample whose major contributions were in applied botany or horticulture were originally trained by their fathers. Pinckney (1722–93) managed her father's plantation at seventeen but continued her plant experiments, particularly on indigo, for many years after her marriage and widowhood. Elizabeth Allston Pringle (1845–1921), whose father had written a treatise on rice culture, experimented with and wrote about rice and other crops. both during her marriage and as a widow. Both to these women lived in South Carolina, but over a century apart.

Of the NAW sample of women, ten were single. The remaining eleven were married, three of them twice. Seven were widowed relatively early in life, a surprising statistic today, but common then. Marriage and children were the norm during this period, and for those women who followed that norm throughout their lives, notable activity in botany is the exception—four out of twenty-one in this sample. Two women who made outstanding contributions, Kate Brandegee and Elizabeth Britton, were married during the period of their major botanical work, but they remained childless.

One pattern, of which Jane Colden Farquahar is the prototype among American women botanists, is the woman who did good botanical work while single but dropped out of botany after the wedding. This pattern was probably quite common in the nineteenth and the first half of the twentieth centuries, but most of the women who fit it did not do work notable enough before their marriages for us to know of them. Jane Colden, as she is generally known, was the daughter of Caldwallader Colden, who was well known as a botanist. Jane Colden was something of a phenomenon, but largely because "her father presented her to an international circle of scholars and collectors," including John and William Bartram, Alexander Garden, and by letter to Linnaeus himself. Her education in botany was conducted entirely in their home. By 1757 she had described and sketched three hundred local plants. By 1759, however, she had married and had shifted to other concerns. In a letter to her sister Katherine, she wrote, "O Cate, if you had a Husband you would feel for every little Sickness or pain he had in a manner you knew nothing of before."[15] She subsequently had one child and did no more botany that we know of. Sadly, in 1766 both she and the child died.

A more recent but similar life pattern of a woman mycologist, Violetta White (1875–1949), was recently discovered.[16] White worked with L. M. Underwood, a Columbia professor, at the New York Botanical Garden from 1900 to 1902. She wrote scholarly monographs of two families of gasteromycetes (puffballs and allies), which were published in 1901 and 1902 in the *Bulletin of the Torrey Botanical Club*. She corresponded with and borrowed specimens from the leading mycologists of the time, including Charles Horton Peck (1833–1915), the state botanist in Albany (a correspondent of and mentor to several of the women discussed in this essay). White discovered a number of new species of fungi. In 1902 she was working on a third monograph, which was never published. Her beautiful illustrations for it were recently found at the New York Botanical Garden. In a letter to Peck she wrote about the fungi: "The more I work on these plants, the more the fascination of them grows upon me; I only wish I could devote all my time to their study and collection."[17] Shortly thereafter, though, Violetta White married prominent lawyer, John Ross Delafield, had two sons and eventually seven grandchildren. We hear no more of mycology after she married.

The single women in the NAW sample were among the most productive botanists of the late nineteenth and the first half of the twentieth centuries. These include E. Lucy Braun (1889–1971), Alice Eastwood (1859–1953), Margaret Ferguson (1863–1951), and Kate Sessions (1857–1940). Two earlier women, Lydia Shattock (1822–89) of Mount Holyoke and Rachel Bodly (1831–88), dean of the Women's Medical College of Pennsylvania, both included in NAW as botanists, were more important as teachers and mentors than for their publications.

It is hard to know whether remaining single was a choice for some of these women; it is a difficult question to ask those single women botanists of this era who are still alive. One woman who studied botany at Hunter, graduating in 1917, told me that all the women in her class hoped to marry, whatever their academic interests. Many of these Hunter women did find employment after graduation and worked until marriage or motherhood. Others never married and managed to support themselves. Several of the women in the NAW sample are described as attractive. Shattock, for example was "singularly attractive with bright blue eyes, a clear complexion, regular features and fair curly hair." Sessions, who entered Berkeley in 1877, was called the "prettiest girl on campus."[18] The only direct word we have is from Winifred Goldring (1888–1971),[19] a paleontologist sometimes included in listings of botanists, who wrote in a report to her Wellesley class in 1919, that she had "as yet discovered no one more attractive than my work."[20] One does know, however, that women were not hired at women's colleges if married and would resign their positions if they later married. Rossiter related the unfortunate story of one woman scientist at Barnard who tried, unsuccessfully, to hold on to her position.[21] The botanists at women's colleges, who also include Susan Hallowell (1835–1911), Margaret Ferguson Clara Cummings (1855–1906), Henrietta Hooker (1851–1929), and Julia Warner Shaw (1863–1927), did have "families" of a sort at their colleges—mentors and protégés. The mentors sometimes trained them, sent them off to teach elsewhere, and then called them back to the family. Hallowell, for example, called former student Ferguson back to the family at Wellesley after she had obtained her Ph.D. at Cornell, Ferguson, thereafter, built up the Wellesley botany department and served as mentor to the next generation of Wellesley botanists. Some of these women did pioneering research in the newer aspects of botany, such as physiology, cytology, and plant genetics. Ferguson, for example, published important work on the cytology of the pine and the genetics of the petunia.

As will be seen from the life histories of the botanical women in the NAW sample who married, were widowed early, and subsequently carried out their important botanical work, it was not only the written and unwritten regulations and prevailing opinions that kept married women out of the few professional positions available to them in the nineteenth and early twentieth centuries. It was nearly logistically impossible for a married woman with children to carry out original research. Much of the research in botany at least in the the nineteenth century (and in cryptogamic botany or lower plants well into the twentieth) was in systematics and often involved exploration for new species.[22] In spite of difficulties and prejudices, a number of single women botanists succeeded. Alice Eastwood, for example, traveled all over California, Utah, and Colorado in search of undescribed species. Goldring, although complaining that her male colleagues rarely invited her on field trips outside the state and wrote to her about the limitations of taking women into the field, invented a bloomer outfit, learned to shoot a revolver, and went into the field alone to collect. Kate Furbish (1834–1931), another unmarried botanist, collected alone in wild, inaccessible places in Maine. Kate Curran, Agnes Meara Chase, and the most amazing botanical explorer of all, Ynes Mexia, also collected but not until they were widows.

Agnes Chase had been married at eighteen to a newspaper editor, but she was widowed in 1889 after one year of marriage. She never acquired any degrees or higher education, but she had an unusual mentor, the Reverend E. J. Hill, a botanist. He encouraged her and helped her find positions, first in the Chicago stockyards and later in the USDA Bureau of Plant Industry and Exploration, where she eventually became principal scientist for agrostology (the study of grasses). She traveled widely into her seventies, published important books, and along the way was jailed for her suffragist activities. One wonders what her life would have been like if William Chase, her husband and editor of the *School Herald,* had not died young.

Kate (Mary Katherine Layne Curran) Brandegee (1844–1920) graduated from a California seminary and married a constable, Hugh Curran, at twenty-one. He died in 1874, before she was thirty. The following year, the widow Kate Curran went to San Francisco and earned an M.D. at the University of California. She practiced medicine, but women doctors were not in great demand in 1880. Patients were sparse, and she became increasingly interested in botany. Her botanical mentor was Dr. Hans Behr, through whose help she joined the California Academy of Sciences. In 1883 she became the curator of the academy's herbarium, one of the best professional positions held by a woman anywhere in America at that time. She collected plants for the herbarium all over California. At forty-five she married Townshend Brandegee, a civil engineer with botanical interests. They walked the five hundred miles from San Diego to San Francisco on their wedding trip, collecting plants all the way! According to Hunter Dupree and Marian Gade, Kate Curran Brandegee "was able to make an even greater imprint on California botany . . . with the financial and moral support of her second husband."[23] In addition to collecting, she published papers and founded and edited, together with her husband, the botanical journal *Zoe.* She remained as curator for five years, after which they continued their joint botanical work in San Diego and later Berkeley. In Kate Curran Brandegee we have a scientifically productive widow who continued her important botanical research while in a companionable and financially supportive second marriage.

Ynes Mexia (1870–1938) grew up in Texas and attended Quaker schools in Philadelphia, Ontario, and Maryland. She married twice, the second time to A. A. de Reygados in Mexico City. In 1908, widowed, but with no children, she moved to San Francisco and eventually entered the University of California in 1921, when she was over fifty. She thus serves as an early role model to today's many returning women students. At fifty-five she went off to Mexico to collect plants with another woman botanist, Roxana S. Ferris. The next year she went back to Mexico alone and returned with one new genus and two new species of plants. The genus, *Mexianthus,* was named for her. In succeeding years she collected in Mount McKinley Park, Alaska, and in Brazil and Peru. The USDA Bureau of Plant Industry and Exploration sent her to Ecuador in search of cinchona (for quinine) and other plants. Even at sixty-five she traveled alone to the wilds of South America. Her expeditions were mostly financed by the sale of specimens—not a career for a wife and mother, hardly a career in the 1920s for any woman, single, married, or widowed.[24]

IN ORDER to look further at the effects of different types of marriage and of widowhood on the botanical work of women, I will consider to some detail the life of Almira Hart Lincoln Phelps (1793–1884), the one important woman in botany who has left us her own views about marriage and her work. When Almira Hart (henceforth referred to as Phelps) was born in 1793, a woman could not even be a "winter," that is, a regular teacher in a district school in Connecticut. She was Lydia Hinsdale Hart's tenth child; altogether there were seventeen children in the family, including seven from her father's first marriage. Once married, a woman was almost certain to expect many and frequent pregnancies and the care of small children over an extended period. In the eighteenth and nineteenth centuries young women often married widowers with children.[25]

Phelps's early life history provides an especially well-documented example of a home in which the mother, apparently a superb organizer, ran an enormous household without any servants (except the children). Religion, though not strictly Calvinist, was devoutly practiced, and this made a lasting impression, as can be seen in Phelps's later writings. There was reading aloud and discussion of literature, as well as of politics and religion, although neither parent had a formal education. We can see Lydia Hart dissecting the chicken and joints of meat and forming a "tolerably correct idea of human anatomy."[26] It was a family in which curiosity and ideas were valued; Almira's first mentors were certainly her parents and siblings. Her sister Emma (later Emma Willard), six years her senior, was her most important early mentor. She was her first teacher (the summer teacher) at a Berlin district school. Emma became, at an early age, the head of a female academy in Middlebury, and at twenty-two she married a prominent Middlebury physician, John Willard, age fifty. Marriage, as noted above, was the only route to a full life for a woman in the early nineteenth century, and position—the husband's position, of course—was an important consideration.

Emma's marriage and her subsequent invitation to Almira to come and live with them enabled Emma to continue as Almira's mentor. There was another mentor in the home, John Willard, the nephew of Emma's husband. This relationship was described in some detail by Almira herself.[27] We learn that she and John, a sophomore at Middlebury College, studied higher mathematics together and that he made "clear by the light of his own mind, difficult problems." The following year three of John's fellow students came to board at the house as well. Almira wrote: "The six students [including Emma] came together for the morning meal with renewed ardor for the studies of the day . . . conversation seldom ever flagged, sometimes argumentative and literary, with sallies of wit and humor." But the young men's "studies of the day" were in the Middlebury classrooms and even included "natural philosophy," complete with apparatus. Almira's studies were at home with Emma. They included mathematics and philosophy, but natural science and especially experimental science with apparatus came considerably later in Phelps's life.

In order to obtain the formal credentials needed for teaching beyond the elementary level, Phelps went to her cousin Nancy Hinsdale's academy in Pittsfield, Massachusetts. Subsequently she had several schools of her own, notably a winter school in the town of Berlin, a position to which she was elected over some opposition, this being the first time a woman had ever been the winter teacher in that town. She later took charge of an academy at Sandy Hill, New York. Here she introduced many new subjects, including geography, using her sister's new methods. But after two years at Sandy Hill, Phelps, as her older sister had done, left teaching for marriage. Also like her sister, she married a man of position, Simeon Lincoln, the editor of a conservative Federalist newspaper in Hartford, Connecticut.

In keeping with the views of society, and very likely her own as well, Phelps became the complete housewife. Emma Bolzau described Phelps's difficulties with housekeeping and servants, and Almira herself recalled giving up "poetry and literature for receipt books, *Domestic Cooking,* and the *Young Housekeeper's Companion.*"[28] The Lincolns were heavily involved in entertaining people of their social stratum in Hartford, but Almira also was able to find time for church work, including visiting the sick and dying. Three pregnancies followed in three years, and there seems to have been no thought of ever returning to teaching. There is also no indication that her husband ever encouraged her to step out of her prescribed role as the wife of a prominent citizen and the mother of many. Her husband's sudden death from yellow fever in 1823 left her with two surviving children aged one and two, and two insolvent estates, her husband's and her father-in-law's. She needed a means of financial support for herself and her children and at first found a teaching position in one of the New Britain, Connecticut, public schools. But the following spring she moved to Troy, New York, and her sister Emma's school, the Troy Female Seminary. There Phelps became both student and teacher. She was finally able to study science (an 1825 seminary announcement offered lectures on natural philosophy, chemistry, botany, and other branches of natural history). Widowhood and financial hardship had thus resulted in academic motivation and opportunity lacking in her marriage.

At this time Phelps's second mentor, Amos Eaton, appeared and her interest in botany apparently began. Eaton, a well-known botanical author and eventual founder of Rensselaer Polytechnic Institute, had, since 1822, given lectures in natural philosophy, including mineralogy, zoology, phrenology, and botany at the Troy Lyceum. These lectures included the use of such apparatus as air pumps and microscopes. They were set up "with a particular view to accommodate" the students at the Troy Female Seminary.[29]

Phelps was at this time teaching elementary courses and educating herself to teach more advanced courses in languages and mathematics, but she soon became involved with the courses in science given by Amos Eaton to the young women of the seminary and she became fascinated with his new methods of instruction—experimentation and demonstration. There are many indications of Eaton's mentorship of Phelps in both botany and chemistry.[30] He encouraged her to use the hands-on methods for teaching science that he had developed. The botanical course that Phelps taught at the Troy Seminary, for example, included getting firsthand knowledge of plants through field excursions. "The region around Troy is rich in its flora and scarcely a dell, ravine or island of the Hudson in its vicinity was not explored in these expeditions," though it appears the gentleman students of Rensselaer often procured the specimens that the Troy Seminary girls studied.[31] There was no suitable textbook for Phelps's botany course at that time, so she prepared detailed lecture notes and, at Eaton's urging, published her *Familiar Lectures on Botany* in 1829. This textbook was reprinted for more than forty years and sold 375,000 copies. The book traveled not only to the Arkansas Territory, but to Europe as well.[32] Phelps acknowledged Eaton's help and encouragement profusely, in and out of print.[33]

It is clear that Eaton thought well of Phelps's abilities. He asked her to prepare a chemical dictionary from a French one by Vauquelin, and he praised her work on it.[34] Published in 1830, it was followed by several other books, notably *Botany for Beginners,* for use in lower schools, and *Chemistry for Beginners.*[35]

By 1834, when *Chemistry for Beginners* was published, Phelps was no longer in Troy. She had remarried, retired from teaching, and was living in Brattleboro, Vermont. Her new husband was the Honorable John Phelps, a state senator and the father of a Troy Seminary student, Lucy Phelps. In addition to Lucy, Almira Phelps acquired five other stepchildren, ranging in age from nine to twenty-six.[36] It is clear that Phelps took her new responsibilities as wife and stepmother seriously. Lucy Phelps wrote to her sister in 1832: "Mother [stepmother] is very kind indeed and seems to take a great interest in our affairs. . . . She does everything for father's comfort and he seems very happy indeed, and his health has never been better." The Phelpses social life kept them busy attending and giving parties, as was also documented by Lucy.[37] As during her earlier marriage in Hartford, Phelps became involved in church work. In a letter to Emma Willard she wrote: "I find my present retirement sufficient occupation and hope that I am enabled [not only] to be useful in my domestic relation, but to do some good in the society in which I am placed. I take great interest in the Sabbath School and find Sunday the busiest and happiest day of the week."[38]

Thus marriage and domestic duties threatened to permanently still the pen of one of the best botanical authors in America. She appears to have completely accepted her "domestic relation" and the conventional type of community contributions made by upper-middle-class women in the society in which she was placed by her marriage to Phelps.

This marriage proved different from her first, however. Before she married she had published her botany book, and it had been well received. John Phelps knew of her accomplishments and appears to have courted her, at least in part, because of them. By all accounts he encouraged her to continue her writing. He had been elected to the state legislature and was absent for long periods during winter, when he lived in Montpelier, the state capital. This provided his wife with time for her writing. The older girls were also away at school in the winter, but there were, on the other hand, two more pregnancies, with the babies born in 1833 and 1836. Whatever her family duties, Phelps managed to write several more books, revise her botany books, contribute to two women's magazines, and deliver several addresses on education. Thus marriage alone, even if it included the care of a family was not enough to prohibit botanical work. The attitudes of spouse and children were crucial determinants.

The extent of the husband's encouragement in this case was extraordinary. John Phelps resigned from his office as state senator in 1837 when he was sixty and his wife was forty-four. She was offered a position as the head of a new seminary to be established in West Chester, Pennsylvania, and accepted at her husband's suggestion.[39] Thereafter, he followed her from West Chester to a school in Rahway, New Jersey, in 1839, and then to Patasco Female Institute in Ellicott's Mills, Maryland, in 1841. Although Bolzau stated that "Mr. and Mrs. Phelps commenced their work at Patasco Female Institute," there is no question but that it was *her* school.[40] Emma Willard wrote in a letter to Lydia Sigourney: "My sister is about to remove her school from Rahway, a large and elegant establishment having been offered her by Bishop Wittingham. . . . I regret her greater distance from me but rejoice in her prosperity and opportunity of increased usefulness."[41]

The institute flourished under Phelps's guidance. The curriculum in science was particularly strong. By 1852 pupils from the preparatory department to the senior year studied botany, chemistry, and geology, much of it from "Lincoln's Botany," as *Familiar Lectures on Botany* was called, and Phelps's other books. Her husband, who acted as the school's business manager, died in 1848, but she continued to run the institute herself until her retirement in 1856.

Almira Phelps was not a botanical researcher. She did collect plants as early as her first acquaintance with Amos Eaton in Troy and as late as her trip to Switzerland in 1854. Her herbarium was presented to the Maryland Academy of Sciences, of which she was the first woman member. She was, as Sally Kohlstedt discusses in this volume, the second woman elected to the American Association for the Advancement of Science (AAAS), following Maria Mitchell. She presented several papers to the AAAS in the years following her election. One, an 1869 defense of the Linnaean system, was presented long after the demise of that paradigm, but the issue for Phelps was pedagogy, not classification per se. The title of the paper was "Popular Science," and in it she voiced dismay at the decline in the teaching of botany at schools and colleges following the abandonment of Linnaean methods. She was herself quite aware of new developments in botany as a science.[42]

There is no question that her botany and other textbooks, and their wide-ranging effects on the teaching of science in secondary schools and later in colleges, were her most important contributions. Although she belongs to what Kohlstedt calls the "second generation," the educators and popularizers,[43] she was probably the most important of these women, particularly in botany.[44]

Despite her pioneering in botanical teaching and writing, Phelps's views on women in society were typical of her generation. As her history of marriages and "retirements" shows, marriage and its attendant childbearing were, in her own view, the preferred state of women, even unusual women capable of supporting themselves. A woman's sphere was different from a man's and was meant to be. Not only do her own writings echo some of the nineteenth century's views of woman's restrictive physiology,[45] but she even took her views back to the Bible and Adam and Eve, at least in some of her later writings: "She [woman] was created to be the companion of man, to cheer his solitude and to assist him in his duties. This very relation implies a different between them. A companion or assistant fills a secondary position." Religion, including a literal belief in God and his creation, was a central part of her life, reflected in her reasons for studying botany, as expounded in her texts.[46]

Her views on the education of women, in science as in other subjects, stressed the often stated need for educated mothers to teach their sons and to be a good moral influence on them. This was not just rhetoric to an outside world ambivalent about the education of women; it was her own opinion. After discussing women's rights in a letter to a friend, she wrote: "Mothers must try to bring up their sons to respect and sustain the weaker sex—wives must seek to gain their proper place and influence by wise management, not by opposition and contention."[47] In her case, as Mrs. Phelps, contention was not necessary. Her second husband encouraged her scientific writing and moved with her from Vermont to Pennsylvania to New Jersey and finally to Maryland as she changed positions, most unusual in the nineteenth or even the twentieth century.

THREE WOMEN who worked in botany and were the wives of botanists are Edith Schwartz Clements and Elizabeth Knight Britton, both of whom studied botany before their marriages, and Eliza Wheeler Sullivant, who began her botanical work at the time of her marriage and was her husband's sole co-worker until her death. These three botanical couples provide a contrast in time, from the first half of the nineteenth century to the early twentieth and in place, from New York City to Ohio and Nebraska. They also show contrasting relationships between marriage and botanical work.

Two other botanical couples have already been mentioned. Wanda Kirkbride Farr and Clifford Farr did joint research on plant cells in his laboratory at the University of Iowa and later at Washington University. Kate Curran Brandegee and Townshend S. Brandegee are probably the best-known botanical couple; their lives and work are already well documented.[48] Kate Curran was already a well-known botanist with an important position before she married Brandegee in 1889.

English botanical wives are discussed in Ann Shteir's "Linnaeus's Daughters: Women and British Botany."[49] Some, Maria Hooker for example, were both botanical wives and daughters. These women were wives of a "dominant male with prior expertise" from whom they received training and encouragement and to whom they provided assistance. These wives served as "Institutional, financial and technical helpmeets," sometimes doing botanical illustrations or editing, usually without much, or in some cases any, recognition. Most of the women discussed by Shteir were born before 1800, and British society in their time essentially prevented women from having professional positions or belonging to the important national scientific socieities, although women were admitted to the newly formed Botanical Society of London in 1836. Scientific publications by these women were unlikely; only one of the thirty-three women who joined the botanical society in its twenty years of existence ever had a paper presented. She and her husband worked together on ferns and presented a long monograph to the society, but one periodical did not believe that a woman could have carried out such work and attributed to solely to "Mr. Riley."[50]

The three American botanical wives lived somewhat later and in a more open society, or societies. Columbus, Ohio, still the frontier when Eliza Wheeler married William Sullivant in 1834, was a far cry from the New York City of the Brittons at the turn of the century. Both were very different from University of Nebraska, where the Clementses met around 1900.

William Starling Sullivant, an early Ohio settler, was a close associate of the two leading botanists of his time, John Torrey and Asa Gray. He was one of the first men from west of the Alleghenies to attend Yale. Before he could complete his studies, he was called home upon the death of his father to attend to the family business. In the 1830s that business involved farmland, mills, a stage coach company, a bank, and a limestone company for the building of the new state capitol at Columbus. To quote his biographer, "His leisure time was devoted to Botany."[51] By 1839 Asa Gray was looking for a microscope and botanical books in Europe to buy for Sullivant.[52]

Sullivant, equipped with microscope, books, journals, and European correspondents, went on to become the leading American cryptogamic botanist of his time, specializing in mosses. But much of the actual work was done by his wife, Eliza, as we learn from his own writings.

Sullivant's first wife had died very young after the birth of their first child. Ten years later he married Eliza Wheeler and brought her back to Ohio. Franklinton, which the Sullivant family founded when the area was only wilderness, later became part of Columbus in the new state of Ohio.

When William and Eliza Sullivant began their joint botanical work there in the mid-1830s, it was an unlikely place for anyone to do scientific work. There was no academic institution anywhere nearby. There was neither a herbarium nor another botanist nearer than Cincinnati or Louisville, neither library nor bookstore from which to obtain botanical books or journals. Even the microscope and references Gray obtained had to wait for the spring thaws, when they could arrive by canal. Only Sullivant's considerable financial resources made it possible for him to obtain the tools needed for their work. Sullivant started working on mosses at about the time of his marriage to Eliza, and he often referred to her as his only "co-laborer."[53] She did microscopic work and prepared sets of mosses with labels to send to Europeans, such as Sir William Hooker and J.F.C. Montagne. In 1841 Gray sent a set of their mosses to Hooker in England, writing: "I must not forget that my package also comprises a set of Ohio Mosses from my friend Sullivant, of whom I have often spoken, and of whom as a botanist we have high hopes, as he has an independence (for this country), talent, and much zeal."[54] Gray did not mention Eliza Sullivant. Andrew Rodgers, however, did give her credit in his biography of Sullivant, particularly in terms of botanical illustrations.[55] Eliza Wheeler Sullivant died of cholera in 1850. Sullivant wrote to Gray:

> Thus ends sixteen years by far the happiest portion of my life. To me she was evry [*sic*] thing. What she was as a wife and as a mother [of five children] I will say nothing; in those relations few equalled her. But there is an other subject on which it is fitting that to you of whom she always had the highest admiration I should say a few words. I mean her devotion to botanical studies. In these she was my constant companion & coworker for the last fourteen years; and I can say with the strictest truth that more than half of whatever I have done in these pursuits is due to her. Such a clear and discriminating judgment, such memory, industry, energy and dispatch as were hers is rarely to be met with. She had acquired of Latin and German and French sufficient to read botanical descriptions with ease. In the mechanical labor of writing she was of vast assistance to me. . . . In the most difficult and delicate dissections and microscopical examination she was astonishingly successful. Her enthusiasm for our studies really required to be checked. Her knowledge of Musci and Hepaticae was very extensive. No female I am sure ever equalled her and few of the other sex excelled her.[56]

A year later he wrote to Gray: "Her attainments amply entitle her to a short notice among the obituaries of those devoted to the Natural Sciences.—From no hand could it come with more propriety than yours. Schimper [a leading European bryologist] who knew well her proficiency in his favorite study has named two species of mosses for her."[57] Gray never wrote the obituary, but wrote in a letter to Charles Wright: "I have just found a letter of Sullivant's dated May 27, 1850. . . . Poor fellow! as I wrote you before, he lost his wife while I was away, and was overwhelmed, as she was everything to him, and as good a muscologist [moss specialist] almost as he."[58] But Eliza Wheeler Sullivant's name cannot be found in Rossiter's book or among the botanists in *Notable American Women* or in *Women in the Scientific Search.*

Edith Schwartz Clements's story is different. She does appear in Rossiter's book as an example of "conjugal collaboration."[59] Her husband, Frederick Clements, an early plant ecologist, was already coauthor of an important book, *The Phytogeography of Nebraska,* by 1898. His career took him from Nebraska to the University of Minnesota to the Carnegie institute, and he did field research in many parts of the country, including pioneering work in the Colorado Rockies. His wife accompanied him on all these travels. Frederick Clements was one of the two most important theorists in American plant ecology for the first forty years of this century.[60] Did Edith Clements have an important but unrecognized role in her husband's accomplishments?

Edith Schwartz probably would have not had a botanical career if she had not married Clements. In that respect she is similar to Eliza Sullivant. But the circumstances were different. By 1900 women could and did earn doctorates in science and could find teaching positions, though largely in women's colleges. About four hundred women published papers in American scientific journals between 1884 and 1900, whereas only three had published papers between 1860 and 1863,[61] more than ten years after Eliza Sullivant's death. Edith Clements's problems were closer to those of William Sullivant—having to terminated her studies to support her family. Her marriage provided financial support, status—her husband's and the opportunity to complete her Ph.D. studies. Antinepotism regulations may have prevented her from obtaining a faculty position, but not from working with her husband.[62]

In her autobiographical *Adventures in Ecology,* Edith Clements recounts her life and work with Frederick Clements.[63] It is clear that she was in many different ways the facilitator of his work, and especially for his travels. The subtitle is *Half a Million Miles . . . from Mud to Macadam,* and much of her contribution was as chauffeur over difficult terrain in the early days of both automobiles and roads in the West. But right from the beginning, as at their ecology lab in Colorado, she was off painting mountain flowers while

Frederick and students, male and female, as well as a woman scientist at the laboratory, were doing ecological research.

Edith Clements did in fact produce four popular flower books of western America, one with her husband, but three on her own. Her husband meanwhile wrote *Plant Succession: Dynamics of Vegetation,* and many other ecology books. There is little indication in her autobiographical book that she had any real involvement in her husband's theoretical work. She was obviously proud of him and happy to be of service, whether as cook or even auto mechanic when needed, but apparently not as scientist.

There is an interesting passage in the book indicative of the way she conceived of her own role as botanical wife. The young daughter of a friend asked her mother what she would have to know "in order to get a job visiting the National Parks and traveling all over the country" the way the Clementses did. Her mother replied that she would have to be able to read scientific papers in half a dozen languages, study geology, physics, chemistry, and physiology, and "of course everything about plants—probably as much as in all those books your 'Uncle Fred' has written." The daughter answered: "Dear me, I believe I'd rather get married." To that Edith Clements replied: "That's the idea, Margaret. . . . It's much easier to marry someone who knows it all and if you know some, yourself, and can draw and paint and take photographs and drive the car, and typewrite . . . you can go along the way I do."[64]

In the context of her book I do not think she was being cynical; that was how she viewed her own role in her husband's science. While this in no way detracts from her accomplishments as a botanical artist and author, we cannot turn her into a botanical researcher who had an important role in her husband's work.

The third husband-and-wife team, Nathaniel Lord Britton and Elizabeth Gertrude Knight Britton, differs in many ways from the Sullivants and the Clementses. Unlike the other two examples of botanical helpmates, Elizabeth Britton was, as noted above, the most important woman botanist in the last quarter of the nineteenth and the first quarter of the twentieth century. Rossiter considered her one of the four most prominent women scientists of the 1890s, together with Cornelia Clapp, Ellen Richards, and Alice Fletcher.[65] Perhaps even more relevant here, in 1893 she was the only woman, among twenty-four men, to be a charter member of the American Botanical Society. Although she had no advanced degree, she had graduated from and taught at the New York Normal School (later Hunter) and published papers before her marriage. She was, also before her marriage, the first woman member of the Torrey Botanical Club in 1879 and was elected the club's curator in 1884 and 1885.

In 1885 Elizabeth Gertrude Knight married Nathaniel Lord Britton and gave up teaching. But this did not mean the end of her botanical studies. As John Hendley Barnhart put it: "No longer a teacher, she was able to devote more time to her favorite science."[66]

She did devote more time to botany. In 1886, 1887, and 1888 she was editor of the *Bulletin of the Torrey Botanical Club* (still extant). Her specialty was bryophytes, mosses, and their relatives, and she wrote over two hundred papers and reviews in this field. She was instrumental in founding a society devoted to that branch of botany, the Sullivant Moss Chapter, named for William, not Eliza, Sullivant. It eventually became the American Bryological and Lichenological Society, which publishes the *Bryologist,* a journal in which Elizabeth Britton also had an important founding and editing role. From 1899 until her death in 1934 she was the curator of bryophytes at the New York Botanical Garden, which both Brittons were instrumental in founding. In 1912 this position was made official: Honorary Curator of Mosses.[67] Although she never received any pay as curator, it was an important position. Her professional activities ranged from negotiations for the purchase of European collections to the identification of specimens sent from around the world, to supervising at least one doctoral dissertation, that of Abel Joel Grout, her student and the leading bryologist of the next generation.

Nathaniel Lord Britton, originally trained in geology at Columbia, became the leading botanist in New York after John Torrey's time. He was the major author of a large work still popularly known as *Britton and Brown,* although long since revised by Henry Gleason.[68] When the New York Botanical Garden was established, Britton became its first director, a position he held almost to the end of his life. Although Elizabeth Britton's position was the most important held by a woman in botany in New York, her husband's was much more important and powerful.

The Brittons both came from well-off families. The Knight family owned a sugar plantation and a furniture factory in Cuba, and Elizabeth spent much of her childhood there. She was, however, sent to live with her grandmother (Compton) while attending the private school run by Dr. Benedict in New York.[69] At a time when almost all the funding for science

---

Opposite: Elizabeth Britton at work at her microscope. Courtesy of the New York Botanical Gardens.

and scientific institutions came from philanthropy, Nathaniel Lord Britton had the right friends, including Cornelius Vanderbilt and judges and philanthropists Charles Daly and Addison Brown, all of whom were instrumental in fund-raising efforts on behalf of the New York Botanical Garden. Elizabeth Britton and Mrs. Charles Daly organized a women's committee of the socially elite. By 1896 the needed $250,000 had been raised, and Nathaniel Britton was appointed director-in-chief. The Board of Managers included Cornelius Vanderbilt, Andrew Carnegie, and J. Pierpont Morgan.[70] The Brittons were not as wealthy as these philanthropists, but their life-style was definitely upper class. They kept a city and a country house, complete with servants, and did a lot of entertaining. They spent their winters in warm places—Cuba, the Bahamas, Puerto Rico, Trinidad, and Jamaica—where they both did research on the indigenous plants. Obviously, the affluent wife of the director did not require a paid position, no matter how valuable her service.

The prejudices of the time probably further limited the ethnic and religious (and of course racial) backgrounds of the women who were even only peripherally employed in botany as recently as the early twentieth century. Anglo-Saxon names predominate and first-generation immigrants are absent almost entirely.[71] Elizabeth Britton shared some of these prejudices. Although Britton died in 1934, several people who knew her are alive and still working at the New York Botanical Garden.[72] One woman told me how, when she was very young, she had been hired as Mrs. Britton's secretary. She was asking nothing about her shorthand or other secretarial skills, only about her religion. She was Episcoplian, as was Elizabeth Britton, and she got the position. The applicant just ahead of her, a Catholic, was turned down. According to another source, a Jewish secretary was hired while the Brittons were on one of their winter trips. On their return, Elizabeth Britton met her and commented that they never had anyone with a name like that working there before.

Elizabeth Britton's marriage was an exceptional one. Both because of and in spite of her husband's importance in botany, she was able to do a great deal of research, to get it published, and to receive credit for it. It is very doubtful that she would have had any position, even "honorary" and unpaid, from which she could have carried on research had she not been Mrs. Britton. If she had remained single she might have continued as a botany teacher, perhaps eventually as a professor at Hunter, but her publications would probably have been few. Her ability to shape events in botany, as she did at the New York Botanical Garden, would have been nonexistent.

Her contemporaries, Susan Hallowell and Emily Gregory, well known as botany professors but not for publications, were not asked to be charter members of the new American Botanical Society, as Britton was. If she had married someone else in her social title, but a nonscientist, as Violetta White did, she probably would not have had a position of influence, or even a place to work. Had she had several children, perhaps that role, surely the primary one for all married women in 1900, would have consumed most of her time and energy. But the Brittons were childless. Still, her husband might have made her his assistant, like Eliza Sullivant earlier and many botanical wives later.[73] He did not. He studied vascular plants and she studied bryophytes, a different group of plants. Although they often worked in the same geographical areas and sometimes produced joint publications, her work was always her own and recognized as such. Nepotism does not seem to have been a problem at the New York Botanical Garden, at least in the absence of a paid position. Her husband even signed her letter of appointment as honorary curator.[74]

I have not been able to find any of Elizabeth Britton's letters to her husband, but several of his to her are in the New York Botanical Garden Archives, and they provide important insight into their marriage. He certainly appears to have treated her as an equal and a highly competent person. For example, on August 10, 1891, in a letter addressed "Dear B," he wrote to her from Geneva that he had been escorted all over the city and introduced to "Casimir De Candolle [a famous botanist] who was tremendously polite and put his herbarium at my disposal." The mosses, he wrote, were not in the best of order, "but I think you will be able to see anything you want. But the thing that will make you squeak is as follows: Barbey has instructed him to offer you the Jaeger Herbarium . . . for 3000 francs. . . It is richest in European species but I noticed in a few covers a considerable number from Tropical and South America." He noted "M. Barbey's great kindness in affording you [Elizabeth Britton] this opportunity to purchase it. So you will please devise a method of extracting the said $6000 from certain of our ornamental members of the Club."[75] She did indeed devise a method, since two years later an article by E. Britton described the Jaeger Moss Herbarium as "the most complete collection of exotic mosses in this country," which had recently been purchased through the "kindness of M. Barbey . . . and the generosity of several friends."[76]

The Britton's marriage was companionable; they obviously shared their experiences even when apart. A later letter from Nathaniel to Elizabeth Britton after he had returned to New York and while she was still in

Jamaica reads "The trip was entirely uneventful except for the fact that they insisted on my sitting at the captain's table with a bunch of English so I had to dress up, a blooming nuisance."[77]

Elizabeth Britton went on long plant-collecting trips in the Adirondack wilderness without her husband, though usually with a guide. Britton's faith in his wife's ability to manage difficult travels on her own are revealed in another letter from Geneva in August 1891, giving her directions to travel alone from Neufchatel to Lausanne. The next day she was to travel from Lausanne to Mortigny, arriving at 10:41 A.M.: "And you can get over to Chamonix the same day very nicely. I remember I walked it in about 9 hours."[78]

The Brittons' was an unusual marriage, and one in which a great deal of important botanical research was done by both partners. Elizabeth Britton wrote a total of 346 papers and reviews,[79] no mean feat for any botanist, male or female, in that age or this. She had great influence in her field of botany bryology, through her writings, her immense correspondence, and her many visitors, American and foreign, to the New York Botanical Garden.

In all three of the botanical marriages discussed above, the wives received both financial support and position from their husbands. Their husbands were three of the most important botanists of their times and fields and their contributions are still important today. Two of their wives are nearly unknown; they did not have positions. In Eliza Sullivant's time there were no positions for women botanists. In the case of the Sullivants, the inspiration and encouragement for botanical work were mutual, as was the work, but not the public credit for it. In Edith Clements's case, a faculty position might have been possible, but would almost certainly have been incompatible with marriage. Her botanical work, it appears, was also limited by her own conception of herself as primarily a helpmate, a common conception of educated wives even at late as the 1950s. Why were Elizabeth Britton's accomplishments within a botanical marriage so different? First, she was already a fairly well established botanist in her own right at the time of her marriage, and she had a husband who saw her as a full partner in their field. Second, although the power structure in botany was almost entirely a male one during this period, Nathaniel Britton had the power and position to enable his wife to use her considerable talents and ambition. No other married woman was so successful as a botanist in the nineteenth century, or much of the twentieth.

THROUGHOUT the eighteenth and nineteenth centuries in the United States, and for most of the twentieth, marriage and children have been considered the ideal of a good life for women, far different from the major goals of men. These were the views not only of society, but of the great majority of women themselves. Until the last quarter of the nineteenth century there were no professional positions for women botanists. Those whose achievements we remember either were supported by fathers until their marriages (e.g., Jane Colden); worked on the research of their husbands, who supported them and the work (e.g., Eliza Sullivant); or were widows of at least some means. Nearly all of them had children, often many children. In the latter two decades of the nineteenth century and the first part of the twentieth, positions for single women were available, especially at women's colleges, but also at government laboratories, and in spite of the negative connotations of spinsterhood,[80] many women chose that path. In 1921 nearly half of the married women scientists with sufficient training and accomplishments to be listed in *American Men of Science* were unemployed. Only 19.5 percent of those listed under botany were married. Even in 1938 only 21 percent were married.[81] A woman with a Ph.D. in botany who did marry usually found herself unemployed or working in her husband's lab. A woman who wanted a lifelong career as a botany professor at a women's college, however, had no choice other than to remain single.

Thus it would appear that the best strategy for recognition in botany for the women I've discussed, all born before the twentieth century, was to remain single. Yet there were a fairly large number, as we have seen, who did marry and were still heard from, that is, published significant botanical work. Their relationships with husband and work varied; some wives were helpmates, others independent researchers supported by the husband, and in the case of the Phelps, one cannot help but feel that he, at least after the age of sixty, was the helpmate. Widows occupy a surprisingly important place in this account. A number of women with at least some previous education were able to turn widowhood into a life of productive botanical work. Some of these widows had means; some had to support their children. They were similar in having had the traditional role of wife, highly prescribed in these times, suddenly removed, thus creating, along with many difficulties, a certain amount of freedom and, with it, new ambitions. Children per se do not seem to have been the major impediment to botanical achievement. Although two of the most successful married botanists, Kate Brandegee and Elizabeth Britton, were childless, several of the widows, including Almira Phelps and Wanda Farr, countinued their work and cared for their young children as well. A similar phenomenon can be seen today in high-achieving divorced women. A study of the relationship between marital status and graduate study showed that the most committed women graduate students were neither single nor married, but divorced. A higher proportion of divorced women students had published papers and were planning research careers. The opposite was true of married versus divorced male students. The author concluded that with divorce "men lose a supportive relationship, while women lose a source of severe role conflict." It is interesting to note that almost 70 percent of these divorced women graduate students had at least one child.[82] For the nineteenth and early twentieth-century widows, as well as for the more recent divorced women in science, it was not the responsibility for children that made scientific work difficult, but the role of the wife and mother in society.

When we look at those women botanists—single, married, or widowed—whose achievements in botany are apparent, we need to remember the hundreds of others in Rudolph's sample who showed an active interest but whose actual achievements were few.[83] They were severely limited by society's views of the proper place of women and, in many cases, their own acquiescence in them.

These views have changed very slowly, as have women's options and strategies in science. For women born before 1900, widowhood sometimes led to unusual achievement, but it cannot really be called a strategy. Even to remain single and teach at a women's college was hardly a strategy, in the absence of other choices. One can speak of strategies in the careers of male scientists in terms of positions and patronage as early as Galileo and Harvey and certainly in the late eighteenth and early nineteenth centuries, as ably shown by Dorinda Outram in the case of Georges Cuvier.[84] It was certainly true for mid-nineteenth-century American elite male scientists, for example, Benjamin Gould.[85] It may have been less true in botany than in astronomy or geology, since botany was only imperfectly professionalized even for men, but there were strategies or career choices open to Asa Gray in the 1840s: Michigan or Harvard, a professorship of botany and zoology or botany alone, albeit at lower pay.[86] No such options existed for lesser male botanists, and certainly not for women.

Even in the 1920s women had few strategies available. Nobel Prize winner Barbara McClintock cannot be included here as a nineteenth-century woman botanist. She was born in 1902 and is considered a geneticist, not a botanist, although she was worked all her life on corn. Genetics was a field separate from botany by the 1920s.[87] McClintock's problems were very similar, however, to those of the women discussed in this essay. She turned down the option of a teaching career at a women's college because she wanted a research career. But beyond that there were few available strategies. She could not obtain a position at a major university, even though her early work was very important and well recognized.[88]

Career obstacles rather than career strategies have been the norm for women botanists even in the past few decades, at both the graduate and the postgraduate levels.[89] Marriage compounds the problems caused by prejudice, since a woman, at least until very recently, tended to follow her husband's career relocations. Even the well-documented increase in academic commuter marriages in the 1980s, while a more equitable, less sexist solution, often puts strains on both career and marriage.[90]

Botany has changed dramatically in the 1980s, however. With the advent of plant recombinant DNA research, botany has become an elite science. A young woman well trained in this area does have options available. She can choose a prestigious postdoctoral fellowship, a good university position, or a highly paid job in industry. Perhaps her husband will follow her to the location of her choice and even transport their child to the day care center. But there remain many problems for women in botany, or any other science, particularly at the upper levels of academic and industry. The problems, however, have changed since the nineteenth century. For this generation of women botanists, unlike for their earlier sisters, to choose to remain single is no longer to be a "spinster," to marry is less likely to confine one to one's husband's laboratory, and widowhood is no longer a "strategy" for achievement.

6

MARILYN BAILEY OGILVIE

Marital Collaboration: An Approach to Science

> Allow me, first of all, to tell you that I am happy to speak today before the Academy of Sciences which has conferred on Mme. Curie and myself the very great honour of awarding us a Nobel Prize.[1]

THUS Pierre Curie, belatedly fulfilling his and his wife's Nobel Prize obligations to the Swedish Academy of Sciences, reported the results of this best-known husband-wife scientific collaboration. The marital-team approach of the Curies has surfaced numerous times throughout the history of science, for collaboration with a mate traditionally has been a way through which women could enter science unobtrusively. Some husband-wife collaborative efforts took the form of the husband-creator, wife-executor mold, in which the wife was highly regarded by the scientific community and by her husband partially because of her willingness to accept a subordinate role. Margaret Huggins (1848–1915) was very comfortable in that position. Although she coauthored papers with her husband, William (1824–1910), they were written from William's perspective. Never overstepping traditional sex-role boundaries, Margaret Huggins was popular with her male contemporaries. Hertha Ayrton (1854–1923), on the other hand, was brash and bold, although never as secure as external appearances indicated. Collaboration in her case provided her an entrée into the world of science. Initially working with her physicist husband, she branched off into independent research, publishing and presenting papers on her own. Audacious

Ayrton was very different from the proper Huggins. Yet she, like Huggins, was able to command the attention, although not always the affection of, her scientific colleagues. A third nineteenth-century woman scientist, Amalie Dietrich (1821–91), would never have gotten involved in science if it had not been for her husband, Wilhelm. Yet after she learned from him initially, she far surpassed anything he did scientifically. Diffident throughout her life, she nevertheless overcame incredible odds to contribute to science.

Of the many women who have collaborated with their husbands in science, Ayrton, Dietrich, and Huggins were selected as the subjects of this essay instead of better-known examples, such as Marie and Pierre Curie and Irène and Frédéric-Joliot Curie, precisely because they are not so well known. They represent three different fields, have made documentable contributions to science, offer an opportunity for further research, and demonstrate vastly different ways in which marital collaboration was productive to science. Ayrton's and Huggins's science reflects the happy fruit of collaboration, whereas Dietrich's is an example of collaboration gone awry. Their scientific contributions, educational backgrounds, relationships with spouses and children, and views about the role of women are explored in order to elucidate what proved to be a viable approach to the alternatives of a career or a family.

Science, of course, is never created in a vacuum. Politics, economics, religion, education, institutional affiliations, and human associations of all kinds intrude on the creative process. For both men and women scientists, family relationships are important in determining the character and quantity of the science produced.[2] Running counter to the typical Victorian family size, Dietrich and Ayrton each had only one child and Huggins was childless; the presence of the single child or the absence of children influenced the course of their scientific achievements. The effect of husbands Wilhelm Dietrich, William Huggins, and W. E. Ayrton on their wives' scientific activities was sometimes liberating, sometimes constraining. Nevertheless, the common denominator, the pervasive presence of a scientific spouse, undergirded the scientific careers of Amalie Dietrich, Margaret Huggins, and Hertha Ayrton. In each case, both husband and wife worked in the same field. The Dietrichs were involved in botanical collecting, the Hugginses in observational astronomy, and the Ayrtons in applied physics.

If science is defined as an attempt to explain natural phenomena, the inclusion of Amalie and Wilhelm Dietrich under the rubric scientist might be questionable. Certainly little explanation was involved in their work as natural history collectors. Wilhelm Dietrich's assessment of himself as a great botanist seemed to be shared by few others. Before their marriage, Wilhelm regaled Amalie with tales of his family's commitment to botany. Although several members of this country botanist (*Bauernbotaniker*) family from Jena in Germany had made significant contributions to systematic botany, Wilhelm's own contributions were meager. In biographical dictionaries that include the Dietrich family, Wilhelm is consistently omitted. Ironically, in view of his condescending attitude toward Amalie, an account of her life, not his, is included in the *Neue deutsche Biographie*.[3]

Wilhelm Dietrich's botanical collections were not unique. Amalie's, on the other hand, were. Her collections added to the understanding of the flora and fauna of Australia and Tonga. Although she did not invent any scientific theories herself, she was familiar, through Wilhelm's tutoring, with the Linnaean system of classification and was aware of the advantages and disadvantages of natural and artificial systems of classification.[4] Through her understanding of this system and by carefully classifying and preparing her collections of botanical, zoological, and ethnological objects, she provided research material for others to interpret. Although she did not publish any research papers, her letters from Australia provide information about the flora and fauna of this continent, as well as her reactions to a variety of situations.[5] Her work was recognized by her contemporaries. Two species of wasps, *Nortonia amaliae* and *Odynerus dietrichianus;* a moss, *Endotrichella dietrichiae;* and two species of algae, *Amansia dietrichiana* and *Sargassum amaliae,* were named for her. In 1867 Amalie Dietrich was named a fellow of the Entomological Society of Stettin (Pomerania Province, Prussia, now Poland).

Born in Siebenlehn, Saxony, Amalie (Nelle) Dietrich's purse maker father and housewife mother were uneducated villagers. After a brief time at the local school, she became an avid reader, but it was not until she met Wilhelm Dietrich, a "gentleman naturalist" who described himself as having "a love for botany in . . . [his] blood," that she became interested in natural history.[6] Wilhelm taught and Amalie absorbed the Latin names for plants and methods of specimen preparation. The relationship between Amalie Dietrich and her husband was dictated to a large degree by Wilhelm's personality. An arrogant tyrant, he found amiable, pliable Amalie, the untutored country girl, an ideal mate. Flattered that a "famous" botanist was interested in her, she promised to help him whenever and however he needed. Prophetically, she promised to "fetch and carry" for him. Nothing, she predicted, "will be too heavy."[7] As it turned out, when she went traipsing after Wilhelm she was the one who usually carried the collecting equipment. She explained:

> I should have to carry the loads about from one land to another. I offered to do so of my own accord: Wilhelm did not demand it of me. The basket used to chafe my back till it was raw, so I suggested that we should try a cart and dog. . . . You ask angrily why Wilhelm does not do the carrying when there is carrying to be done? Have I not told you often enough, that Wilhelm is a refined scholar of gentle birth and delicate physique? Do try and take that in. He simply can't do it. I can, and why shouldn't I. . . ."[8]

During her labors Amalie Dietrich learned to identify plants and to prepare the collections. She later left Wilhelm, became successful on her own, and reversed the situation. Much later, she wrote to Wilhelm and invited him to be her assistant. He declined, but confessed to her that "the scales fell from my eyes when you took leave of me, when I saw that others recognized your true worth."[9]

The conflicts that precipitated Amalie's independent stand stemmed from Wilhelm's treatment of their daughter, Charitas Concordia Sophia. He was incensed, first, because their child was a girl and, second, because her presence intruded on Amalie's time with him. Wilhelm Dietrich insisted that she leave the child with her own parents so that she could continue helping him in the field. Amalie's mother died when Charitas was four years old, however, leaving Amalie with the child and unprepared to manage a household. Wilhelm expected perfection in the operation of the household, as well as expert assistance in his work. He flew into a rage when Amalie burned his collar; he complained when she forgot to change the papers in the plant press. Amalie meekly submitted to his demands until she realized he was having an affair with the hired girl. Then she took Charitas and traveled to Bucharest to live with her brother and his wife. They agreed to care for Charitas, and Amalie accepted a position as a housekeeper. But unwilling to spend the rest of her life as a maid, she accepted what seemed her only other option, to return to Wilhelm. Pleased to find his excellent help returned, Wilhelm now insisted that she accompany him on long collecting trips, leaving Charitas with any stranger who agreed to take her. "You see what a nuisance the girl is. It is a tragedy that she is not a boy—a little Gottlieb Dietrich."[10] In a letter to her brother, Karl, Amalie wrote: "You know, every time I have to give Charitas to strangers I ask myself again and again: Which duty comes first? Should I be the helpmate of my husband, or a mother to my child? But in any case, whatever way my heart may decide, necessity leaves no choice. Wilhelm says—and I quite see his point—we must travel; that is essential to our profession."[11]

The situation, however, continued to deteriorate. The final blow occurred when Wilhelm commanded her to go to Holland and Belgium to collect algae. On this trip she became very ill, spent weeks in the hospital, and upon returning home found that Wilhelm had become a tutor to a count's son and had sent Charitas away to be a household servant. This time she left him for good. Doing the unthinkable, she went to Hamburg by herself to try to sell her collections. She met a Herr Walther, who wanted to buy the specimens but did not have enough money. In exchange for her moss collection, however, he provided an introduction to Dr. H. A. Meyer, a businessman who was interested in plants. If the arrangement did not work to Dietrich's advantage, Walther promised to return the mosses.

This bargain marked the turning point in Dietrich's career. Not only did Meyer buy the collection, but he befriended Amalie as well. Meyer and his wife became almost foster parents to Charitas. Free to travel now that her child was well cared for, Amalie could pursue her profession. During her long absences, the Meyers served as Charitas's surrogate parents, seeing to her education and preparing her to earn a living by teaching kindergarten. Amalie began to realize that the daughter she had left so many years ago had developed a system of values different from her own. In 1870 she wrote rather petulantly: "You are taken up with all manner of things, yet it seems to me that for natural history in particular you have little genuine interest. . . . You write about every possible thing, but I have never heard if you have visited the Zoological Garden in London, or if you have been to the Aquarium."[12]

Amalie and Charitas were reunited in 1873, ten years after Amalie had left for her collecting trip. Gradually, mother and daughter became reacquainted. Amalie had expected her relationship with her daughter to resume as if the ten-year separation had never happened. She was disappointed to hear that Charitas was to be married and that her daughter had a life of her own. Eventually, however, they were reconciled.

Amalie Dietrich's involvement in women's rights was dictated by necessity rather than through self-conscious decision. During her marriage to Wilhelm, she accepted his role as ruler and her role as subject. When Amalie left Wilhelm for the first time, her alternatives were so bleak that she returned; life was even more dismal than before, if that was possible. When she left him for the second and last time, she decided to continue collecting independently of Wilhelm. While not an explicit statement on women's rights, Amalie Dietrich's independent stand reflects a radical feminism of its own.

Initially dictated by desperation, Dietrich pioneered a career as a natural history collector. Through Meyer's help, she was employed by Caesar Godeffroy, who was establishing a museum for the geography, natural science, and ethnology of the South Pacific. Originally skeptical about employing a woman, Godeffroy was finally convinced of Dietrich's suitability by testimonials of competence from associates and Meyer's recommendation. Her first assignment was in Australia. Before leaving, she returned to her home in Siebenlehn to pay bills, to say good-bye to her father, and to visit Wilhelm once again. Lashing out at her good luck, he expressed indignation that she, a woman, and not he, "who comes of the famous old botanical stock," got the offer.

From this point on, Dietrich's self-confidence and self-sufficiency gradually began to grow. Conservative in her first venture at calculating the projected cost of her supplies, she won Godeffroy's confidence in her financial responsibility. After an eighty-one-day voyage to Australia, she undertook her first collecting trip on that continent. Glorying in her new independence, she soon expanded her collecting beyond plants.

> No one circumscribes my zeal. I stride across the wide plains, wander through the virgin forest. I have trees felled in order to collect different kinds of wood, blossoms, and fruit. I cross rivers and lakes in a small canoe, visit islands and collect—collect—collect. I speedily forget the discomforts of heat and mosquitoes in the unbounded feeling of joy that animates me when, at every step, I light upon treasures that no one has secured before me. I have no fear of not fulfilling the expectations that Godeffroy's have placed in me. When I wander over great spaces without let or hindrance, I think no king can feel so happy and so free as I. It is just as if Herr Godeffroy had made me a present of this vast continent.[13]

Throughout her collecting experiences she underwent many hardships, all of which made her more confident in her own abilities as a woman and as a competent collector. After she returned to Germany, she became a popular guest at some of the best-known Hamburg houses. One of her last triumphs occurred when she attended an anthropological conference in Berlin at which a paper was read on Australia. Since women were not admitted, the doorkeeper would not allow her to enter. Finally, she received permission to listen from a corner of the gallery. After the doorkeeper explained to the privy councillor who she was, he came out himself, "brought the old lady in, conducted her past all the rows of the audience, and introduced her to the committee."[14]

ASIDE FROM the fact that Amalie Dietrich and Margaret Huggins both worked in scientific disciplines and both had scientific spouses, they appear, on the surface, to have little in common. Different national origins, socioeconomic backgrounds, subject interests, and relationships with their spouse dominate their lives. Yet both illustrate the importance of marital collaboration in science.

Margaret Murray, born in Dublin, Ireland, became interested in science before she met her husband. Her mother died when Margaret was quite young, and after her father remarried she was left alone much of the time. Her grandfather, an amateur astronomer, spent evenings teaching Margaret to identify the constellations. From this early exposure, she became interested in studying the heavens with home-made instruments. After reading a "how to" article in the magazine *Good Words*, she constructed a spectroscope. William Huggins, already fascinated by stellar spectroscopy, was impressed by Margaret Murray's interests and abilities. They were married in 1875. Of this union he said: "[I] had the great happiness of having secured an able and enthusiastic assistant."[15]

Both Margaret and William Huggins stressed observation over theory. Although they assumed that explanation eventually would emerge from persistent, patient observation, they were less concerned with theoretical conclusions than with recording the phenomena themselves. Demanding precise observational accuracy, they abhorred the idea that a preconceived theory might influence the course of observation. William Huggins had established his observatory in 1856, in connection with "my private dwelling-house at 90, Upper Tulse Hill, London," and opened "up a passage from the house, and raised it so as to command an uninterrupted view of the sky except on the north side."[16] Because William was twenty-six years older than Margaret, he already provided insight into the potential of several new observational techniques before his wife became involved. After studying Gustav Robert Kirchhoff's (1824–87) ideas that the lines in the solar spectrum showed the chemical composition of its atmosphere, William concluded: "Here at last presented itself the very order of work for which in an indefinite way I was looking—namely to extend his novel methods of research upon the sun to the other heavenly bodies."[17]

Huggins's spectroscopic observations rescued the nebular hypothesis for the origin of the stars from obscurity. Once popular, this theory, which postulated that the stars originated from a gaseous nebular material, had fallen into disfavor after William Parson's (1800–1867) powerful telescope resolved many "gaseous" nebulae into component stars, causing most astronomers to assume that all nebulae could be so resolved. William Huggins's spectroscopic observations, however, implied two distinct types of nebulae: those resolvable into stars and those composed of an unresolvable luminous gas. The latter category, he proposed, could be the formative material of future stars, thus reinstating the credibility of the nebular hypothesis.[18] He also speculated that the Doppler effect, combined with spectral analysis, might be useful in determining the motions of stars in the line of sight. In a paper of 1869, he explained: "If the stars were moving towards or from the earth, their motion, compounded with the earth's motion, would alter to an observer on the earth."[19] His method provided a technique for determining stellar motions, the rotation of the sun, of "Saturn and his Rings, and Jupiter as well as a means of separating double-stars."[20]

When she married William Huggins, Margaret stepped into a ready-made set of astronomical problems. She had no formal training, although as mentioned she had as a child exhibited a passionate interest in astronomy, examining the sun during the day with an instrument of her own construction and the stars at night by "means of a star atlas and dark lantern."[21] She began the systematic observation of sunspots at the age of ten, taught herself the principles of photography (eventually becoming an excellent photographer), and manufactured a crude spectroscope. In 1875, the year of their marriage, the feasibility of photographing stellar spectra occurred to both Margaret and William Huggins. Her skill as a photographer, as well as "her sensitiveness of eyesight and extreme accuracy in measurement," became very important to the collaboration.[22]

The common interest in spectroscopy that initially brought Margaret and William Huggins together persisted throughout their careers. Although Margaret often is characterized as William's assistant, good at photographic manipulations and visual observations, he increasingly respected her work. The nature of their collaboration can be understood by analyzing their published papers. In their early joint efforts, William wrote in the first person singular.[23] For example, he used this form in a paper on the Orion Nebula, noting: "I have added the name of Mrs. Huggins to the title of the papers, because she has not only assisted generally in the work, but has repeated independently the delicate observations made by the eye."[24] Later memoirs that issued from the Tulse-hill Observatory bore both their names; because

---

(Overleaf)

William Huggins, about 1900, from *The Scientific Papers of Sir William Huggins, K.C.B., O.M.* (London, 1909). Courtesy of the History of Science Collections, University of Oklahoma.

Margaret Huggins, about 1905, from *The Scientific Papers of Sir. William Huggins, K.C.B., O.M.* Courtesy of the History of Science Collections, University of Oklahoma.

SIR
K·C·B · O·M
ÆTAT

of the nature of these publications, it is difficult to isolate individual contributions. When the papers were written under both names, William always was listed as the senior author and the communication was always worded from his perspective. Although in the later papers he wrote in the first person plural, he made it apparent that he was the one who did the writing.[25]

William and Margaret Huggins produced joint papers on the planets, the spectrum of the star Nova Aurigae, and chemical spectra. By examining the dates on these papers and the person and number in which they are written, Margaret's increased participation in the process is implied. Of two papers on the planets issued under both their names, the first one, published in 1889, was written in the first person singular, whereas a second, published in 1895, was written in the first person plural.[26] The three papers on Nova Aurigae were written in 1892 and 1893 in the first person plural.[27] Six papers on chemical spectra published jointly from 1897 to 1905 were all written in the first person plural. Although William evidently still considered himself the chief author, in the later papers he recognized his wife's importance in the collaboration.[28]

In addition to numerous papers, most of which were communicated to the Royal Society, Sir William and Lady Margaret Huggins brought out two volumes of the publications of the Tulse-hill Observatory. Margaret contributed pencil sketches and "charming initial letters which beautify this stately scientific treatise."[29] In evaluating Margaret Huggins's contributions to astronomy, there is little doubt that she added to the bulk of astronomical data and worked well as a part of a team with her husband. Yet an assessment of her originality must wait for a detailed investigation of primary source materials.

Margaret and William Huggins shared common interests in many subjects other than their work. They were both interested in music. William loved to play his Stradivarius violin while Margaret accompanied him on the piano or organ. Both were interested in violins and their history; Margaret published an account of the life and work of a Brescian master violin maker, Giovanni Paolo Maggini. A fine artist as well as a musician, she painted during their holidays and transformed watercolor sketches of their vacations into Christmas cards. Her interest in art and history was reflected in her hobby of locating and restoring antiques. Known for her excellent letters, Margaret Huggins documented the multifaceted life of a nineteenth-century "Renaissance woman."

Margaret Huggins was typical of the best of the intelligent, sensitive, diligent nineteenth-century intellectual women. She appreciated her position as a woman and resented any insinuation that men discriminated against women scientists. "I find that men welcome women scientists provided they have the proper knowledge," she asserted. "It is absurd to suppose that anyone can have useful knowledge of any subject without a great deal of study." Women, she said, often have not engaged in the study necessary for excellence in science. "When women have really taken the pains to fit themselves to assist or to do original work, scientific men are willing to treat them as equals." She denied that men "wish to throw hindrances in the way of women who wish to pursue science." She agreed that "the lady doctors had a great fight, it is true, but that is old history now, and there were special ancient prejudices involved."[30] Interestingly, her own experience seemed to contradict her pronouncements, for since women were not allowed to become regular fellows of the Royal Astronomical Society, she was forced to be content with an honorary membership in May 1903.

HERTHA AYRTON had little in common with Amalie Dietrich or Margaret Huggins. Sources available on her life and work include a biography by Evelyn Sharp, numerous scientific publications, and unpublished references at Girton College, Cambridge. Although Hertha and her husband, W. E. Ayrton, explored the explanatory dimensions of science, they remained more interested in practical problems than in abstract theory. W. E. Ayrton, elected a fellow of the Royal Society in 1881, was a pioneer in electrical engineering, a staunch advocate of technical education, and a supporter of women's rights. During his lifetime he published 151 papers and a book on practical electricity (1887). Hertha Ayrton entered science through invention. Based on an idea of her cousin Ansel Lee, she created a device designed to divide a line into any number of equal parts. The instrument, she explained, was useful to artists, decorators, engineers, and ship's navigators.[31] After working for a year on the model, she obtained a patent in 1884. The favorable reception accorded this invention by architects, artists, and engineers encouraged her to consider scientific research as a profession.[32]

(Overleaf)

W. E. Ayrton, from Rollo Appleyard, *The History of the Institution of Electrical Engineers, 1871–1931* (London, 1939). Courtesy of the History of Science Collections, University of Oklahoma.

Hertha Ayrton, from *The History of the Institution of Electrical Engineers, 1871–1931*. Courtesy of the History of Science Collections, University of Oklahoma.

Hertha Ayrton was born Phoebe Sarah Marks in Portsea, England, the third child of Alice Theresa and Levi Marks. Her father was a Polish-Jewish refugee who constantly struggled for solvency in his clockmaking-jewelry trade in Portsea. After Levi's death in 1861, Alice Theresa attempted to support the family by her needlework, no easy task, for there were eight children in the family. Influenced by her mother's determination to support her family as well as her liberal attitudes toward child rearing, Sarah Marks became self-assured and self-sufficient at an early age, sometimes to the extent of appearing abrasive. One of her sisters was an invalid, so Marks helped care for her siblings and learned to sew, cook, and keep house, allowing her mother to earn money by her needlework. Independent, freethinking, and stubborn, Marks was able to get an education only because she had an aunt who ran a school in London. As an expression of independence, she rejected her given name and adopted a new one, Hertha, suggested by her friend Ottilie Blind.

A devout Jew in her younger days, Marks became a skeptic after her association with her cousin Marcus Hertog, a freethinking graduate of Cambridge. Nevertheless, throughout her life she continued to take pride in her Jewishness. The largess of Barbara Bodichon, an eccentric philanthropist interested in women's causes and one of the founders of Girton College, Cambridge, made it possible for Marks to take advantage of new opportunities for women to obtain a university education in England. In the early 1860s Emily Davies had begun organizing women teachers into schoolmistresses' associations. After convincing Cambridge authorities to open its local examinations to girls in 1863, she made plans to establish a women's college at Hitchin. This college opened in 1869 with five girls and moved to the village of Girton, outside of Cambridge, in 1873. Davies insisted that girls must study the same subjects and take the same examinations as boys.

If it had not been for Barbara Bodichon, Marks's financial position would have made it impossible for her to attend the new institution. Even though she was reared in her aunt's home, she considered herself under an obligation not only to support herself, but to send money home to her family as well. After six years of working as a governess, thus providing for herself and contributing to her family's needs, Marks explored ways to continue her education at the newly established Girton College. An introduction to Bodichon in 1873 opened the door to further education.[33] Because of her family requirements she did not enter Cambridge immediately, but passed the Cambridge University Examination for Women in 1874 with honors in English and mathematics. She took the scholarship examinations in 1876, but failed to win either of the two openings. Her mediocre showing on the examination was characteristic of her future examination performance.

Though unsuccessful in the scholarship attempt, her friends, especially Barbara Bodichon, scraped together enough money to allow her to enter Girton as a student in October 1876. A short time after Marks entered Girton, she became ill and was forced to rest. Again, her friends were loyal and made it possible for her to return. Although during the time she was in attendance students at Girton College could sit for examinations, they had to do so unofficially in a room in their own college. After the papers were marked, the mistress received the names of the successful women. Their names, however, were not printed with the list of successful male candidates. Marks took her tripos examination in 1880, again with disappointing results. Apologetically writing to Barbara Bodichon, she noted that she was only fifteenth in the Third Class (lowest third): "I am so sorry. I am afraid you will be very disappointed."[34]

Marks's first foray into the world of work after leaving Girton was as a mathematics teacher at Kensington High School. Uncomfortable in a classroom situation because she was unable to give scope to individual differences, she resigned her position and accepted private students.

During her teaching career, Marks met her future husband; she remarked in a letter to Barbara Bodichon in 1882 that she had "had a great deal of conversation with a Professor Ayrton an electrician."[35] She was, therefore, acquainted with him before she invented her line divider in 1883–84 and decided to pursue a career in research.

The strongly independent personalities of both Ayrtons made their relationship more of a mutually supportive one than an actual partnership. When Hertha Marks announced to her family in 1885 that "yesterday evening, Professor Ayrton asked me to marry him, and I accepted him," there was some family disapproval because he was not Jewish.[36] Ayrton, a widower with one daughter, Edith, believed in equality of opportunity between men and women. After he became ill in 1905, he increasingly relied on his wife to complete his scientific commissions. Ayrton obviously was proud of his wife and made it clear that he wanted her to get the credit she deserved. On the fourth report he insisted that Hertha's name be included as the author of the publication. As Hertha Ayrton became more involved in the militant wing of the suffrage movement, Professor Ayrton staunchly supported her stand.

The birth of a daughter to the Ayrtons evoked very different emotions in her parents than did the birth of Charitas to the Dietrichs. Both parents were delighted with Barbara Bodichon Ayrton, named after Hertha's benefactress. Ayrton cheerfully devoted her early married life to caring for the child. Because of her husband's attitude, she was able to continue with activities outside of the home, including science and social concerns. The Ayrtons even involved Barbara in their research, much to her delight. In 1898 W. E. Ayrton produced a paper, "The Science of Smell," in which he enlisted the help of both his wife and daughter.[37] As Barbara grew older, she exhibited a typical teen-aged complaint about her parents, insisting that she wished "Mother had a boudoir, all filled with yellow satin furniture, instead of a laboratory—like the mothers of other girls!"[38]

Hertha Ayrton consciously supported women's causes. Her fierce independence, her educational experiences, her husband's egalitarian attitudes, and her success at penetrating male-dominated institutions nurtured her inclinations. When the Institution of Electrical Engineers broke tradition and invited Ayrton to read her paper "The Hissing of the Electric Arc" on March 23, 1899, the presentation attracted considerable notice. Although the newspapers were less concerned with the subject matter of her paper than the sex of the presenter, the scientists who were present took her seriously and she soon became known as an authority on the arc.

Hertha Ayrton's lifelong interest in the electric arc had begun in 1893. During that year, W. E. Ayrton went to Chicago to read a paper. Trusting his wife to complete his unfinished experiments, he arranged for his laboratory facilities at Central Technical College, Exhibition Road, Kensington, to be made available to her. She forwarded the experimental results to her husband in the United States, and her results were incorporated into his Chicago paper. This paper included experimental results from both of their efforts. The experiments encouraged her to continue on her own in order to "solve the whole mystery of the arc from the beginning to the end."[39] Unfamiliar with the accepted explanations and lacking the formal background of most workers on the subject, Ayrton produced novel interpretations of the experimental results. Her very ignorance of the "establishment" point of view may have allowed her the freedom to be creative in her interpretation. Others recognized her expertise, for in 1895 she was invited to write a series of articles for the *Electrician*.

In 1897 one of Hertha Ayrton's papers was read by proxy before the British Association for the Advancement of Science at Toronto. If she had not been physically exhausted and unequal to the long journey, she would have read the paper herself.[40] For the next meeting, in 1898 at Bristol, she read her own paper, in which she described her search for a satisfactory insulating material that would both withstand the heat of the arc and not disturb it by its thickness.[41] W. E. Ayrton was president of the Mathematical and Physical Science Section of the meeting at which Hertha presented this paper. In 1899 she presented a paper to the Institution of Electrical Engineers, reporting the results of an experiment she had conducted. She attempted to explain why the arc lamp was initially silent, then hummed and hissed, and finally produced a range of brilliant colors with a marked reduction of steadiness of illumination.[42] Her experiments indicated that "there can be no shadow of doubt that the phenomena accompanying the hissing of the open arc are due to the oxygen in the air getting directly at the crater and combining with the carbon at its surface."[43] A practical application resulted from the theory, for by protecting the arc from direct contact with air, the current could be increased without producing the ill effects of hissing or loss in light.[44] This paper, in addition to her articles in the *Electrician*, made her recognized as a leading authority on the electric arc. (The summary of her efforts in this area later appeared in her book *The Electric Arc.*[45])

The president of the Institution of Electrical Engineers, Silvanus Phillips Thompson, FRS, found Ayrton's paper especially commendable, since Ayrton was not a member of the society. The excellence of the paper encouraged him to support her for membership. He "was unable to say whether there was any legal disability against a lady being elected," but "expressed the hope that she would soon be numbered among them." The hope was fulfilled, for "appropriately, on Lady Day [May 25], 1899—she was elected a full Member." The president was then able to add to his summary of the session a report that "the list of members includes the name of the first lady elected to the Institution. I am very glad to have been President during the Session that has seen that innovation."[46] During the same year, Ayrton presided over the physical science section of the International Congress of Women in London. In 1900 she spoke at the International Electrical Congress in Paris.

Even though the Royal Society was not yet ready to allow a woman to read a paper, it compromised by inducing John Perry, FRS, to read Ayrton's paper "The Mechanism of the Electric Arc."[47] In November 1901, after W. E. Ayrton received a Royal Society Medal, an attempt was made to secure Hertha Ayrton's election to the society. Because of the provisions of the charter, a married woman could not be admitted. Although it was hoped that by bringing the issue before the group a supplemental charter would be issued, the attempt was unsuccessful. Henry E. Armstrong, writing in *Nature*, expressed the views of some of her male associates when he said: "Most of us thought, at the time, that they [John Perry and W. E. Ayrton] were ill-advised in preferring her claim to the Royal Society."[48]

After 1901 Ayrton changed the focus of her research. For the sake of

W. E. Ayrton's health, the family visited the seaside community of Margate. Without a laboratory, Hertha Ayrton could not continue her work on the arc. Rather than brood over the loss, she became intrigued with a problem in her new environment. The investigation, involving the formation of sand ripples, extended from 1901 to 1905. She observed that the sandy shore was "covered with innumerable ridges and furrows, as if combed with a giant comb."[49] Two kinds of sand ripples were produced by waves. The small ripples, she postulated, were formed by water flowing over irregularities on the seashore, causing sand to be formed on either side of the primary ridge. Large ripples appeared when stationary waves flowed constantly to and fro over the same spot. As the oscillation continued, the ripple-marked sand was pushed into two mounds between the crests of the waves. She pointed to the chains of sandbanks in the sea as well as sand dunes on the shore to illustrate the second kind of ripple formation.[50] Moving from theory to practice, she suggested that it was not "altogether beyond the bounds of engineering science . . . to imagine some way of altering the trend of the sea, so as to keep it from forming the stationary waves which, I believe, heap up the dangerous Goodwin Sands."[51]

In addition to the electric arc and the sand ripple problem, Ayrton was involved in several other areas of research. After her husband became ill, Ayrton completed several projects assigned to him by the admiralty. One such effort involved determining specifications for carbon to burn successfully in searchlight projectors. Invention rather than science marked her last research endeavor. When England became increasingly involved in World War I, Ayrton applied her inventiveness to a fan designed to make it possible "for our men to drive off poisonous gases and bring in fresh air from behind by simply giving impulses to the air with hand fans."[52]

During the years 1906 to 1908, W. E. Ayrton's health continued to fail, and Hertha became involved in consolidating much of his research. At the same time, she was becoming more of a political activist for the militant wing of the suffrage movement. Her involvement, however, did not destroy her scientific pursuits. After her husband died in 1908, Hertha Ayrton continued in her roles of scientist and suffragist. She took part in the most militant demonstrations. Her actions in the demonstrations of November 1910 even led to the postponement of her paper "On Some New Facts Connected with the Motion of Oscillating Water." Although she had read the paper to the Royal Society in 1911, it was not accepted for publication. She was convinced that the referee had rejected it because of her suffrage activities.[53]

The last time she would participate in a demonstration was in 1911. But she passed her feminist ideas down to her daughter, Barbie, who became very active and eventually was jailed for the cause. This event prompted Ayrton to write to her stepdaughter, Edie: "I am *very* proud of her."[54]

Although Ayrton never became a fellow of the Royal Society, she was awarded the Hughes Medal for original research. She received notification on November 1, 1906, that she had achieved this honor for her experimental investigations on the electric arc and her work on sand ripples. Favorable press notices, such as the one in the *Times* engineering supplement, proclaimed: "Our countrymen have been slow to realize that women possess brains; we are too apt to regard them as the mere helpmeets of men. . . . It seems that the time has now come when woman should be permitted to take her place in the ranks of all our learned bodies."[55]

MARGARET HUGGINS and Hertha Ayrton were interested in science before they met their husbands, whereas Amalie Dietrich probably would never have considered a scientific career if it had not been for her marriage. Yet because of their circumstances, collaboration with their spouses enhanced the probability of their success in science.

An uneducated German country girl, Amalie Dietrich originally was totally dependent on her husband for guidance. She had no knowledge of scientific possibilities, and if it had not been for her husband's ungentle tutoring, she never would have become interested in science. His instruction in the Linnaean system, techniques of specimen preparation and record keeping, and the honor involved in a scientific career made possible her later independent efforts. After an initial period when she was only a servant and unpaid assistant to her husband, she, through necessity, began to recognize her own potential. Necessity sparked her emancipation, yet once it was attained, she demonstrated her self-reliance, and her achievements far outshone those of her husband. Diffidence because of her humble origins and lack of formal education remained a part of her personality, yet she developed an almost passionate devotion to her career in natural history. Her self-effacing image posed no threat to her male associates; they could afford to congratulate her on her successes for she was not in direct competition with them. They found economic necessity easier to understand as a female motive for entering a male-dominated sphere than mere intellectual curiosity.

Whereas Ayrton and Dietrich branched out on independent courses after an initial period of collaboration with their spouses, Huggins's career remained welded to that of her husband. Like Ayrton, Huggins was interested in science before she met her husband, but unlike her, she did not receive a university education. Collaboration gave her an opportunity to work successfully in a field that had previously intrigued her. In the case of Margaret Huggins, it is difficult to distinguish individual contributions. Although William Huggins wrote the papers, acknowledging the contributions of his wife, he did not specify the exact nature of these contributions.

More than the others, Ayrton had the potential to succeed on her own. Aggressive and ambitious, she had every intention of exploiting her talents to their fullest. Her association with radical feminists, her flirtation with liberal ideas and causes, and her determination to obtain the best possible education suggest that she was unlikely to occupy a subordinate role. Yet these were the very characteristics that raised the hackles of her scientific associates. Her interest in science predated her association with W. E. Ayrton, but as she herself no doubt recognized, her chances of scientific success were enhanced by her marriage. Professor Ayrton's liberal views about the capabilities of women and his tolerance for eccentricity and creativity in his wife made him an ideal spouse for Hertha. As an established scientist, his name allowed her ingress to circles that normally would have been closed to her. His professorial positions provided her with laboratory facilities; his familiarity with the subject matter introduced her to fields of research that she otherwise might not have considered. From Hertha Ayrton's early tentative ventures as an assistant to her husband and through a period of collaboration with him, she emerged as a competent investigator on her own. By the time his health had failed, she was confident in her own abilities and was able to strike out in original directions.

Hertha Ayrton's outspoken views alienated her from some of her male colleagues. For example, H.C.H. Carpenter, in his obituary account in *Nature,* compared her unfavorably to W. E. Ayrton's first wife by using operatic allusions. Wife number one, although also a feminist, was soft-spoken, a "Melisande," whereas about Hertha he could only say: "I should not quite know where to place the second musically but it would be near to Brunhilde, as she had much of the vigour of Wotan's masterful daughter and, at least, aspired to be an active companion of scientific heroes."[56] Carpenter was irked by Hertha Ayrton's self-confidence, acquired through overcoming early economic struggles and her acceptance of the possibility of success. Such an aggressive woman, he asserted, was not a proper companion for W. E. Ayrton.

> I often told him that he and his wife were an ill-assorted couple: being both enthusiastic and having cognate interests, they constantly worried each other

> about the work they were doing. He should have had a humdrum wife, "an active, useful sort of person," such as Lady Catherine recommended Mr. Collins to marry, who would have put him into carpet-slippers when he came home, fed him well and led him not to worry either himself or other people, especially other people; then he would have lived a longer and happier life and done far more effective work, I believe.[57]

Carpenter's biases made it impossible for him to see much of value in her research. Anything original must have derived from her husband: "Indeed, I always thought that she was far more subject to her husband's lead than either he or she imagined . . . though a capable worker, she was a complete specialist and had neither the extent nor depth of knowledge, the penetrative faculty, required to give her entire grasp of her subject." He concluded: "She was a good woman, despite of being tinged with the scientific afflatus."[58]

From a study of these three women, professional collaboration with a spouse emerges as one avenue through which nineteenth-century women entered science. Different personality types, different economic and social backgrounds, and different types of spouses caused different manifestations of marital collaboration. Amalie Dietrich—self-effacing, self-educated, economically insecure, responsible for a child, and wife to a tyrant—entered a science-related occupation through necessity. Yet as her successes accumulated, she became totally committed to her profession as a collector and classifier. Although limited by her background, she built on the training given her by her husband, learned to inspire confidence in the males upon whom she was dependent, and became respected in the field. Margaret Huggins—affable, cosmopolitan, economically secure, childless, and with a symbiotic relationship to her spouse—found collaboration a way to pursue her early interest in astronomy; she was accepted as a part of a team, albeit the subordinate partner. Hertha Ayrton—opinionated, ambitious, university educated, a vocal exponent of women's rights, mother of one child, and wife of a sympathetic spouse—polarized her colleagues. It would be difficult to find three more diverse women scientists. Yet all three illustrate the importance of marital collaboration in science.

TWO
Biographical Studies

7

SALLY GREGORY KOHLSTEDT

Maria Mitchell and the Advancement of Women in Science

ASTRONOMER Maria Mitchell (1818–89) was well known among her contemporaries as the first woman to win an international medal, the first American woman elected to the American Academy of Arts and Sciences, the first woman member of the American Association for the Advancement of Science, and the first woman professor of astronomy. Her successes serve to highlight the paucity of recognized women scientists in the nineteenth century. The barriers to women in science were high, limiting even those women who worked on the periphery of the scientific community as illustrators, textbook writers, and herbarium owners.[1] Thomas Henry Huxley, for example, argued that since women were ipso facto amateurs, they were not worthy of membership in learned societies devoted to serious discussion and debate.[2] Similarly, access to institutions offering advanced scientific education was negligible until the end of the century. Other more subtle factors also prevented women from realizing their scientific skills and promise. Most women internalized the gender stereotypes of Victorian society and thus lacked confidence levels and aspirations that reached as high as their potential. It is significant that Maria Mitchell, having to some extent surmounted these obstacles, chose to work directly with women rather than to defy male hegemony. This decision appears to be based on her perception that women's self-definition and experience contributed to their limited participation in science; that is, she believed that women had a fundamental problem with personal self-confidence as well as public encouragement. Her initiatives regarding women in science, unanalyzed by her biographers, offer an illuminating perspective on a period typically viewed as quiescent on women's rights.[3]

Maria Mitchell became a symbol to her contemporaries, men and women alike, of the contributions women were able to make in science. Her discovery of a comet in 1847 and her calculations of its exact position at the time of discovery brought her a gold medal from the king of Denmark and led to her membership in the American Academy of Arts and Science, "in spite of being a woman."[4] One result of these honors was Lucretia Mott's citation of Mitchell's achievement at the Seneca Falls Women's Rights Convention in 1848 as evidence of women's capability in all occupations.[5] Thirty years old at the time, Mitchell was an independent, hard-working librarian and astronomical observer on Nantucket Island in Massachusetts and far removed from reform activism.[6] Public recognition, however, gave her an opportunity to travel, not only to annual meetings of the American Association for the Advancement of Science, but also to Europe, where in 1857 she met leading astronomers, including the Herschel family and Mary Somerville, and visited their observatories.[7] Shortly after her return, a group of women led by Elizabeth Peabody raised money for an Alvan Clark equatorial telescope to enable Mitchell to make more precise observations.[8] Appointment as field researcher and computer for the *Nautical Almanac*—not coincidentally she was asked to compute the position of the planet Venus—permitted her more time for astronomical study. In short, Mitchell was perhaps the only American woman to have self-supporting scientific employment and international recognition in the 1850s.

Mitchell knew from experience that self-help was slow, tedious, and sometimes unrewarding. Her father, William Mitchell, was an active observer and calculator on Nantucket. Through her father's influence she won astronomical friends at Harvard. As the first woman in a number of male societies, she never felt the force of absolute exclusion; but she learned that nominal membership meant little in terms of collegiality. At some annual meetings of the American Association for the Advancement of Science, she attracted more curiosity than professional interaction.[9] Such experience helps explain why she grew less attentive to the problems of discrimination and more concerned with the aspirations of young women.

Mitchell had two stated incentives for involving women in science. Like many contemporaries, she viewed science as a way of thinking. Concerned about the "half-educated, loose and inaccurate ways" in which women seemed to think and act, Mitchell hoped that science education would provide a systematic and rational way of problem solving. She also believed that science offered a unique intellectual challenge and could help women escape the narrowness typical of their lives. Her crusade to involve women in science thus embodied both a concern for social progress and a commitment to expand opportunities for individual women.

Her decision to work with women is not surprising. As Carroll Smith-Rosenberg and others have noted, the period was one during which spheres of public and private activity were largely identified by sex.[10] Coming from Nantucket, where Quaker women were often responsible for civic and economic matters while their husbands were away at sea, Mitchell appreciated the effectiveness of women's networks. Although aspiring professionals might ultimately challenge the boundary of the sphere traditionally assigned to women, they had been shaped by its requirements and used its sources of strength. Mitchell's opportunity came with an offer to teach at Vassar. She agreed with a trustee who felt students should be "fed from living springs rather than a reservoir," and teachers should be a fount of knowledge and inspiration.[11]

Her course in astronomy proved challenging. No student was admitted who had not passed an examination in mathematics. Juniors were given extensive reading assignments, but each Monday Mitchell counterbalanced this work with lectures on the history and meaning of science reputed to be "racy and interesting" and which had a "more literary than mathematical" connection with astronomy.[12] Mitchell challenged but put no distance between herself and her students (their respect was evident in the fact that she was the only faculty member not given a nickname); she opened her advanced class with the comment "We are women studying together."[13] Mitchell took students on trips to New York City and into the countryside to make astronomical observations. She also invited prominent friends, including Julia Ward Howe and Mary Livermore, to the campus. Although Livermore avoided the controversial topic of suffrage, a group discussion led to mutual agreement on the proposition that "every girl should have some aim and some means by which to support herself in time of need."[14]

Vassar was exhilarating, but there were also frustrations. Perhaps jealousy prompted one suggestion that relationships might be too close between the teacher and certain astronomy students.[15] Although such rumors

(Overleaf)

Maria Mitchell studied with her father, William, who taught her to use a telescope at their home on Nantucket Island and introduced her to his astronomer colleagues in Cambridge. Courtesty of Special Collections, Vassar College Library.

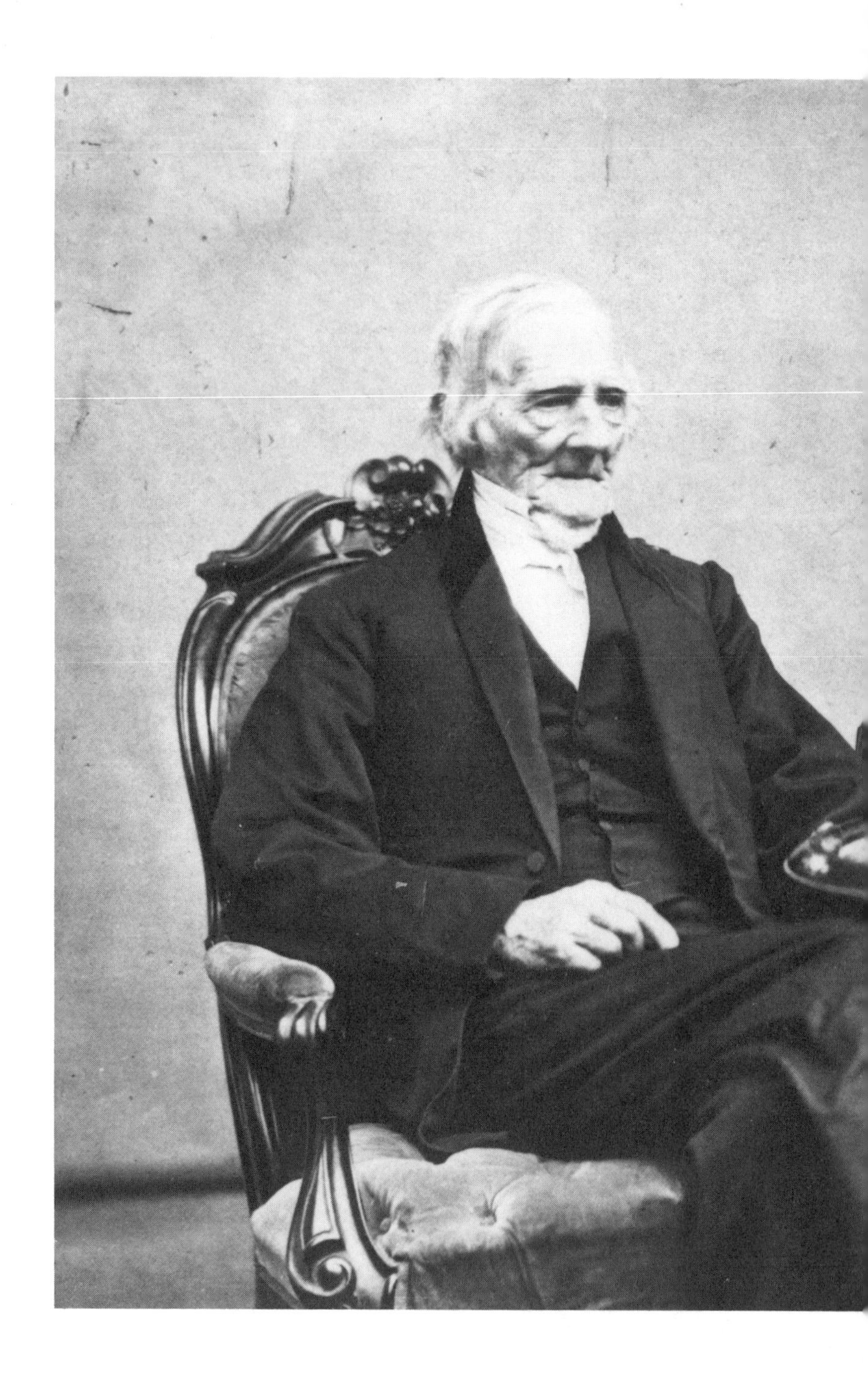

gained little attention, their existence indicates how easily suspicions arose about an unmarried woman scholar. Pressures were various, but student anxiety was a constant source of concern to Mitchell. Julia Pease, for example, whose early college letters frequently mentioned Mitchell's classes, was discouraged from selecting advanced courses in astronomy by disapproving parents.[16] Even those who remained in the astronomy class were confronted with the question of what they would do after graduation. One student reflected: "I have Miss Mitchell and all these grand instruments, and nobody makes fun of it all here. But when I get home no one there will take any interest in astronomy. I shall have no telescope at first, and there will be no one there to help me on. Do you think I shall be brave enough then to hold on tight to what I have begun? When I think of it, I get discouraged."[17]

The college problems of her students reinforced Mitchell's conviction that offering women an opportunity to study was necessary but hardly sufficient to insure their participation in science. Within a year after coming to Vassar she resolved: "In case of my outliving father & being in good health, to give my effort to the intellectual culture of women without regard to salary."[18] Women scientists needed to be recruited in greater numbers and society needed to be made ready for them.

Englishwoman Frances Bacon Cobbe observed; "As women grew older, if they had led independent lives, they tended to be women's rights women."[19] This was certainly true for Mitchell. Maturity brought confidence in her own intellectual and leadership abilities. More significant, she became committed to women's rights. Once settled in Vassar, she went often to Boston, visiting friends and attending lectures on a wide range of social issues.[20] Her travel journals for 1857 and 1873 reveal a remarkable change in her concerns. On her second trip to Europe in 1873, she again visited scientific friends and astronomical observatories; but now her primary concern was women's higher education.[21] She sought out persons working for women's education, and constantly discussed women's issues with fellow passengers.[22] On her return from Europe Mitchell for the first time actively joined a national women's movement. Vassar had shown her the need to raise the aspirations of women, and she was ready to contribute her energy to a women's organization whose initial goals seemed broadly

---

Opposite: Maria Mitchell trained Mary Watson Whitney (*right*), later her assistant and successor, at the observatory at Vassar College. Courtesy of Special Collections, Vassar College Library.

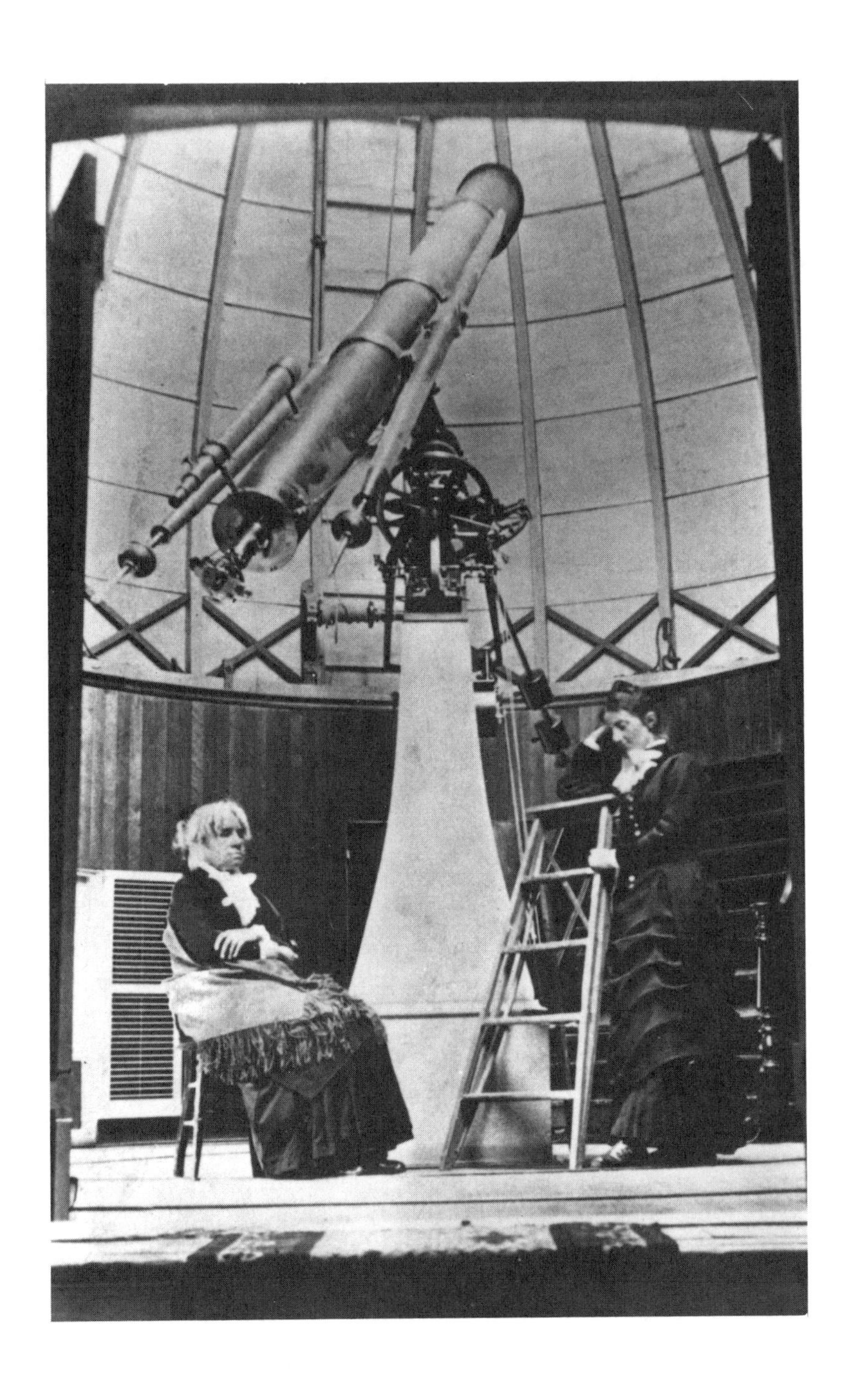

defined to include professional advancement. She had chosen not to deal directly with the male scientific community, but to work with women. Her strategy was positive: to encourage public awareness and to prepare the coming generation psychologically and practically for acceptance as professional peers. It was to these ends that she directed her efforts in the Association for the Advancement of Women (AAW).

For a number of years there had been discussion about forming some umbrella organization to focus on women's club activities on a larger than local scale. The successful initiative came from Sorosis, a group in New York City. In 1869 several Boston women had visited the newly founded society to discuss a women's "parliament," without results.[23] Four years later, however, Sorosis decided to issue a publication called the *Messenger* to leading women involved in educational reform and in established women's clubs. Its stated intention was to unite women working in isolation from one another in order to undertake a "careful consideration [of] the more important questions that affect our women's lives."[24] Maria Mitchell was among the persons signing a "call" for a meeting in September 1873.[25] Specifically invited were women preachers, teachers, professors, artists, lawyers, businesswomen, editors, authors, and "practical philanthropists."[26]

The AAW won approval from leading mid-century reformers. A wide range of women responded to the prearranged theme of education: Elizabeth Peabody discussed coeducation at Boston University; Catharine Beecher stressed the need for endowments of women's colleges; and Elizabeth Cady Stanton submitted thoughts on coeducation. Attendance by several hundred women revealed a measure of broad interest. Some papers were published, the participants were enthusiastic, and Sorosis itself gained more local members.[27] A midyear conference of officers resolved to hold the second annual meeting in Chicago, thus setting a precedent for peripatetic sessions. The western meeting was intended to attract new affiliates and serve as a stimulus to women in other parts of the country; subsequent meeting places included Syracuse, Philadelphia (in the year of the centennial celebration), Cleveland, and Providence.

Public opinion had changed only slowly about the acceptable leisure-time activities for women outside the home. Local groups organized for religious or charitable work were acceptable because they seemed an extension of women's domestic and moral concerns. Involvement in mixed reform groups and women's rights associations had come under steady public criticism. The opposition, however, began to erode as prominent upper- and middle-class community leaders spoke out. The new association, with its highly visible leadership, found the press supportive of "learned ladies," and several newspapers went out of their way to explain that these women were practical and hardworking, "not mere talkers" or radicals. Having anticipated criticism, one AAW member recorded to her satisfaction that the "croakers and fogies handsomely acknowledged that the deliberations were hightoned."[28]

The formation of the AAW gave women with Mitchell's vision a valuable public forum in which to explain and promote their ideas. The binding commitment was to education, formal and informal. In 1874 Mitchell's address "The Higher Education of Women" traced the initiatives of Emily Davies and others who had founded Girton College at Cambridge, England. While Mitchell deplored the unwillingness of Cambridge University to grant degrees to the graduates, she approved its intention to establish courses for women equal to the best male education. But where to start? Mitchell was practical, and her conclusion was a plea to every women's club to investigate its local situation and to work for the education of women by campaigning for positions on local school boards and by supporting women's colleges.[29]

Mitchell had substantial hopes for the AAW as a vehicle for reaching a wide audience of women. She actively recruited new members and solicited papers from friends and former pupils for the annual meetings.[30] With Mitchell on the program committee, it is not surprising that several papers dealt with women and science. One was by former student Ellen Swallow (later Richards), whose attention had turned from astronomy to applied chemistry. In her paper, "What Practical Science Is Open to Women?," Swallow pointed toward her own initiative on the subject of nutrition. Another read a paper called "The Value of Natural Science for Women." Probably Mitchell's own address, her efforts to build an effective program, and her personal prestige led to her election as second president of the AAW for 1875 and her reelection in 1876.

As president Mitchell planned the annual meeting, she organized the AAW for more comprehensive action.[31] Although surprised by the administrative effort required and reluctant at first to address large crowds, she proved to be a successful presiding officer.[32] Her presidential address reviewed the association's origins and offered suggestions for the future. She praised the enthusiasm and dedication shown by women at the first two congresses, but she argued that it was essential to create more stability and to develop new programs. Her recommendations reflected her own approach to life in emphasizing that permanent changes in the condition of women could come only by "combined, continuous, and systematic action." Simply by virtue of annual meetings that brought together dedicated women from around the country and awakened local women to "definite views, higher ambitions or braver efforts," the women's congresses made a difference. Nonetheless Mitchell wanted to produce more concrete results, calling for statistics on women in education, industry, and such special circumstances as prison, poverty, or chronic ill health. Accurate statistics would, she argued, enable activists to indicate the number of girls who suffered from "aimless lives" as well as those who had studied too hard, the number of working women at all levels of employment, and the success of asylums for the poor and criminal. Statistics, too, could be a "formidable opponent to the flourish of rhetoric" when bandied about with regard to women's presumed characteristics. Her presentation was tart, eloquent, and explicit when she demanded "fairness for the individual woman."[33]

Mitchell's special concern remained that of aspiring professionals and she tried to make the audience, composed primarily of women whose work focused on the home and voluntary activities, understand the problems confronted by aspirants. On the one hand, she noted, society underestimated women and thus women rated themselves too low. At the same time, "with our ideas of women, we demand too much of a woman. We are generous toward the species but severe upon the specimen." She cited the example of the president of a coeducational school who indicated that he would hire a woman professor if he could find the equal of Mary Somerville. Mitchell argued that this demand was "enormous," and that if each male appointment had to be the equal of Pierre Laplace, then college chairs "must be vacant." Laplace was internationally famous for mathematical techniques he developed in celestial mechanics, a term he coined in his major work, *Mécanique Céleste* (Paris, 1795–1805). Mary Somerville became well known in Britain in 1848 as a translator and expositor of some of Laplace's work. While applauding contemporary pioneers, Mitchell expressed concern over the single-mindedness of the most dedicated and suggested, "There is something almost painful in the seriousness of the best girl graduates from our colleges." Witty and down-to-earth, Mitchell was quoted at length in newspapers, which commended her for her "sturdy common sense, her scientific knowledge, her ready talk, her abundant good nature, her prompt application of rules, and her steady presence."[34]

At the centennial celebration in Philadelphia, Mitchell redirected her attention toward the "need for women in science." Earlier she had argued that women needed the discipline of effective science teaching to improve their natural facilities. The AAW address, written in response to a debate between Thomas Wentworth Higginson, Mary Livermore, and Louis Agassiz on the capability of women at a social science meeting in Boston, approached the situation from a different perspective.[35] She argued that not only could women use scientific training, but that science also could benefit from women in the field. Childhood training developed in girls certain skills and characteristics that made women meticulous observers of natural phenomena. Moreover, she pointed out, the increasing need for science education in the primary schools could only be solved by better preparation of the women who taught at this level. While Mitchell preferred to deemphasize sex differences, she was pragmatic enough to mention women's socialized skills as an encouragement to women to engage in scientific studies.

Anticipating the question of why so few women had contributed to science in the past, she gave an exaggerated comparison to make her point. Danish astronomer Tycho Brahe had received a royal grant of an entire island, instruments, and assistants in order to pursue his research; this was the "ideal" intellectual life. In contrast, Mitchell cited excerpts from astronomer Caroline Hershel's diary; at fourteen Herschel attended church and school, but at home was so busy in the scullery that she could rarely join in family discussions; at twenty she was knitting socks for her famous astronomer brother; and only at thirty-seven did she gain an annual income of 30 pounds as assistant to her brother, which she said was "the first money I ever in all my life thought myself to be at liberty to spend to my own liking." Mitchell's conclusion, like that of Virginia Woolf's *A Room of One's Own* a half century later, was simple: "The laws of nature are not discovered by accidents; theories do not come by chance, even to the greatest minds; they are not born of the hurry and worry of daily toil; they are diligently sought, they are patiently waited for, they are received with cautious reserve, they are accepted with reverance and awe. And until able women have given their lives to investigation, it is idle to discuss their capacity for original work."[36]

Mitchell was all too aware of the cost of multiple responsibilities. Two years as president of the AAW, with its tedious administrative details and distracting controversies over personality and presentations, were enough.[37] Skipping the Cleveland meeting in 1877, she returned the following year to present a scientific paper on the solar eclipse observed by her Vassar students. Public education, however, was only a first step for Mitchell.[38]

In order to support women with talent and incentive for advanced work in science, Mitchell suggested that the AAW sponsor a scientific society for women. Although a growing number of "ordinary" scientific societies were open to women, she pointed out that "there is really little room for papers by women and we have so long been accustomed to listen in silence, and not to speak, to receive views and not to advance them" that few women were yet ready to speak to mixed audiences as "equally unmixed as they are in science."[39] Her women's scientific society would be, in effect, a board of experienced researchers who could encourage younger women in laboratories and museums to present the results of their research. Mitchell envisioned a far-reaching society that would eventually establish courses, found schools, and develop local organizations. Women, once attracted to science, needed ongoing encouragement to offset the negative public response. Mitchell wanted to create an environment in which aspiring young women found concrete advice, encouragement, and the ways and means to pursue research. Dr. Mary Putnam Jacobi, already committed to the professional tendency toward specialization, responded skeptically to the idea: "I confess I do not see where you are to find enough women knowing enough about science to write about it, to form a committee,—and when you have,—who will be the audience to listen to them? or how would Prof. Mitchell and Mrs. Treat [a botanist] meet on common ground?—As for myself, I cannot more than any practicing physician, claim to be a 'scientist.'"[40] Mitchell very shortly was forced to agree that there were not yet the critical numbers of women in the sciences required to form a workable organization.

No society was ever formed, but in 1876 the AAW created several committees, including one on science. Until her death in 1889 Mitchell served as its head. The purpose of the Committee on Science was to stimulate interest in science by annual reports "showing the fitness of women students to pursue investigations in different fields and their gradual assumption of responsibilities as professors and teachers."[41] Mitchell solicited information on the status and opportunities for women in science and medicine through an open-ended questionnaire. Specifically, she asked:

1. Are there women in your community engaged in scientific research, or in original investigation?
2. Are there women who make a specialty of any branch of physical science, chemistry, or natural history, as teachers or as writers?
3. Are there women physicians, and have they a successful practice?
4. Are there scientific schools open to women?
5. Are there women who are giving pecuniary aid to such scientific schools?[42]

In 1876 Mitchell sent the questionnaire to friends, former students, and institutions in various parts of the country, asking that the circulars be returned even if every answer was negative. The exact number mailed is unrecorded, but by the time Mitchell made her report she had sixty-five replies, and fourteen arrived later.[43]

There were more negative responses than positive, and the net results demonstrated Jacobi's prediction regarding the dearth of women engaging in scientific research. Mitchell tabulated the results, readily admitting that the numbers so casually acquired were only useful for general observations: seven questionnaires were totally affirmative, forty-seven were partially affirmative, and nineteen were totally negative.[44] The questions hinted at some relationship between scientific study and medical practice, but the two activities proved to be basically unrelated. In Albany, for example, the Dana Natural History Society (an amateur group of educated women) was reported as "large and reasonably prosperous," while the four women physicians in town had limited practices because, its reporter felt, "Albany is desperately conservative."[45] Even more common was no reported activity in science but a number of successful physicians. The early nineteenth-century nexus between physician and scientist was nearly over, with a few notable exceptions such as the enterprising Sarah R. M. Dolley of Rochester.[46]

It proved easier to deal with the question of physicians, who formed a distinct occupational group, than to identify women scientists. Who should be included? The scientists were themselves in a process of distinguishing amateurs from professionals and provided no clear standards. Respondents showed understandable confusion, and their answers varied from an uninspired yes or no to discussion of the interests of specific local women. Question 1, with the requirement of "original investigation," elicited only seventeen positive replies out of seventy-nine; twenty-nine persons were identified in answer to the second question. Altogether fifty-one women were named, most of whom were local schoolteachers and members of natural history societies. The list was incomplete, lacking such names as Almira Phelps, who had written numerous textbooks in natural science; Emily Gregory, then studying at Cornell; and astronomy student Mary Whitney, who had graduated from Vassar and would eventually become Mitchell's successor there. Because Mitchell had relied on her own network of students and friends, there was a distinct regional bias; only Kentucky and Texas represented the South.

The questionnaire left no doubt that women were capable of scientific study. Women teachers in elementary and high schools, normal schools, and women's colleges taught every subject from natural history to chemistry and astronomy. Often, however, there was a note of apology. Emma Morrill, who belonged to an active botanical society in Wilmington, Delaware, wrote somewhat defensively: "Women are not idle, they are good workers, many of them are efficient and capable, but their work is of a kind that cannot be put down in a schedule."[47] Women teachers had often "stumbled into the subject" of science, and many, like young Mary Reed of Philadelphia, were assigned to several areas with little opportunity to develop any special expertise. During the summer, however, Reed conscientiously read and attended summer school at Cornell in an effort to learn more about the subjects she taught.[48] The circulars also indicated efforts to secure opportunities for women seeking careers in science. Ellen Swallow Richards, who had become the first woman to study chemistry at the Massachusetts Institute of Technology, for example, wrote her former teacher about a proposal for women to study applied chemistry at MIT. A joint committee composed of MIT faculty and members of the Boston Women's Educational Association approved the proposal, and efforts by the latter were begun to raise financial support. Pleased with recent progress, Richards wrote confidently, if naively with regard to public opinion: "There is no longer any doubt as to the ability of women to carry out any work of this kind."[49]

Indeed, the list of women from the Boston area was impressive, with twelve scientists and ten physicians.[50] Moreover, local women were gaining recognition. At the Boston Society of Natural History (BSNH), where women were still only "patrons," the question of membership had come under serious discussion. Lucretia Crocker, a former science teacher and member of the Boston school committee responsible for developing a Saturday program for teachers at the BSNH's museum, reported the sequence of events. Once women began studying science at the museum, the "inevitable" had happened: "At the close of a course of lectures or lessons by a Harvard Assistant Professor of Botany it was proposed to form a Botanical Section, when the dilemma arose—How can women join a section if they are not eligible as members of the Society?"[51] In less than a year several women had gained at least partial recognition as members in the newly created category of associate members of the BSNH.[52]

Mitchell found such news fascinating, although she never made use of the names she gathered to form a scientific club. The data became the basis for some annual reports, and in 1880 she circulated the same questionnaire but identified Henrietta Walcott and Antoinette Brown Blackwell as new committee members.[53] This survey was apparently smaller and more select; Mitchell's manuscripts reveal only nineteen replies. Her report for 1881 stressed the progress made in five years. The best results again related to women physicians, whose numbers appeared to be rapidly increasing.[54]

Positive reports from Lucretia Crocker in Boston and Grace Anna Lewis from Philadelphia suggested to Mitchell that if in 1876 a scientific institution had been partly open to women, by 1881 it was wholly so.[55] Still she paused in her report to ask: "At what time did scientific associations close to women?" Princess Catherine of Russia had been chosen a member of the American Philosophical Society in 1789, and the American Academy of Arts and Sciences in Boston had elected fifteen members in 1780 without apparent regard to sex.[56] Yet both organizations had since basically closed their doors to women—with the exception of Mitchell's own election to the academy in 1848 (the last woman until the twentieth century). Mitchell's efforts led her to a more sophisticated understanding of history and the realization that progress could not be assumed.

Thus success appeared to her to be dependent on women working cooperatively to advance the situation of women. On the negative side, she cited the paltry response to question 5, concerning women's financial support of other women, and gave explicit credit to five persons, including Elizabeth Thompson, who contributed to the cause of women in science.[57] More positively, she pointed out the research activity of faculty at such women's colleges as Wellesley, Mount Holyoke, and Vassar.[58] She stressed the expansion and vitality of women's groups in Syracuse, Chicago, and other cities where the AAW had held annual meetings. Privately, however, she subscribed as well to a "great woman" theory, recording in her diary: "Given one wide awake woman in town . . . you can rouse the whole."[59]

Astronomer Mitchell was reluctantly forced by the questionnaire results to conclude that "those sciences which touch life and health" were most attractive to women. The defection of Mitchell's promising student Ellen Swallow to applied chemistry (analysis of purity of foods and water) served to underscore her observation that the "abstract sciences"—mathematics, astronomy, and physics—attracted only a small proportion of those women entering science. Social consciousness helped explain the emphasis —a significant challenge to women's expanding rights came from arguments regarding health.[60] Opportunity was the other side of the incentive. The life sciences led to medical schools, which were admitting more women to study, but graduate education was more difficult to obtain.[61] Although most medical school graduates opened private practice, several respondents reported women working as superintendents in state and charitable institutions, medical directors in gymnasiums, members of city or county boards of physicians, medical missionaries, and delegates to medical conventions.[62] Such facts seemed to indicate that women were being integrated into regular medical circles, and in the 1880s most women expected the pattern to continue.

The emphasis of Mitchell's reports was not on those who enjoyed science as an avocation, but on those who took it up as a "serious and life-long occupation."[63] She described the work of women who presented their results at exhibitions, who published in *Popular Science Monthly* and other journals, who attended and presented papers at professional meetings, and who held paid positions in science. Employment was an evident problem and Mitchell's reports began to document the variety of possibilities, mentioning women as naturalists and assayists with mining companies and government surveys, as computers at observatories, as pharmacists, as assistants at the Smithsonian Institution and the Museum of Comparative Zoology in Cambridge, and as teachers on every level. Gains in science were less dramatic than those in medicine, and cautioning against her own impatience, Mitchell insisted: "Measured by a year's interval, the scientific study of women shows little change. If we take half a century, the growth is astonishing."[64]

Mitchell allowed the science reports to dwindle in content throughout the 1880s.[65] A partial explanation may have been fatigue, since Mitchell was in her late sixties. In addition, she found the AAW less personally rewarding and professionally significant than she had hoped.[66] The tone and substance of annual meetings shifted as the leadership of the AAW became dominated by philanthropic leaders with leisure time. The Committee on Science, as conducted by Mitchell, seemed an anomaly representing professional aspirations in the midst of the voluntarism and moral philanthropy advocated by Julia Ward Howe, who served last and longest as AAW president from 1880 to 1900. In retrospect, the AAW met its stated purpose of disseminating information and inspiring local women at annual meeting places into literary or reform-minded groups.[67] Yet, ironically, the association's deliberate effort to be a stimulus meant that some of the very groups it helped initiate eventually joined the historically more visible General Federation of Women's Clubs, founded in 1889.

The AAW initially had stressed working activists. The Sorosis *Messenger* (1873) expressed the hope of uniting foremost writers and thinkers, women working in either traditional or nontraditional areas, and practical philanthropists.[68] Perhaps the needs of these various groups were too diverse to be united, although representatives of each undoubtedly joined the AAW in order to secure a national base. By the time the AAW had sufficient support for a regular publication in 1882, the group was beginning to lose its membership to alternative organizations. Mitchell's own former students, including Christine Ladd-Franklin, Ellen Swallow Richards, and Martha Talbot—women who might have been expected to play a larger role in the AAW—instead helped to organize women college graduates into the Association of Collegiate Alumnae, directed particularly toward women who were career-minded or ready to support others who were.[69] Since Mitchell had expressed some dismay over the quality of mind of AAW leadership as early as 1875, it is not surprising that she gave full support to the new collegiate association.[70]

Mitchell's own prodigious activity and the vigor of the early years of the AAW belie any assertion that feminism was quiescent in the 1870s. Although the General Federation of Women's Clubs and the reunited suffrage movement have made the 1890s appear to be a take-off period for women's activism, the 1870s and 1880s were critical for raising the consciousness of numerous women and laying the essential groundwork for later coordinated activism. The AAW took up many topics—the advancement of higher education for women, the expansion of women's interests outside the home, the strengthening of responsibility within the home, health, physical development, and sensible dressing—thus introducing significant discussion to groups of women not apt to subscribe to a feminist journal or to have much contact with women in other geographical regions. The leaders of the AAW were, as newspapers often observed, eminently practical. The AAW's low profile may well have helped obscure the group's considerable influence, as viewed by historians as well as many contemporaries.

The influence was indirect. Mitchell's "statistical investigations" were never applied to problems. The Committee on Science, like the AAW, simply supplied information to arouse women's awareness, but left to other groups any specific action. In the case of a science society, Mitchell's hopes were premature; the critical number of women involved in research did not yet exist. Without such groups, women had little leverage within professional circles. In 1879, for example, Mitchell recorded in her diary: "I wanted to nominate some women in some of the comm. of the Association [for the Advancement of Science] but my friends assured me that I should do more harm than good."[71] The slow, persistent work of the Committee on Science and Mitchell could only lay groundwork for the later activism by encouraging segmented efforts of isolated women to enter scientific careers and to continue their private study.[72]

The foundation laid by Mitchell and others has been hidden because it was built outside the scientific community and in conjunction with an organization that flourished only briefly. Mitchell helped invigorate the early years of the AAW. She lent her name and shared her vision from the presidential podium and then stepped down to head the Committee on Science, whose questionnaire and annual reports became a model among the AAW committees. Her own work was presented in sufficiently popular form to win attention from local audiences, which frequently numbered over a hundred persons, and she actively solicited papers from others in science and medicine to demonstrate the quality of women's research. Her public posture was deliberate. On the one hand, she spoke out for women's education and represented a fulfilled single career woman. On the other hand, she avoided public statements about suffrage and opposed presentations that might offend.[73] It is evident that her own astronomical work suffered from her commitment to help other women overcome barriers.

The AAW questionnaire indicated a substantial number of women interested in science and hinted at their rising aspirations. Although other women, including such friends of Mitchell as Grace Anna Lewis, Antoinette Brown Blackwell, and Lucretia Crocker, were influential in this movement, one cannot exaggerate Mitchell's key role in the 1870s. Influence cannot be measured in quantitative terms, but after the 1876 meeting Phoebe Hanaford placed Mitchell at the "head of the list of scientific women in America."[74] Mitchell held a vision and pursued strategies that coincided with changing but cautious views of women's clubs; that is, she sought and utilized the psychological and practical reinforcement found in the sphere of a female world. Mitchell, in her later years at least, found homosocial society congenial and a more useful vehicle for change than any direct challenge to male bastions. She believed that the integration of women into professional circles could come only after women had gained, within a supportive environment, both confidence and expertise. Her own strategies were to teach women at Vassar, encourage any woman engaged in science, and publicize successes. The remarkable increase of women in science by the turn of the century suggests that indeed her students and younger friends had begun to realize the results of her long-term strategies involving women's collective action and professional participation.

8

JOY HARVEY

"Strangers to Each Other": Male and Female Relationships in the Life and Work of Clémence Royer

> Up until now, science like law, made exclusively by men, has too often considered woman as an absolutely passive being, without instincts or passions or her own interests; as a purely plastic material capable of taking any form given to her without resistance; a being without the inner resources to react against the education she receives or against the discipline to which she submits as part of law, custom or opinion. Woman is not made like this.
>
> CLÉMENCE ROYER,
> "Sur la natalité" (suppressed communication read before the Société d'Anthropologie de Paris, July 1874)

THERE ARE FEW nineteenth-century women, outside of major literary figures, who produced writings so closely connected to their female roles as Clémence Royer, best known for her controversial French translation of Charles Darwin's *Origin of Species*.[1] While the preface to this translation has been recently examined for its eugenic and social evolutionary claims, this represents only a very small part of her output, which includes articles and books on evolution, social evolution, economics, race theories, and archeology, and even a novel. In her writing, she questioned the same female stereotypes that she challenged in her personal life. She felt, as the quotation at the beginning of this essay illustrates, that male science had poorly understood women. "Woman is the one animal in all creation about which man knows the least," she insisted.[2] From her first reading of Darwin, she had come to believe that there had been a continuous selection of women for the wrong qualities, resulting in extreme biological as well as social consequences.[3]

Clémence Royer's judgment that men and women held false or uneasy relationships with each other came out of her own life experiences. Her biography points up with a certain poignancy the problems of even a self-aware and articulate woman in the nineteenth century. She had reacted against the restrictions placed on her during her childhood under the thumb of a loving but tyrannical father and a strong-minded mother. Her rebellion against a Catholic and Royalist upbringing resulted in her development into a broadly educated freethinking republican who came to see in science an answer to many of the problems that concerned her.[4] She became the first woman in France to be elected to a scientific society (the Société d'Anthropologie de Paris). Out of her struggles to educate herself, she produced a new, original, and controversial philosophy of science that suggested the possibility of the flowering of a "female genius" in science.[5] She also became a strong feminist in her later years, extolled by the middle-class feminists around her as "the greatest living French woman scientist."[6]

As a woman Royer came to reject marriage as an option for herself, regarding it as "a dangerous lottery," partly in reaction to the parental quarrels she had witnessed throughout her childhood.[7] Yet she formed a lifelong alliance with a politician and publicist in exile, Pascal Duprat (1815–85), who fled in 1850 from the Second Empire to Brussels and then Switzerland, returning to public life in France only for the last years of his life. She bore his only son, René Duprat, and played the part of colleague, mistress, and hostess for him in Paris. Although the relationship was not always an easy one, either personally or financially, she regarded Duprat as the "one love of my life" and saw herself as having helped Duprat hold at a distance the heavy "chains" of his unhappy marriage.[8]

Clémence Royer was born in Nantes, Brittany, in 1830, out of wedlock to a mother who embodied the energy and practicality of many Breton women. Her maternal grandfather was a ship captain who had taught his daughter a great deal about the sea. Her father, from the province of Maine,

---

Opposite: Clémence Royer, as caricatured for the series *Les hommes d'aujourd'hui* (Men of Today), 4 no. 170 (1881). Royer is shown writing her latest book, *Le bien et le loi morale* (Welfare and the Moral Law). Around her are stacked other works, identified by subject rather than by title. Courtesy of Houghton Library, Harvard University.

4e volume. — No 170. 10 c. Un an : 6 fr.

LES HOMMES D'AUJOURD'HUI

DESSINS DE DEMARE

BUREAUX : 48, RUE MONSIEUR-LE-PRINCE, PARIS

CLÉMENCE ROYER

was an army captain trained as an engineer, who regularized his marriage only six months after Clémence's birth. An ardent Catholic and Royalist, Augustin-René Royer was involved in the 1832 conspiracy to put Henri V on the throne. As a result he was exiled from France for the first years of Clémence's life, and she grew up along Lake Lucerne in Switzerland, to which she returned during a deep emotional and intellectual crisis in her late twenties.[9]

Clémence Royer's relationship with her father was a complex one. He had taught her mathematics, in which she always delighted and presumably gave her a taste for the abstract thought and scientific knowledge that dominated her life. "Like a true pagan," as she described herself at the age of eleven, she had prayed for an Aladdin's lamp with which to turn herself into another Joan of Arc so that she could bring Henri V to the throne and vindicate her father, whom she was later to describe as "quixotic" and a "hero born out of his time."[10] Royer's father rejected the adolescent asceticism she adopted in her early teens during a short mystical and religious period, but found it amusing when she adopted republican sentiments after reading Lamartine's account of the French Revolution, laughing at the fears of an old female relative who thought she was "turning into a Red."[11] Her father, however, became paranoid in his belief that his wife, family, and business associates were conspiring against him and left Paris to retire to his natal village. This allowed Clémence to develop under the practical eye of her bright and lively mother. In 1849 her father voted for Louis-Napoléon during the short-lived Second Republic, knowing an empire would be the consequence. He died of a stroke in that year of social change before his prediction could come true. His death left Clémence with a personal and financial freedom she had never had before.[12] Perhaps her most direct comment on his role was her depiction of paternal tyranny as the source of all autocratic and tyrannical government, endured by its subjects for the sake of limited protection.[13]

Clémence took her small inheritance, which was meant to be her dowry, but which she saw as insufficient to attract the kind of husband she might have preferred. The uneasy relationship between her father and mother had made her wary of marriage. Instead she prepared herself to teach languages and music, obtaining the necessary certificates within three years. It came as a shock to her, however, that she had no knowledge of the structure of her own language, and she realized she had been badly taught.[14] Soon after, she left for Wales, where she taught in a girls' school for a little over a year and mastered English well enough to be recommended for translations. She came away from England with a positive impression of children's nurseries and the consequent improvement of English family life, later reporting on the differences between France and England to the Société d'Anthropologie.[15]

Royer's encounter in Britain with Protestant, especially Unitarian, beliefs had a profound effect upon her religious ideas. When she encountered the writings of such Enlightenment figures as Voltaire, Rousseau, and Diderot upon her return to France, around 1855, her reaction to this new knowledge was overwhelming. She experienced another mystical crisis, stemming from what she termed "the last echoes of my monastic training."[16] Again she saw herself as a Joan of Arc, this time with the task of liberating human beings from the oppression of the Church which she believed had "darkened her childhood."[17] In her rage at getting no satisfactory answers to her doubts from any priest, she left all her personal possessions in Paris, took what was left of her inheritance, and returned to Switzerland.

In a small farmhouse a few miles from Lausanne, Royer lived a frugal life, dressing and eating like a peasant and reading her way slowly through the local library. Her mother feared for her sanity and sent one of the local governmental officers to speak to her. Royer succeeded in convincing him that she was completely in possession of her reason, and he reassured her family.[18] Reeducating herself in both philosophy and science, she came in contact for the first time with Pascal Duprat, the man who was to change the course of her life, markedly affecting it "for good or ill."[19]

At twenty-seven years old, Royer was a small, curiously dressed, bright young woman with piercing blue eyes, which remained unusually lively even in extreme old age. The account she gave many years later of her first meeting with Pascal Duprat in 1858 is unusual in that it describes in more detail his wife and his house than the man himself.[20] Yet she recorded elsewhere a vivid description of him in the course of a biographical memoir of his friend the scientist François Arago: "In both of them there was the same incisive mind, brilliant and yet ironic, the same marvelous memory, the same natural eloquence even more persuasive in intimate conversation than in public."[21]

Royer had gone to visit Duprat to thank him for the support he had given her in his journal, advertising the philosophical lectures for women she had been inspired to give after hearing the Swedish novelist Frederika Brenner speak in Lausanne that year. Duprat was then teaching political science at the University of Lausanne. He had been a deputy of some distinction under the short-lived Second Republic (1848–49), and then had gone into exile following Louis-Napoléon's rise to power. In Lausanne he edited two journals—one, *Le nouvel economiste,* emphasized the new field of social science, the other, *La recherche libre,* published in Brussels, was intended for a more general audience.[22] He immediately suggested that Royer write a novel for the second journal. Impressed by Duprat, who was sixteen years her senior, she immediately acted on his suggestion. In three months she returned with the first draft of her one and only novel, *Les jumeaux d'Hellas* (Hellas's Twins).[23] One can only speculate whether this draft contained the same passages extolling free love and free thought that the novel later displayed. One of her early biographers claimed that Duprat's reading of the novel led to the love relationship between the two.[24] Royer herself denied to a friend that the novel played any part in the growing mutual interest between them, since his journal had gone into receivership and she was convinced he did not read the novel until it appeared in print in 1864.[25]

Reticent about the course of her relationship to Duprat, Royer described Duprat's wife with great rancor. Meeting her for the first and only time on the day of her first encounter with Duprat, she took an immediate dislike to her, describing her as "very pretty," a "constant whirlwind," a woman seductive in her dealings with men but rude and brusque with other women.[26] Duprat had married her ten years before, following the birth of their daughter. One wonders what echoes such a story had in Royer's own life. The wife's extravagance, which Royer described in some detail, proved to be a real and continuing problem once Royer and Duprat became lovers. Duprat's salary was seized more than once to support his wife and child; divorce was never an option, since it was legalized only shortly after Duprat's death.[27] The financial struggles with Duprat's wife may have led Royer to write some years later of the danger, if economic reforms were not enacted, of "war declared between the two halves of humanity who would be engaged in tricking and exploiting each other alternately and reciprocally."[28]

Whatever the nature of her early relationship between Pascal Duprat and Clémence Royer, her undeniable response to her meeting with him was the release of a remarkable flood of creativity under the stimulus of Duprat and his circle of freethinking, republican friends. She began to colloborate and assist him on his remaining journal, *Le nouvel économiste,* and to attend and report on social science conferences in the pages of his journal and in *Journal des économistes.* Her lectures in philosophy for women were extended to the topic of science in 1859 and were an immediate success. The opening lecture (printed by Duprat's journal in pamphlet form) invited women to learn and absorb science not only for their own benefit, but for the benefit of the family: "It is greatly to be desired that women devote themselves to science, that they do so with pleasure, with taste, with love as well with philosophy. The difference of language, ideas and opinions between the two sexes renders them in some degree, strangers to each other, dividing and disuniting them not only in society but in the family."[29] Clémence Royer attributed this barrier to the lack of knowledge of Greek and Latin, which made scientific terms seem "like scarecrows set in the fields to frighten the birds." She urged women to realize that their fear of the language of science was a similar illusion. "Approach, touch, it won't bite you."[30] Although Clémence Royer does not make the parallel, her description of the female fear of science is not too different from her later descriptions of young women's fear of sexuality and maternity.

Above all, Clémence Royer stressed in her women's lectures that women could contribute something unique to science, as well as to literature. Her concept of the "special genius" of women included a vision of a high degree of synthesis and a capacity to breathe life into the "cold marble" of male science.[31] Women, by introducing science into the family, would serve as the real popularizers of science. She later carried her concept of "female genius" further, to suggest a possible reversal of sex roles, which might produce a considerable improvement in the human economy. In such a situation there was the chance for "some human variety capable of forming and prospering in a manner now realized only at the top of the lowest branches in the scale of animals, those marvels of female genius which in no way astonish us among the insects and that we find completely natural among the bees and ants."[32]

In 1861 the canton of Vaud in Switzerland set up a prize competition on the theory of taxation. A conference sponsored by Duprat's journal discussed the questions involved. Clémence Royer, with books suggested and provided by Duprat, set out to write an essay on the topic. Not surprisingly, she captured the second prize of 300 francs. (Pierre-Joseph Proudhon, the great French philosopher of socialism, won the first prize.) Her essay was published the following year (1862) and won her the first attention from a wider public.[33]

Her lectures for women, however, lost their enthusiastic middle-class audience when Royer, who had been reading Lamarck, put forward a nonreligious and evolutionary explanation for human origin and development.[34] She broadened her audience to men, who at that time could more easily accept freethinking conclusions, and spoke in Geneva and Brussels. Soon after when she was offered the opportunity to translate Darwin's *Origin of Species,* she saw in this new evolutionary theory a successful combination of biology and Malthusian economic theory applied to the natural world.[35] She completed her strong and vivid translation of Darwin in 1862 (with the assistance of Edouard Claparède, who provided the scientific background she needed).[36] Her startling and controversial preface to this translation begins with a direct challenge to revealed religion, more a response to the audience of her philosophical lectures than an introduction to Darwin's evolutionary ideas: "Yes, I believe in revelation, but it is the revelation of man to himself by himself in a rational revelation which is the simple result of science and contemporary revelations."[37]

Clémence Royer's application of the concept of natural selection (or as she translated the term, "élection naturelle") to the human condition was a shock to some and an inspiration to others. She was, perhaps, the first writer to publicly draw eugenic conclusions from evolutionary theory. More significant for the purposes of this paper are the implications she drew from what she believed was the continuous selection of men for intellect and women for beauty. As she expressed this concern: "One must conclude that in order to hasten the progress of the race in all senses, it would become necessary to ask for woman a part of what has been asked for man, that is strength united to beauty, for intelligence united to gentleness, and for man, a little idealism united to vigor of mind and body."[38]

Darwin was amused rather than annoyed by this preface, commenting on Clémence Royer's "Deism," and concluding that she was the "oddest and cleverest woman in France."[39] It was only later that author and translator were to come to a parting of the ways.

The Darwin translation earned Royer an enduring reputation in the French-speaking world. Although her eugenic and social evolutionary claims provided a tinted glass through which to read Darwin, it had an enormous impact on readers of her generation. Some indication of the enthusiasm with which this translation was hailed can be gleaned from the comments, many years later, of social evolutionist Charles Letourneau, an ardent scientific materialist and a colleague of Clémence Royer's in the Société d'Anthropologie. He recalled that on his first reading of this preface, he felt it was "shattering the windows" of establishment thinking.[40] In its challenging language, he thought he discerned a man hiding from the Empire under a woman's name. Letourneau's many volumes on the evolution of social institutions owed more than a little to Clémence Royer, as well as to Spencer. Many years later the first French woman graduate in medicine, Dr. Madeleine Brès, recalled one of her professors, Dr. Gavarret, respectfully ushering in a woman scientist, Clémence Royer:

> When Gavarret who had (as I am certain you have not forgotten) a profound admiration for you, your character and your remarkable work, pronounced your name, everyone in the room stood up to cheer you. For it is true that young

> people of great and generous aspirations, health, and accurate judgement render homage to all that is good, beautiful, and true even when it concerns a woman. That meeting I have never forgotten. It raised in my mind the same breath of enthusiasm. The next day I read the *Origin of Species* and learned of the immortal work of Darwin.[41]

Clémence Royer endorsed in her preface and in her later lectures a view of Darwinism as the successful reworking of the earlier Lamarckian theory of evolution, in which progress became a biological law. She immediately saw the implications for a new social science to be based on evolutionary doctrines: "It is especially in its moral and humanitarian consequences that the theory of Darwin is so fecund. Never has anything so vast been conceived of in natural history. One could say that this is the universal synthesis of economic laws, the social science par excellence, the code of living beings for all races and all times.[42]

When Clémence Royer came to interpret Darwin's use of Malthus, she saw the implications for human beings as markedly different from the conclusions Malthus himself had drawn: "It is the exuberance of a species which leads to its perfectibility. To stop this exuberance is to put an obstacle in the way of progress."[43] She expressed in her initial reaction to Darwin a concern about the effects of lowered population growth that many French scientists came to share. Between 1870 and 1890 in the forum of the Société d'Anthropologie, she was to renew these concerns. Darwin's concept of nature, as she read it, acting continuously to individualize and diversify species, was a blow against the concept of "ideal types." The incessant rivalries that occur between competing individuals she termed "concurrence vitale" (vital competition), which she derived from the language of economics. Her preference for this term was hailed by one of her biographers at the time of World War I as a far more humane concept than the Germanic rendition of Darwin's borrowed phrase "struggle for existence" as a "battle for life."[44]

Royer had begun to make her own living, following the publication of her major books, by writing for the *Journal des économistes* and lecturing on Darwinism in Switzerland, Belgium, and Italy. Her audience had increased greatly since her original philosophical lectures. Duprat, meanwhile, had decided to edit his journal in Geneva, but then, fleeing his extravagant wife and his creditors, he left Switzerland for Italy. His wife followed him to Italy, but then returned to Paris, where she remained. She urged Duprat to make his peace with the Empire, but he rejected both the government of which he disapproved and the wife with whom he could no longer live. Instead, by 1865 he and Clémence Royer were planning a life together. At one of their regular meetings during the International Social Science Conference in Berne, they decided to make a final commitment to each other. To their embarrassment, Duprat was arrested for his wife's debts, which his political enemies had bought up. Although Clémence Royer felt severely compromised by having her baggage impounded along with Duprat's, she appealed to their mutual friends, who rallied to their support and had Duprat released. Since this left the two lovers without funds and unable to travel to Italy as they had planned, they returned to Paris, where Duprat lived in Clémence Royer's rooms, hiding from the Empire (and probably his wife). Both of them wrote books to earn money for their new life together.[45]

It was at this point that Clémence Royer completed her second edition of Darwin (which appeared in 1866), incorporating many corrections Darwin had requested.[46] For example, she dropped her term "élection naturelle," which she had preferred to the non-French term "selection," and adopted the term "natural selection" to conform to worldwide usage. Yet in her discussions she continued to prefer the term "election," partly because it implied an elitist, hierarchical component. She also adopted "lutte pour l'existence" for "struggle for existence," although she retained "concurrence vitale" for sections in which natural rivalries were more explicitly indicated. The subtitle of the book was corrected, eliminating the "laws of progress" and substituting "natural selection."[47] Darwin was then on cordial terms with her, corresponding on corrections and additions, as both he and she explained at the time of their formal disagreement some years later.[48]

Clémence Royer sent two copies of this new edition to Paul Broca, secretary-general of the Société d'Anthropologie de Paris. She seems to have already met him, almost certainly through Pascal Duprat, who had been a member of the earlier, by then defunct, Société Ethnologique. Broca shared many of Duprat's political and social attitudes, and like him was to become a significant political figure in the Third Republic. The letter accompanying the Darwin edition expressed Royer's unhappiness that the rules of the society excluded her, "trapped as I am in the body of a woman," as she put it.[49] The new foreword to the second edition (added without eliminating her first preface) hailed the Société d'Anthropologie (formed in 1859), "so young and so active." Here, she claimed was a place where "the theory of Charles Darwin reigns uncontested among the most influential and competent members."[50] Paul Broca was embarrassed by this statement, although there was more than a little truth in the claim when one examines the later positions of the leading members of the society on the issue of evolution. Yet at this time, with the Empire only beginning to ease its repression, he insisted publicly and privately that "as a polygenist I cannot be a Darwinist," and added privately to Carl Vogt, who had developed a polygenist Darwinism, that only about nineteen members of the Société d'Anthropologie were Darwinists, "in spite of what Mme. Royer has said."[51] He did not add that this number included most of the members of the Central Committee. Later he encouraged the opening of a debate on Darwinism that began at the end of 1868 and went through 1870. By 1870 he had given his assent to a reformulated (polygenist) reading of Darwinism and evolutionary theory.[52]

Finally having saved enough to move to Italy, Clémence Royer and Pascal Duprat set up their household in Florence, where Duprat also edited a new journal and taught. Here their son, René, was born, legally recognized by his father outside of marriage, as Italian law allowed at the time.[53] Royer showed her practicality and cleverness by dividing their one enormous room into many smaller pleasant ones by skillfully hanging curtains.[54] She continued to write for French as well as Italian scholarly journals, publishing three long essays on Lamarck for Emile Littré's positivist journal *La philosophie positive,* which helped reintroduce Lamarck's evolutionary ideas to a French audience.[55] She insisted that Lamarck would have had all the necessary ingredients for a truly effective theory had he but known of the work of contemporary English economists, an interesting though questionable claim. She also began her detailed application of Darwinism to human society with her book *L'origine de l'homme et des sociétés.*[56] Here she recorded many interesting insights into the role of women, including a speculation that the change in the status of women from equals of men to a role "fatally subordinate" can be traced to the acquisition of goods and children, which must be guarded by the women.[57] Does this comment reflect a perception of the change in her own life, the woman in crinolines waiting in a Victorian train station with child and steamer trunks?

At the end of the Empire, Clémence Royer returned to Paris with Duprat and their four-year-old child. Her book on the evolution of human society was published in 1870, but public attention and sales were damaged by the Franco-Prussian War and the events of the siege of Paris and the Commune. It was not a success, and Clémence Royer complained that in part this was due to the misinterpretation of her book as another version of Darwin's own book on human evolution, *Descent of Man,* published the following year.[58] However, she was accepted as a full member of the Société d'Anthropologie, and she entered the Darwinist debates of 1870 as a very active participant.[59] She was the only woman member of the society for the next fifteen years (and the only notable woman of the society for the rest of the nineteenth century), eventually becoming a member of the Central Committee. When her name was proposed, it met with some opposition, since her name, as Letourneau put it, "rang like a revolutionary bell," in fact, "doubly revolutionary"—as a woman and as Darwin's translator.[60]

With the coming of the Third Republic, Pascal Duprat reentered political life as a deputy, first appointed to the National Assembly and later elected from the seventeenth arrondissement. He came to national attention through his active support of laws to ensure secular education and proposals for the more democratic election of the Senate, measures strongly promoted by republican leaders Léon Gambetta and Jules Ferry.[61] Clémence Royer served as Duprat's hostess, which gave her a social as well as a political visibility she may not have appreciated until after Duprat's death. Nor should it be overlooked that his politics paralleled that of the leading members of both the Société d'Anthropologie and the Association Française pour l'Avancement des Sciences, of which she was also an active member.[62] Most certainly through these links, Clémence Royer came to contribute unsigned science articles to Gambetta's newspaper *La République Française,* under the editorship of Paul Bert, whose science column, written by him and others, was the only such feature in any French daily of the time.[63]

But with this recognition came some checks into Royer's authority as a Darwin translator. Darwin, annoyed at the condemnation of his hereditary theory of pangenesis in the forward of Royer's third edition (1870) of the *Origin,* withdrew his authorization.[64] Pangenesis was a sore point with Darwin, for more than one friend who had warmly supported him on other controversial issues expressed doubts about this new theory. Moreover, this theory of heredity appeared not in the *Origin,* but in Darwin's later *Variation of Plants and Animals under Domestication,* and therefore it was inappropriate for a translator to discuss it in a new edition of the *Origin.* Nor could he overlook Royer's failure to correct her translation according to his new editions or her disregard for his notes. For financial reasons Clémence Royer had been in a hurry to get out a new edition and had simply issued her previous edition with a new foreword. "She abused me like a pickpocket for pangenesis," Darwin complained, and he asked his publishers for a new translator.[65]

Some reviewers felt the new translation of the *Origin* by J. J. Moulinié (who had also translated the *Descent of Man* into French) was poorly written and continued to defend and to prefer Clémence Royer's edition.[66] When scientist Charles Lévèque attacked Clémence Royer for having "betrayed" Darwin in her translation, she denied the accusation, citing letters from

Darwin in support of her first two editions. The resulting debate within the Académie des Sciences Morales et Politiques (part of the Institut de France) was decided in Clémence Royer's favor, and the vice-president of the academy apologized to her.[67] This was the beginning of a long relationship with this prestigious body, which periodically awarded her prizes in competitions, but because she was a woman, never offered her membership. Royer continued to issue her own editions of Darwin with another publisher well into the 1880s.

In the Société d'Anthropologie, Clémence Royer had proven herself to be an articulate and prolific participant in debates. Although she was not trained in medicine or one of the other professions, as were most of the members of the society, Royer felt at no disadvantage, since she believed any realm of thought could yield to logical analysis. Nor did her participation in the society end with these debates. She extended her ideas of evolution to models of atavism, and to a new "dynamogenesis" (which she proposed as a hereditary theory to replace pangenesis).[68] Her views on prehistoric lakes, the origin of Aryan races, and the evolution of the mind were all duly recorded in the bulletins of the society, as were her lively commentaries in discussion.[69]

The one topic on which she could be said to have authoritative knowledge far beyond her colleagues—woman's role in society—was the only one suppressed by the society. The story behind her suppressed manuscript deserves to be explored at some length.

After the election of 1873 and the rise to power of the monarchist party led by Marshall MacMahon, Léon Gambetta, editor of *La République Française,* cautioned his readers: "We have discovered the politic method and the method of patience and solitude."[70] The political setback for the liberal republicans had also resulted in the suppression of feminist lectures and journals. In this context the failure to print Clémence Royer's contributions on women becomes more understandable, although she took the circumspection very unkindly and personally. Patience was not one of her strong points.

She had begun a heated debate on the topic of the causes of French depopulation with Louis-Adophe Bertillon, statistician for the city of Paris and the head of the publication committee of the Société d'Anthropologie. In 1873, at the Lyon conference of the Association Francaise pour l'Avancement des Sciences, Royer had insisted that the decline in French population could not be explained in statistical or economic terms alone.[71] Women were involved in their own fertility, she argued, and it is the nature of their decisions that should be examined.

The French scientists in the Société d'Anthropologie continued this debate on the drop in the birthrate in July 1874. When Clémence Royer joined in the discussion,[72] it was to point out that women were more directly involved in their own fertility than most of the male discussants had been willing to concede.[72] Since men did not understand how women thought in the first place, the evident desire of women to conceal their real attitudes from men should be recognized. Most women were unable to analyze their motives through lack of proper instincts or education. Royer offered her own special knowledge as a woman to investigate this problem. First, she argued, woman was not the passive creature so often depicted by men. "Woman is not made like this," she insisted, detailing her own observations. She believed that, especially in the "cultivated classes," there had been a real weakening of both sexual and maternal drives: Only the residual instincts, which Clémence Royer applauded, kept society from extinction. In young women, "the first child is generally welcome because at his birth a weakened residue of maternal love is awakened in her as a result of a life absolutely chaste until then which predisposed the young girl to become a mother."[73] For subsequent children, however, the picture changes. "The second pregnancy is looked on with unease, the third with hesitation and the fourth avoided." As a consequence, "women suffer the infidelity of their husbands almost gladly," since they can avoid further pregnancies this way. She blamed the Catholic church as the source of this attitude because of representation of procreation "as a shameful affliction of the race." The ambivalent attitude of women was most focused on the experience of pregnancy and childbirth, partly because the bearing of children was seen as something "excusable only on condition that it be preceded by a sort of anticipatory pardon by the Church." (Not having had such a "pardon," Royer may have felt this attitude more keenly.) She continued with a comment that seems even more emphatically to come out of her own experience:

> Maternity? Do women even know what it is before they are wives? It is for them up to the day of delivery a terrible unknown, frightening since they know it might result simply in their deaths. How could they desire it? They only feel fear. Marriage is for the most part a sad surprise, a terrible disappointment, a plunge into reality from the high ideals which imagination has held up to the young girl. . . . First surprised by motherhood, she refuses to nurse her child herself and takes care subsequently to prevent pregnancy.[74]

It must have been startling to these male scientists to hear such an impassioned speech from a woman on such a sensitive subject.

Royer went on to insist that women were well accustomed to using methods to control their fertility, "learning to be mistresses so they do not have to be mothers." She reported that a woman doctor from America had informed her that some physicians had large practices among middle-class women because they had the ability to "skillfully kill off the fruit without injuring the tree."[75] This recourse to abortion occurred in spite of the expanded opportunities in America for economic prosperity for additional offspring and called into question the economic motives usually given as a reason for the low French birthrate. Endorsing the colonialism of her day (and of the republican leadership), she depicted maternity as "an obligation" a woman "owed to her race, to produce intelligent and robust citizens, audacious and healthy enough to venture to wrest new lands from other inferior races."[76] In return for insisting on this obligation, women should be protected and ensured adequate legal rights. They should be allowed to bear children outside marriage if they intend to take care of them, and the law, in turn, should ensure those children's inheritance rights. They should be educated equally with men. Finally, the family itself should be rearranged to allow the young couple to live with their respective parents, who would help raise the children and allow the wife as well as the husband to hold professional positions.[77] Needless to say, many of these descriptions of womens' positions reflected Clémence Royer's own situation and her pressing concerns.

Clémence Royer did not hesitate to draw eugenic conclusions, as she had done twelve years before in her Darwin preface. Whereas in 1862 she had warned of "blind and imprudent charity," which would "aggravate and multiply the weaknesses of the human race," now she insisted on the need to "impose a single life" on any individual who could not expect to bear healthy children. In compensation, however, such individuals might live in communities that would allow "all license without restriction as long as no germ of birth be allowed there." Only under such conditions would abortion and even infanticide be permitted.[78]

In this same manuscript, Clémence Royal hinted at a possible matriarchal change in society, which would permit an entirely new civil code to be created: "(Woman) will put aside what her husband has written and effaced ten times over in a century, his ephemeral institutions and codes always copied on ancient codified errors. She will engrave new customs on the solid rock of hereditary instinct, giving to the world sons instructed by her, who will themselves transcribe (new) customs and laws."[79] This statement embodies many recurring themes in her writings: the importance of instinct to correct social errors, the possibility that women could more adequately create a biologically based society, and the belief that an alliance between the mothers and their sons (rather than daughters) might accomplish this. References to matriarchy as a preferred form of society were present in her earlier writings, although more cautiously. In the *Origin of Man and Societies* she had commented (with some disclaimers) about the usefulness of such a plan if it could be shown to be of economic benefit. A reversal of sex roles might result in economic gains in which a more "fertile" use of time would result from less expended effort, coupled with less capital expense.[80] Even earlier, as noted above, she had hoped for the blossoming of a "female science."

Until the end of her life Royer retained a belief in a future matriarchal state, expressing this once more in a letter addressed to the Société d'Anthropologie in 1895, again on the subject of depopulation. In this she recommended that the family be extended to include the grandparents, that primitive matriarchy be restored and a maternal "gens" be created so that name and inheritance could pass through the women. With a comment that echoed her experience, she recommended that mothers and daughters "share a common domestic life" so that they might pursue professions and "protect their children more effectively than do their fathers."[81] (Her mother, until her death in 1876, had formed part of the household during her son's early childhood, and this helped free Royer for professional work.) These themes, expressed so strongly in her anthropological writings, are muted in her other publications, so much so that one recent biographer has questioned her belief in primitive matriarchy.[82]

Given such strong recommendations about the future human society, as well as her emphasis on the "uneasy bargains" men and women had struck with each other, it is not surprising that Royer's male colleagues hesitated to print her remarks. It should be noted that, on the whole, it was not her anticlerical comments that disturbed them, since many of them shared this attitude. Her eugenic comments also might have been difficult to print without response, since Paul Broca had attacked similar remarks in her Darwin preface, insisting that "society must grant protection to all its members or return to the savage state."[83] Twenty years later, however, the Société d'Anthropologie had little difficulty in printing her strong proposals for the reformulation of social codes. But in 1875 they were in the process of delicate negotiations with the Faculty of Medicine, the Municipal Council, and various individual subscribers (including four Rothschilds) to create a School of Anthropology.[84] This was to be the high point of Broca's long-desired Institute of Anthropology, which would unite the facilities of the society, the laboratory, the museum, and the school. Many of the men who had supported the cause of evolution in the transformist debates (among them Louis-Adolphe Bertillon, Gabriel de Mortillet, Eugène Dally, Paul Topinard, Broca himself, and Charles Letourneau) acquired positions in the School of Anthropology, already under attack by the Catholic press as a projected "school of free-thought."[85] To publish Clémence Royer's controversial remarks at such a time may have well seemed an unnecessary risk.

The publication committee handed back the manuscript version of Royer's statements in 1875, just before the bulletin of the society was to go to press, asking her to make major revisions for publication. Broca apparently suggested that the committee feared her attack on the civil (Napoleonic) code might bring the government down on their heads. Her response was one of anger and annoyance. "Have you read my communication?" she asked Broca. "I do not believe it or you would see that nowhere have I attacked the civil code."[86] Yes, she added, perhaps some of her statements were strong, but "it is not my fault that this is the way things are!" Referring to Bertillon sarcastically as "our Cato of Demography" (he had recently created a demographic society), she characterized him, perhaps unwisely, as "more scrupulous in words than in deeds." She claimed to see behind his apparent prudence "a trace of that ambiguous and equivocal morality which rules so strongly today." She went on to explain that she had shown the article to Duprat, "whom I have never known to be mistaken about the measures to be taken on behalf of courageous speech." He had recommended that she make the requested changes and cuts in order to have the piece published. Unfortunately, Royer balked at this solution, for in rereading it she felt she was accurate in everything she said. If she changed it in a few places, it was simply to add notes of justification and explanation. Her appeal to Broca ended with the plea to let the question of birthrate (and of woman) be discussed "at least once by a woman who believes herself as capable as any man to pretend to the title philosopher. . . . Just because she is a woman, she is free from certain prejudices characteristic of men on these delicate questions."[87]

The publication committee, including Bertillon, must have seen this letter along with Clémence Royer's lengthy additions. The manuscript was set into final page proofs, witness to the society's intention to publish it, but then withdrawn and placed in the archives with the laconic note that the author had agreed to its suppression, since she was unwilling to make the necessary changes. No indication of her discussion exists in the published bulletin.

Clémence Royer's remarks were in fact a direct challenge to the civil code, although she was not aware of this.[88] She had questioned the legal obedience of wife to husband and advocated the acceptance of illegitimate births and the use of eugenic abortion. It is a pity that this manuscript was not published even in modified form, since Royer never published comments on women's social roles in quite this form anywhere else, though her letter to Broca indicated her intention to do so. Even when she wrote on similar questions for the society in the 1890s or for the middle-class feminist press, much of the heat and originality was absent in her discussion of these questions. She had viewed these issues with passion and feeling in the 1870s, when she was a young mother with a child born out of wedlock and a privately but not publicly recognized social position as wife and mother.

In a scientific society dedicated to "positive science," as Broca constantly reminded the membership, Clémence Royer's willingness to construct elaborate theories for which she had little "hard" evidence had annoyed some of her colleagues. They may have regarded her statements on women to be similarly unsupported. For example, the young scientist Louis Lartet, son of the renowned paleontologist Edward Lartet, complained in 1874 to his friend Ernst Hamy, a young assistant secretary of the society: "What inexcusable theories Mlle. [sic] Royer permits herself. Fortunately there were no geologists present."[89] And a year later he made it a more general complaint: "I am sick to death of our Société d'Anthropologie with its scientific claims, its encyclopedic pretensions, its sterile and impertinent verbiage. I am not so flattered as you might believe to have my prose in the company of Pictet, Garrigou, Clémence Royer and *tutti quanti* so that something can act as a counter weight."[90]

Yet Clémence Royer saw herself as a philosopher interested in science rather than as a working scientist. She made some attempts to train herself in the practical aspect of science in 1876, enrolling in the courses given at the Laboratory of Anthropology, where one fellow scientist remembered her as the only member who never wore a laboratory coat.[91] Here she would have learned how to preserve brains, examine and measure skulls, and become acquainted with the developing battery of instruments of physical anthropology. She never found this methodology very sympathetic, later speaking disparagingly of the scientist's fascination with "the kitchen apparatus of science" to the neglect of conceptual analysis.[92] Toward the end of her life she described her working style quite accurately when she remarked of her years associated with the Société d'Anthropologie: "My colleagues gave me facts, I simply turned them into ideas."[93] Herbert Spencer, of course, had worked in a similar manner. Even Charles Letourneau, a social evolutionist, professor of sociology at the Ecole d'Anthropologie, and for the last decade of the century, general-secretary of the Société d'Anthropologie, had produced works of synthesis, writing book after book on the evolution of social institutions, often less vigorous in style than those of Clémence Royer.[94]

Although there are numerous examples of Clémence Royer's willingness to speculate beyond the evidence, and the continuous cautions expressed by her otherwise vigorous supporters—Mortillet, Quatrefages, Broca, and Letourneau—the basic cordiality of the members is evident.[95] A few letters survive from the 1870s and the 1880s to give us a hint of the manner in which she balanced her personal and professional life, giving advice on archeological and anthropological matters, exchanging mathematical solutions, and exchanging social visits with other families.[96] In these exchanges, both Duprat and their child are regularly mentioned.[97]

One of the greatest virtues and weaknesses of Clémence Royer's personality was her unassailable confidence in her own intelligence and abilities. Philosopher Ernest Renan called her "almost a man of genius," the underlying edge of which was heard in the nineteenth century as clearly as it is today.[98] Royer was aware of the quality of arrogance in herself and remarked in one of her autobiographical statements that she had been warned as a girl at school that "pride would be her undoing."[99] At the same time, she must have known it was this same pride that gave her the courage to overcome obstacles and reverses. At one point Darwin remarked of Clémence Royer, "What conceited people there are in the world," but, then, he possessed an almost exaggerated humility.[100] One of her early biographers, A. Moufflet, who seems to have known both Royer and Duprat well, hinted that her sharp replies extended to the domestic sphere as well: "As great as was her intelligence, she did not know how to render justice to that of others, but liked to cut people down to size. Even Duprat himself did not always find grace in her eyes."[101] She tolerated contradiction poorly and found it difficult to admit her errors, which, according to Moufflet, made the discussions in which she participated in learned societies "painful."[102] It may well be that the sharp exchanges common among men in the scientific societies of the day sounded different to the ear when they came from a strong-minded woman.

Some of the prejudice against her in public forums may have been due to the male fears of the "man-woman" as caricatured by such writers as Alexandre Dumas fils.[103] A political reaction to the growing demands for women's rights sought justification from the scientists. Gustave Le Bon, a colleague in the Société d'Anthropologie, had worked in Broca's laboratory in 1877 and 1878 on the relative differences in brain and skull size between men and women.[104] Having shown that the differences existed from culture to culture, but were more exaggerated among the civilized, he concluded that this proved the natural inferiority of women's brains, which had limited their social roles in the past and should continue to do so in the future. There were women of superior intelligence, but they were abnormal cases—monstrosities. Most women should be kept in the traditional roles for which they were best suited, said Le Bon, for the pressure to compete with men would result in perversion and decadence.[105]

Although the Société d'Anthropologie awarded this paper the Godard Prize for its careful calculations, in the Central Committee meeting Paul Broca insisted that the award include a statement (read by Letourneau) declaring that the society distanced itself from Le Bon's social conclusions.[106] Broca thought that the smaller female brain size reflected the inferior education women had long received.[107] Royer, although she systematically opposed Le Bon on many other questions following his work on women's brains, chose not to oppose him directly on this issue. She shared the common belief that an intelligent woman must have a "man's brain." Léonce Manouvrier, a young craniologist who was Broca's last student, refigured and reanalyzed Le Bon's data a few years later and published his own studies showing that women's smaller size and lower body rate directly correlated to lower brain and skull size and indicated nothing about intelligence. He remarked in 1903 that he was surprised Royer had mistakenly adopted this "myth of male science," which claimed that her brain must be larger than that of most women.[108]

By the early 1880s at the time she had become involved in the International Congresses on Women's Rights, Royer found it difficult to find a publisher for her philosophical and scientific books. She had written *Le bien et la loi morale* in response to Herbert Spencer's *Data of Ethics*.[109] Unlike Thomas Huxley, who denied that evolutionary theory provided a basis for social ethics, she agreed with Spencer that morality in the wider sense could always be traced back to some biological advantage to the species. This slight volume reached only a limited audience, which she attributed to the "retrograde philosophy" that had followed the Franco-Prussian War and the Paris Commune, and the effects of positivism "discouraging questions outside its study" and suppressing all speculations.[110]

In 1885 Pascal Duprat died. He had been under political attack at home and had chosen to accept an offer to go to Chile as ministerial envoy. At the age of seventy, his sudden death on the return voyage could not have been a total surprise, but to Clémence Royer and her colleagues at the Société d'Anthropologie, the unpleasant political attacks on Duprat were responsible for his death.[111]

Soon Royer found herself in straitened circumstances. She was in her early fifties, her son was a young man studying at the École Polytechnique, and neither she nor her son had any legal claim on Duprat's estate. She successfully petitioned for a small pension from the government, and Paul Bert, then minister of education, paid her an amount that declined every year after his death in 1886 (when he was governor of Indochina), until it disappeared completely with the fall of the Jules Ferry government in 1889.[112] Royer had tried to make a living through public lectures, but the attendance was not large enough to provide an income. As a last resort, she appealed for the traditional widow's right to sell tobacco and stamps, but it is not clear whether this was granted her.[113] By this time René Duprat was a lieutenant in the army, but until he became a captain his pay was not sufficient to provide for her. Clémence Royer was never free from financial worries, even during Pascal Duprat's lifetime (she spoke of the "twenty years of poverty she had married" through her relationship with Duprat),[114] since his income was often seized to support his other household; the apartment leases were always kept in her name. Yet she did not experience real deprivation until five years after his death. In 1889, all other solutions having failed her, she accepted the offer of "free" room and board in a retirement home, the Maison Galignani, which offered a few places to "indigent men and women of letters" and where she remained until her death in 1902.[115]

These reversals, however, did not prevent her from continuing her writing and lecturing. In the early 1880s she had attempted to found a scientific society that would study moral philosophy, in imitation of the Académie des Sciences Morales et Politiques. In 1882 her fledgling society discussed her monist theories of the world, "L'unité de la substance du monde," on which she was to base a later work.[116] In 1883 it debated socialism, and in 1884 Léonce Manouvrier spoke before it. The academy did not long survive Duprat's death.

Royer continued to submit manuscripts for prizes offered by the Institut de France, and two of these were awarded monetary prizes by the Académie des Sciences Morales et Politiques.[117] In 1891 she appealed to Armand de Quatrefages, a longtime member of the Institut de France and her former colleague in the Société d'Anthropologie, to support her for a prize in the Académie des Sciences, arguing that she had anticipated Darwin on the subject of man, had discussed monism before Haeckel, and had raised many issues before Spencer. The prize she saw as a sort of recompense for the "many (academic) chairs, prizes and awards not available to me as a woman."[118] This appeal, which has struck some modern readers as self-promoting, was common practice in the nineteenth century, when all candidates for the various academies were required to spend much time soliciting the support of patrons and friends.[119]

When the Société d'Anthropologie established its yearly lectures on evolution, Clémence Royer was asked to give one on "mental evolution" in 1887.[120] She spoke on the development of arithmetical reasoning in primates and man. The lecture was published in a number of issues of *Revue scientifique* and reprinted in English in *Popular Science Monthly*.[121] This also served to give her an audience in the United States. Age, illness, and a great increase in weight, however, made it difficult for her to climb the stairs to the Société d'Anthropologie. She ceased to attend meetings, although she was made an honorary member of the society and then a member of the Central Bureau.[122] Despite her longtime membership in the society, they called only once upon her expertise, according to the archives, to give an opinion on the best and most economical method of recovering the tables of the meeting room.[123] As a good needlewoman she submitted her report in great detail, apparently not taking the request amiss. Her communications continued to appear in the bulletins, although now in the form of letters.[124]

Although Clémence Royer attended the International Congresses on Women's Rights in 1878 and 1888 in Paris, she did not become actively involved with the French feminists until the end of her life. She had joined a "mixed" masonic lodge for men and women along with the women's rights activist Maria Deraismes in 1893,[125] and it may have been this association with the freemasons that led to her rediscovery by the French feminists in that year. Through the interest of Mary Léopold Lacour and her husband and Marguerite Durand, Clémence Royer's submissions and applications to the Institut de France first became generally known.[126] In 1897, with the formation of Marguerite Durand's daily feminist newspaper for middle-class women, *La fronde,* Clémence Royer became a regular correspondent, writing on science, pacifism, women's education, and many other topics.[127] That same year the French feminists organized a banquet in her honor, which for the first time gave her contact with the younger generation and the women of her time.[128] Her response was ecstatic and charming. She spoke of her perception of herself as abandoned and alone, suddenly discovering a world of new friends. She thanked the learned men who attended this banquet in her honor, but it was to the women she particularly addressed her words:

> I thought myself isolated, vanquished, saddened to have worked in vain; I expected nothing more of life; I believed that my thought, rendered sterile would

> never reach that rare elite intelligence which alone can comprehend and judge it. I believed myself forgotten by my generation, unknown by the current generation. Suddenly I see coming towards me young and old. These remembered me, these discovered me . . . such a great crowd hold out their hands to me to encourage and resuscitate me.[129]

Given the many difficulties Clémence Royer had experienced during the previous seven years, it must have been with a sense of irony that she listened to Charles Letourneau praise her at the banquet as someone who had given her life to science without hope of monetary gain.[130]

Her son, René, who had helped her when he was a teen-ager by copying the manuscript of her book *Le bien et la loi morale,* had by this time become a colonial officer and engineer, laying railroads in Indochina.[131] He had not had an easy life, indeed Clémence Royer had confided to a friend that she thought René's problems had been even more difficult than her own. His status as the illegitimate son of a former politician did not help him in a world so highly dependent on patronage. More significant, he had the same intolerance for authority as his mother had, not a useful trait for an army officer. Royer had had a picture taken of him just before he left for the colonies by her friend the photographer Nadar, and she was impressed by how handsome he looked in his uniform.[132] He was to die in the colonies from a bilious fever only a few months after his mother's death.[133]

IN 1900 Clémence Royer was awarded the Legion of Honor.[134] She was too ill to leave the Maison Galignani, even when she was made honorary co-chairman with Marguerite Durand of the Congress for the Rights of Women. Instead, a delegation led by her former colleague in the Société d'Anthropologie Dr. Blanche Edwards-Pilliet and her friend the feminist Avril de Saint-Croix came from the congress and greeted her as the "greatest French woman scientist."[135] Georges Clémenceau, later war hero and president of the republic, visited her at the Maison Galignani. He described the meeting for the popular weekly *L'illustration:*

> This woman who I knew for so long through her writings, I saw in her modest asylum full of peace and serenity. I found her happy in the joy of elevated thought which escapes the attention of the wicked world. Her features remain fine, her great blue eyes singularly lively, her manner of speaking simple and easy, her clear language testifying to her precise, firm thought, evoking in me

the memory of the highest female minds which were the honor and patrimony of our great eighteenth century.[136]

Yet Clémenceau did not overlook the quality that embodied Clémence Royer's last years, the attempt to work in the manner of the eighteenth-century encyclopedists. She had earlier questioned Newton's laws, returning to the earlier concept of plentitude by imagining elastic atoms whose "edges" touched one another, ruling out action at a distance, and presuming force moving toward a center.[137] In 1900 she had just published a philosophical physics *La constitution du monde,* in which she set forth her monist physical theories complete with an attempt to reformulate a new cosmology in 800 pages.[138] One reviewer in *Science* dismissed this work with the comment that it showed a "lamentable lack of scientific training and spirit."[139] More than one reviewer was shocked by her attack on currently accepted physical laws combated by her own philosophical reasoning. She had extended the manner in which she had opposed Newton's laws of attraction, and her objections to Darwin's pangenesis to astronomical, chemical, and physical science, often attacking one aspect of a theory while accepting another without question. To Clémenceau this style of theorizing indicated that at least one woman was willing to be concerned with such problems, although his attitude had a condescending ring to it. He observed that Clémence Royer's theories "await criticism along with those which they have the pretension of supplanting."[140] It is unfortunate that this was her final legacy, coming as it did out of work done in isolation.

When she died in 1902, having achieved at least a limited fame, it was still her Darwin preface that was cited and remembered. Although the French feminists continued to extol her social philosophy and her science, turning her into a cult figure some twenty-eight years later,[141] a Russian writer deplored her inability to understand the problems of class as well as of sex: "Mme. Clémence Royer, who had energetically struggled for the rights of women and showed by her own example the strength and flexibility of the feminine mind, at the same time put forward the most banal arguments of bourgeois thinkers in opposition to the possibility of doing away with class inequality."[142]

Her biographer and hagiographer, Albert Millice, extolled Clémence Royer for precisely the mixture of elitism combined with pacifism and women's rights. She had developed these ideas directly out of her views on natural selection, which had taken "election" to imply elite.[143] She believed that the great underclass, with its greater population growth, provided the motor of progress that drove some individuals upward in the social hierarchy, while the great mass of society remained at the base. It is in this light that one must read her most highly rewarded contribution to the Académie des Sciences Morales et Politiques, a study of public assistance to the poor with the legend "The poor you will always have with you."[144]

Clémence Royer was neither a great scientist nor a humble handmaiden of science. She was not a profound philosopher. She is best remembered for one piece of work, her Darwin translation, which she came to resent as the basis of her fame and which she tried vainly to surpass in her own original writings.[145] Yet she remains a fascinating individual—a woman struggling through her fragmentary education into the closed world of masculine science, a self-educated woman who was impatient of authority, a believer in an older synthetic style of working in a time of experimentation and specialization. In her attempts to reconcile the beliefs and philosophies of the middle-class men and women around her, she expressed a lifelong desire to make them less "strangers to each other."

9

ANN HIBNER KOBLITZ

Career and Home Life in the 1880s: The Choices of Mathematician Sofia Kovalevskaia

THE RUSSIAN mathematician Sofia Kovalevskaia (1850–91) was the first woman outside of eighteenth-century Italy to hold a chair at the university level in Europe. During her lifetime, Kovalevskaia was regarded as one of the most eminent mathematical analysts in the world. She was awarded the Prix Bordin of the French Academy of Sciences, the Oscar Prize of the Swedish Academy of Sciences, and corresponding membership in the Russian Academy of Sciences. Her dissertation work is considered basic to the theory of partial differential equations. Her classic memoir on the revolution of a solid body about a fixed point (the "Kovalevskaia top") has been described as "one of the most famous works of mathematical physics in the 19th century, linking two main theories of the 19th century mathematics—analytical mechanics and complex function theory—in a beautiful way."[1]

Kovalevskaia was a writer as well as a mathematician. Her literary productions included two plays (written in collaboration with the Swedish writer Anna Carlotta Leffler), a memoir of her childhood, a novella, and numerous essays and poems. Moreover, Kovalevskaia was an active campaigner for women's right to higher education and participated in a small way in the revolutionary movements of her native Russia and Western Europe.[2]

But this dry listing of Kovalevskaia's accomplishments tells only part of her life story. For the last eight years of her life—the time, incidentally, of her greatest mathematical and literary activity—Kovalevskaia was a widow struggling to support herself and her daughter on her own.

How did Sofia Kovalevskaia manage to juggle her career and family? Were her colleagues, friends, and relatives helpful to her? What was the reaction of society to her attempt to balance home and professional life? By describing Kovalevskaia's problems, decisions, and compromises, I hope to elucidate the general dilemmas faced by women scientists in the nineteenth century and today.

SOFIA KOVALEVSKAIA (born Korvin-Krukovskaia) had the good fortune to grow up in the 1860s, at a time when the educated classes of her native Russia were engaged in multifaceted debate on social and political reform. The humiliating defeat of tsarist troops in the Crimean War in 1856 had caused turmoil in Russia; many people took the defeat as a sign that Russia needed to change, to modernize. Progressive nobles and intellectuals agitated for reform of the government and educational systems and called for the emancipation of the serfs. Some younger members of these groups went further: they declared that the entire social structure of tsarist society was morally bankrupt, and that Russia would have to be completely restructured along egalitarian, progressive lines. These people called themselves nihilists.

The nihilists believed in the inevitable triumph of a peaceful social revolution in Russia. They had enormous faith in the power of the natural sciences, and in the efficacy of education as an antidote to backwardness. Moreover, they taught that they had a duty to help the masses of uneducated peasants that formed the vast majority of the Russian population. That is why so many of them became scientists, doctors, agronomists, veterinarians, and village schoolteachers (and why so many of them were arrested for conducting educational and political propaganda in the factories and villages).

In line with their beliefs in social revolution and equality for all, nihilists were staunch advocates of women's rights. They held that women of the nobility and the intelligentsia had a right and even a duty to educate themselves and then proceed to educate their less fortunate peasant sisters. Moreover, the nihilists insisted that women should be treated as equals in marriage as well as education, and that the patriarchal foundations of traditional family relationships should be abandoned.

These teachings were not empty verbiage on the part of nihilist men. Most adherents of the movement lived according to these beliefs. Some, like nihilist theorists Nikolai Chernyshevskii and Nikolai Shelgunov, even felt that their wives should be allowed complete freedom in sexual matters, while they themselves should remain faithful. This was, in Chernyshevskii's words, in order to "bend the rod the other way" and give women the choices they had been denied for countless centuries.[3]

Although most nihilists did not go as far as Chernyshevskii and Shelgunov, nevertheless many of them were genuinely eager to have loving, equal, comradely relationships with women. On the whole, they treated women with respect, viewing them as people in their own right rather than as decorative dolls or childbearing machines. The women of the movement responded in kind, accepting the men, just as the men accepted them, as partners in a common struggle.

Kovalevskaia absorbed nihilist ideas of politics and society from the time she was a child through the medium of her adored elder sister, Aniuta, who had a tremendous influence on her. Aniuta had been introduced to nihilism by the son of the village priest. The young man, who was studying the natural sciences in Saint Petersburg, brought Aniuta word of the nihilist discussion circles being organized in the capital and told her tales of the great revolutionary propagandists of the time—Chernyshevskii, Shelgunov, and others. Aniuta listened with great interest and passed on her vague conceptions of the "new ideas" (as nihilism was sometimes delicately called) to her little sister.

Kovalevskaia eagerly accepted nihilism. The philosophy was congenial to her, with its belief in women's equality, its faith in the natural sciences, and its conviction that a social revolution would soon free Russia from centuries of backwardness. Kovalevskaia espoused nihilism early and remained an adherent of the movement until her death.

With such a background, it perhaps becomes less surprising that Kovalevskaia was able to break with tradition and develop her natural talents to their full potential. She and her fellow nihilists were philosophically committed to trying new life-styles, new careers. Kovalevskaia was a conscious pioneer, aware of the uniqueness of her path and determined to open new possibilities to women.

Opposite: Sofia Kovalevskaia (*left*) and her friend Swedish novelist Anna Carlotta Leffler. Photo taken during collaboration (1886–87) on their play *The Struggle for Happiness,* staged in Saint Petersburg in 1895.

By the time Sofia Kovalevskaia was eighteen, she had decided that she wished to enter the university and study either medicine or one of the natural sciences. Officially, universities in Russia were closed to women, as they were almost everywhere in Europe. (The university and polytechnical school in Zürich were the only exceptions; they accepted women from 1865.) But the nihilist scientists who made up a significant portion of the faculties of the university and Medical-Surgical Academy in Saint Petersburg were quite willing to take women students as unofficial auditors.

Sofia wanted to matriculate officially, however, and obtain her degree from a university abroad. She resolved that in order to circumvent what she assumed would be the violent opposition of her father, she needed to contract a so-called "fictitious marriage." Fictitious marriages were a device favored by some early nihilists as a means of liberating young women from the yoke of parental tyranny. Women in tsarist Russia did not have separate legal identities. A woman was listed on either her father's or her husband's internal passport and could not live apart from one or the other without his express written permission. Since nihilist women usually wanted this permission so they could move to Moscow or Saint Petersburg and audit courses at the university or the Medical-Surgical Academy, or so they could travel to Zürich to pursue a formal degree, their parents were generally reluctant to agree.

Kovalevskaia's situation was different from that of other women of her generation in that her father was relatively tolerant of her desire to learn and had even engaged a tutor for her so that she could study university-level mathematics from the time she was fifteen. For some reason, however, Kovalevskaia decided that she, too, needed a fictitious marriage. Perhaps she was carried away by the romance or the clandestine nature of the idea; perhaps she genuinely felt that her father, while agreeable to a tutor, would not want her to leave home to continue her education. Whatever the reason, Kovalevskaia came to the conclusion that in order to pursue her studies, she would have to find a man who would marry her, give her permission to live apart from him, and then disappear from her life.

Through her mathematics tutor, who was also a nihilist and knew many progressive young men of the 1860s, Kovalevskaia was introduced to nihilist book publisher and sometime amateur paleontologist Vladimir Kovalevskii (1842–83). Kovalevskii was a sincere "man of the sixties," as the nihilists often called themselves, and was eager to help Kovalevskaia escape from the "tyranny" of her family. Unfortunately, although Vladimir had some real political commitment and was later to become a famous evolutionary paleontologist, he was also psychologically unstable. His peculiarities increased as he grew older, and in the last years of his life he was a constant source of anxiety for Kovalevskaia.

Moreover, Vladimir had a different idea of what a fictitious marriage entailed than did Kovalevskaia. He was eight years older than she and had been influenced by the marital ideas of the great Chernyshevskii. In his phenomenally popular novel *What Is to Be Done?* (1863), Chernyshevskii had described the ideal fictitious marriage as one in which the partners started out as platonic friends and then gradually, gracefully, naturally drifted into a sexual relationship of two people who are comrades and equals as well as lovers.

Like many in his age group, Vladimir had been deeply affected by the ideal marriage portrayed in Chernyshevskii's novel. He had participated in the nihilist discussion circles of the early 1860s, in which the details of the novel were endlessly debated. Moreover, Vladimir knew personally not only the leading progressive publicists of the day, but also the people on whom Chernyshevskii was rumored to have based his heroine and heroes—Maria Bokova-Sechenova, Ivan Sechenov, and Petr Bokov.

Kovalevskaia was too young, and too far from the centers of nihilist discussion, to have been deeply influenced by Chernyshevskii's novel, which she does not seem to have read before her marriage to Vladimir. *She* thought of fictitious marriage as merely a legal expedient that would transfer her from her father's jurisdiction to her husband's; at first, she does not seem to have had any notion that Vladimir expected to share her life, if not her bed.

These conflicting views of fictitious marriage, plus Vladimir's increasingly unstable mental state, led to problems for Vladimir and Kovalevskaia. Most of these do not concern us here. For the purposes of this essay it is sufficient to say that for about six years the Kovalevskii union was a nonsexual one. During this period, the couple studied at various universities in Western Europe and received their doctoral degrees—Kovalevskaia from Göttingen University in 1874, and Vladimir from Jena University in 1873. The pair met occasionally abroad, spent some of their summer vacations together, and engaged in correspondence of varying degrees of warmth.

Kovalevskaia's main support during her years abroad came from a group of Russian women who were also in search of higher education in Western Europe, and her Berlin adviser, the great mathematical analyst Karl Theodore Weierstrass (1815–97). Her closest friend and constant companion was Iulia Lermontova (1846–1919). Lermontova studied with Kovalevskaia in Heidelberg and Berlin and, like her, received her doctorate (in chemistry) from Göttingen University in 1874.

Lermontova and Kovalevskaia formed the nucleus of what they called their "women's commune." The membership in this group was fluid. It included at various times Kovalevskaia's sister Anna Korvin-Krukovskaia (1843–87), a member of the International Working Man's Association and a participant in the Paris Commune of 1871; the sisters' cousin Anna Evreinova (1844–1919), the first woman to receive her doctorate in jurisprudence and a lecturer on ethnography and women's rights; and the revolutionary Natalia Armfeldt (1852–87), who became a propagandist in the Chaikovtsy circle in Saint Petersburg and died in detention in Siberia. Other women, such as Lermontova's cousin Olga and Evreinova's sister Sofia, visited the commune for various periods.

These women formed Kovalevskaia's "family." They were each other's main comfort and support during their years abroad, even at those times when most of them were not living together. Kovalevskaia, Lermontova, Korvin-Krukovskaia, and Evreinova would remain in close contact throughout their lives. This banding together of preprofessional and professional women into support networks was common in the Russian nihilist intelligentsia and later in the revolutionary underground. Kovalevskaia's Heidelberg women's commune had its analogs among the Russian women students in Zürich, Paris, Geneva, and elsewhere, and among the women revolutionaries in Saint Petersburg and other large cities of the Russian empire.[4]

Perhaps less typical was Kovalevskaia's and Lermontova's close relationship with Kovalevskaia's Berlin adviser. Almost as soon as the two Russian women arrived in the fall of 1870, Weierstrass and his two spinster sisters, with whom he lived, took Kovalevskaia and Lermontova under their wing. They regularly had them over for dinner, arranged Christmas festivities for them, saw to it that the two young women did not neglect themselves too badly in favor of their studies. In addition, Weierstrass acted as a sort of scientific intermediary between his protégées and their mathematical and chemical colleagues: he was instrumental in arranging for Göttingen University to agree to grant Kovalevskaia and Lermontova doctoral degrees.

Though he kept in touch with both women throughout his life, Weierstrass's closer tie was with his student Kovalevskaia. He and his sisters treated her as a beloved daughter. They assiduously looked after her interests, occasionally gently admonished her for tactlessness or overexertion, and above all provided love and, in the case of Weierstrass, professional assistance. These warm personal and mathematical ties lasted throughout Kovalevskaia's life.

In 1874 Kovalevskaia and her husband returned to Russia to try to find employment at the university level. For reasons that are not entirely clear, around this time they decided to change the basis of their relationship and set up house together as real man and wife. This move was nothing short of a disaster. Kovalevskaia and Vladimir were incompatible, they apparently brought out the worst traits in each other's personalities, and Vladimir was showing signs of the instability that would later destroy him.

When neither Vladimir nor Kovalevskaia was able to obtain a university-level position and the two turned to financial speculation and various publishing ventures to support themselves, Kovalevskaia assumed what many of her friends considered a facade. For almost six years—from 1874 to 1880—Kovalevskaia tried to play the role of a more or less traditional wife. She became a typical "society lady," was seen at the opera and the theater, had a literary salon, and virtually abandoned her mathematics.

This six-year period was the only time in Kovalevskaia's life that she subjugated her own career interests to the desires of society and her husband and put aside her experiments in unconventional life-styles. The foray into social conformity was a failure, and Kovalevskaia grew capricious, resentful, and increasingly restless.

Kovalevskaia engaged in some literary work during these years in Russia, participated in the organization of the first women's university-level courses in Saint Petersburg, and maintained contact with the rising generation of young populist revolutionaries. But these activities were not enough to satisfy her, especially since they offered no possibility of doing scientific work. She began to refer to her life as "the soft slime of bourgeois existence."[5] Later she would say that the birth of her daughter, Sofia (nicknamed Fufa), in November 1878 was the only good thing to come out of these years in Russia, the years of her real marriage with Vladimir and her submission to convention and society. Everything else was worthless.[6]

It is only in the years after Kovalevskaia returned to her mathematical work (around 1880) that we can begin to speak of her actually juggling her family and career. Before that time Kovalevskaia was either exclusively engaged in study (1869–74) or almost entirely involved with her family and social obligations (1874–80).

In 1880 the Kovalevskii family moved to Moscow from Saint Petersburg. Partly this was to escape the scene of their financial troubles. Vladimir's rage for speculation and his tendency to try to enrich himself on borrowed capital made him squander the money Kovalevskaia's father had left her. These failed get-rich-quick schemes had the inevitable outcome: the Kovalevskiis had to declare bankruptcy and allow their possessions to be sold at public auction.

Despite the humiliation, Kovalevskaia seems to have viewed this end to the couple's financial dreams as a relief. She saw the bankruptcy as a sign that it was time for her to stop pretending to be what she was not and get back to serious science. She appears to have greeted the move to Moscow as an excuse for a clean break from her recent "bourgeois" past. She resumed her correspondence with Karl Weierstrass, gave a talk at the Congress of Natural Scientists in early 1880, and became a member of the Moscow Mathematical Society.

In Moscow the Kovalevskiis combined their household with that of Sofia's best friend (and Fufa's godmother), Iulia Lermontova, and her sister. Lermontova worked in the laboratory of the eminent chemist V. V. Markovnikov in Moscow and reviewed articles and wrote notices for the journal of the Russian Chemical Society. In addition, she was experimenting with processes for petroleum distillation, which she hoped would be of eventual commercial use.

Kovalevskaia was preparing herself to take examinations for a Russian master's degree, which would have formally qualified her to teach at the university level, and she had become interested in experiments in electricity and magnetism. Iulia also was very busy, so the domestic tasks of the Kovalevskii-Lermontova household appear to have fallen almost entirely to Lermontova's sister Sonia. Sonia good-naturedly handled the marketing, cooking, overseeing of the household's one servant, and the care of little Fufa when her babble proved too trying on her mother's and godmother's nerves.

Fufa seems to have spent a good amount of time with her mother and with Iulia. A friend of Kovalevskaia, the mathematician and pedagogue Elizaveta Litvinova, described how tiny Fufa, who at this time could barely walk or talk, would take a pencil and paper, draw a few squiggles, and announce to all who would listen to her that this was her "appalat" (her mispronounced baby version of "apparatus"; *apparat* in Russian). Clearly, she had often listened in on Sofia's and Iulia's scientific conversations.[7]

Meanwhile, Vladimir was attempting to negotiate a position for himself at Moscow University and was simultaneously trying one last time to get rich by becoming involved in a shady oil operation. Sofia and Vladimir were not getting along very well—she felt that he should have been concentrating on science rather than dividing his attention between paleontology and commerce. Nevertheless, Kovalevskaia subordinated her professional interests to his for what would be the last time.

As Kovalevskaia prepared to take the master's examinations, she had the support of several influential members of the Moscow University mathematics department, as well as that of the university rector. Moreover, she had unpublished work from her student years that she easily could have revised for presentation as a master's dissertation. (Kovalevskaia had written *three* dissertations for submission to the Göttingen faculty in 1874; she and Weierstrass had felt that her case would have to be especially strong, since she was the first woman to apply for a degree in mathematics.)

Kovalevskaia withdrew her petition for permission to take the examinations, however. Apparently, Vladimir's colleagues had told him that negotiations for his position at the university would be adversely affected if it became known that his wife was requesting admission to an examination hitherto given only to men. Kovalevskaia was bitter about having to postpone her plans and resentful that the geology faculty expected her "naturally" to put Vladimir's professional interests before her own. But she submitted and waited until Vladimir's appointment was confirmed.

Ironically, it turned out that Kovalevskaia need not have bothered. Vladimir managed to alienate the university authorities almost immediately with his unexplained absences and frenetic, disorganized lecture style. He held his position for barely two years before he killed himself. In any case, when Kovalevskaia did apply to take the examinations, not even the combined efforts of the entire Moscow mathematics faculty and the rector himself were enough to convince the reactionary tsarist minister of education. Kovalevskaia was told that both she and her daughter would grow old before a woman would be allowed to teach in a Russian university.[8]

After this experience, Kovalevskaia began to think about seeking a university position abroad. She knew that the only way she would have even the slightest chance of obtaining such a post was if she concentrated on producing as many mathematical works as possible. For this, she felt she had to return to Berlin to reestablish contacts with the Western European scientific world. Vladimir, whose early commitment to nihilist ideas of equality between the sexes was disappearing as he grew older and more eccentric, maintained that Kovalevskaia had studied enough and that it was time to devote herself to her family. He opposed Kovalevskaia's trip to Berlin, but she overruled him. She left Fufa in the care of Iulia and especially Sonia Lermontova, and spent the last two months of 1880 working with Weierstrass in Berlin.

This separation seems to have marked the beginning of the end for the Kovalevskii marriage. Relations between Kovalevskaia and Vladimir became increasingly tense. For Fufa's sake, they maintained an appearance of amity, but they took every opportunity to live apart. Vladimir took frequent trips abroad for his oil company, while Kovalevskaia stayed in

Moscow or journeyed to Berlin. In March 1881 Kovalevskaia took Fufa with her to Western Europe, where she would remain, except for short visits to Russia, until her death in 1891.

Kovalevskaia's financial circumstances at this point were shaky, and to some extent would remain so for the rest of her life. But Vladimir was able to send a small monthly sum for Fufa's maintenance, and there are indications that Iulia Lermontova and Vladimir's brother Aleksander occasionally gave monetary assistance as well. Kovalevskaia was thus able to afford a nursemaid for Fufa, although she scrimped and saved and recorded every penny of her expenses in order to do so.[9]

Fufa and her mother lived a frugal life while Kovalevskaia attempted to get back into the mathematical mainstream. Kovalevskaia seems to have known that she and Vladimir would never be able to settle their differences. In any case, she probably would not have wanted to return to her old relationship with him unless he was willing to support her professional interests. By this time it was clear that Kovalevskaia needed mathematics in order to feel completely alive and would not tolerate another immersion in the "soft slime" of her former life.

Kovalevskaia set about searching for a position at the university level. The obstacles were enormous, and it would take almost three years and the death of her husband, not to mention the determined efforts of Weierstrass and Swedish mathematician Gösta Mittag-Leffler (1846–1927), before she would obtain a university post.

Meanwhile, however, the vagabond existence Fufa was forced to lead seemed to be having bad effects on her health. Kovalevskaia had brought the child with her from Berlin to Paris when she moved there in late autumn 1881. Then, in early 1882, Fufa barely survived a bout with something the doctor feared could have been cholera. Both Iulia Lermontova and Kovalevskaia's brother-in-law, the embryologist Aleksander Kovalevskii, repeatedly begged Kovalevskaia to send Fufa to one of them to be raised. Traveling was not good for the child, Aleksander and Iulia insisted. A four-year-old needs stability, which Kovalevskaia could ill provide, they warned.

Kovalevskaia resisted the entreaties of Iulia and Aleksander for several months. She initially felt that Fufa needed her mother and was reluctant to part with her. Eventually, however, Kovalevskaia decided that stability and reasonable material circumstances were more important than a maternal presence, and so in March 1882, the girl was sent to Russia, where she would remain, spending part of the year with Aleksander in Odessa and part with Iulia in Moscow. When Kovalevskaia was firmly established in

Stockholm, in the autumn of 1886, she brought her daughter to live with her.

Kovalevskaia's decision to send Fufa back to Russia had been an agonizing one, as comments in her diaries and letters make clear. In ultimately choosing to part with her daughter, Kovalevskaia was behaving in a typical manner for Russians of her class and time. Members of the Russian gentry and educated classes believed firmly that a comfortable financial situation and stable family structure were crucial to the happiness and well-being of their children. Poorer parents often split up their children among their more well-to-do relatives.[10] Kovalevskaia was convinced that in sending Fufa to Aleksander and Iulia, she was acting in the best interests of her child.

Kovalevskaia remained in France through 1882 and much of 1883. She corresponded with Vladimir and saw him once or twice when he came through Paris. But by this time Vladimir's behavior was so erratic that misunderstandings and quarrels were the inevitable result of any contact they had with each other. Vladimir alternated between periods of deep depression and feverish excitement and optimism. He would respond favorably to some gesture of friendship on Kovalevskaia's part, and then without any warning go out of his way to insult or hurt her. She, meanwhile, had become so immersed in her mathematics, and in her contacts with Russian and Polish émigré revolutionaries, that she had neither the time nor the interest to deal with Vladimir's peculiarities. She seems to have written Vladimir off and maintained contact with him only through the intermediary of Aleksander Kovalevskii. Instead, she concentrated on writing up her paper on the refraction of light in a crystalline medium, and on encouraging the efforts of her colleagues to arrange a university-level position for her.

Gösta Mittag-Leffler, founder of the journal *Acta Mathematica,* was a staunch internationalist and early feminist. His love, respect, and admiration for his teacher Weierstrass meant that he was totally devoted to the propagandizing of Weierstrassian analysis. It was therefore to his advantage to have one of the foremost exponents of the discipline, Kovalevskaia, at the same university with him. Kovalevskaia was known as an excellent expositor, and Mittag-Leffler felt that she would make a stimulating, communicative colleague. The fact that she was a woman and would have difficulty breaking into the all-male world of university teaching only added to Kovalevskaia's appeal for Mittag-Leffler. He had long been a champion of women's access to institutions of higher education, and he seems to have been stimulated by the challenge of helping Kovalevskaia.

Mittag-Leffler had been attempting to find a university post for Kovalevskaia since early 1881. His negotiations had come to nothing in Helsinki, where he had been a professor until 1882, but he was hopeful that he could obtain an appointment for Kovalevskaia at his new university in Stockholm. Mittag-Leffler was having trouble because Kovalevskaia was a woman and a vocal advocate of women's rights and radical causes. These attributes did not upset most mathematicians—Kovalevskaia's colleagues tended to be relatively tolerant of political diversity. In fact, throughout her career, Kovalevskaia met with support and encouragement from fellow mathematicians, with only two exceptions.[11] Unfortunately, Mittag-Leffler was not dealing with mathematicians, since he was the only mathematician on the faculty at the time. And to professors in other disciplines, Kovalevskaia's gender and politics *were* factors in their decision.

One of the main obstacles to Kovalevskaia's appointment at Stockholm appears to have been her ambiguous marital status. Her evident desire to carve out a career for herself apart from Vladimir was something that the staid Swedes strongly objected to, or at least were eager to use as an added argument against her. A woman who lived separately from her husband could not be quite respectable, they insinuated. Mittag-Leffler was stymied, at least for the moment.

At this point, however, fate, in the guise of the hapless Vladimir, took a hand in the proceedings. Vladimir had involved himself in the machinations of his oil company and was facing criminal charges arising from an investigation of the firm. Moreover, his mental faculties were waning; he was losing his powers of concentration and his ability to do original science. It took him several hours of agonizing effort to prepare a single page of class notes, and he found himself incapable of writing up his research findings. The combined misfortunes were too much for him: he committed suicide in late April 1883.

Kovalevskaia was distraught when she heard the news. She had never loved Vladimir, and their marriage had brought her little happiness. Nevertheless, she did feel a certain responsibility for his suicide and blamed herself for not realizing how desperate he was. Vladimir's death, however, simplified Kovalevskaia's life considerably and opened up to her the possibility of a professional career. For purposes of obtaining a university position, nothing could have been better for Kovalevskaia than recent widowhood. In late nineteenth-century Europe, widowhood was eminently respectable. A widow enjoyed a freedom of movement and decision-making power not allowed a single or married woman; a widow was in control of her own fate. Mittag-Leffler pounced with almost indecent haste and by mid-August 1883 had secured a temporary post at Stockholm University for Kovalevskaia. She took up her teaching duties in January 1884, was appointed to a five-year "extraordinary professorship" in June of that year, and in June 1889 was given a permanent "ordinary professorship" (in modern parlance, tenure).

Even after Kovalevskaia moved to Stockholm in the fall of 1883, she did not bring Fufa to live with her. She was convinced that she needed to establish herself in her new home, learn the language, discover whether the position had any long-range prospects, before she subjected Fufa to what might very well be just the first in a series of moves from university to university. Iulia Lermontova and Aleksander Kovalevskii fully endorsed this decision, as did Gösta Mittag-Leffler and his family. But much of Swedish society disapproved. Kovalevskaia was constantly forced to defend her disposition of her daughter to her Swedish acquaintances. Her stock answer was that she felt herself to be the sole guardian of Fufa's happiness, and she was not going to needlessly disrupt the child's life simply for fear of what people would say.[12]

In later life, Sofia Vladimirovna (Fufa) applauded her mother's decision to leave her in Russia until such time as Kovalevskaia was settled in Stockholm.[13] But the separation undeniably had some effect on the relationship of mother and daughter. Kovalevskaia visited Fufa for a month or two each summer, and often for a few days at Christmas time, and sent presents and affectionate letters. Not surprisingly, however, Fufa tended to regard Kovalevskaia more as a glamorous visitor or an indulgent aunt than as her mother. Her closest tie was with Iulia Lermontova (whose sister Sonia had died when Fufa was five); in fact, she called Iulia "Mama Iulia."

Kovalevskaia recognized and regretted the fact that Fufa was closer to Iulia than to herself, but still felt that a stable home for Fufa with her godmother or uncle was better for her in the long run than a bohemian existence with her mother. Only in autumn 1886 did Kovalevskaia consider herself sufficiently settled to bring Fufa back with her to Stockholm. And even then she had to contend with the vociferous objections of Iulia and Aleksander Kovalevskii, both of whom thought that Fufa should remain in Russia with them.

Sofia Vladimirovna later recalled that her mother had to devote much energy to reconciling her little girl to her change of circumstances—everything seemed so strange to Fufa in Sweden. But Kovalevskaia was patient with her daughter, taking her to the market and visiting with her, so that soon Stockholm, the Swedish language, and Sofia's friends no longer frightened Fufa.

The arrival of Fufa made Kovalevskaia change her residence from two furnished rooms to a five-room apartment and compelled her to hire a cook-nursemaid to care for Fufa and help with the housekeeping. Kovalevskaia's attitude toward domestic concerns had always been slapdash. If left to herself, she would wear the same dress for years on end, grab ice cream or a meat pie for dinner, and tidy her room only when one of her inevitable piles of books and papers toppled off its precarious perch on a chair or windowsill. Still, there were certain tasks that had to be performed, and she once complained to a friend: "All these stupid but unpostponable everyday affairs are a serious test of my patience, and I begin to understand why men treasure good, practical housewives so highly. Were I a man, I'd choose myself a beautiful little housewife who'd free me from all this."[14]

Since Kovalevskaia could neither acquire a "housewife" nor afford a retinue of servants, she coped with the details of daily life as she saw fit. She sent to Russia for her share of the decayed grandeur of her old family home (long since sold to pay her brother Fedia's gambling debts), which consisted of cracked mirrors and sofas with missing legs and torn upholstery. She propped up the furniture with books when necessary, threw antimacassars and unfinished pieces of embroidery over the worn spots, and blithely considered herself ready to entertain anyone who came to call.

Swedish society was not impressed with Kovalevskaia's attempts at housekeeping, and her female acquaintances there complained about her incessantly. The Swedish educational writer and feminist Ellen Key recalled with sympathy how puzzled Kovalevskaia was by her friends' attitudes. Key observed that Kovalevskaia had been used to moving in very different circles in Saint Petersburg and Paris—circles in which one was judged by intellectual attainments and political commitment rather than by the elegance of one's drawing room, and where one's personal life was not criticized by one's friends.[15] Kovalevskaia found it difficult to cope with the fact that the Swedes expected her to be a brilliant, productive mathematician and a meticulous housewife in the Swedish manner at the same time.

Kovalevskaia had not expected to encounter such stuffiness among the Swedes. She had reasoned that since Stockholm University had taken a pioneering step by giving her a position, then Swedish society would be free of prejudice and worship of conventionality. This turned out to be far from true. Eventually, the divergence between Kovalevskaia's social views and those of Swedish intellectuals led to a coldness and estrangement between her and most of her Swedish acquaintances. As time went on, Kovalevskaia sought to escape what she increasingly came to consider the oppressive atmosphere of Stockholm by taking frequent trips to visit her more tolerant mathematical colleagues and freethinking friends in Paris, Berlin, and Saint Petersburg.

After Kovalevskaia brought Fufa to Sweden to live with her, she tended to leave Fufa with others when she traveled. Mittag-Leffler and his wife, Signe, often took the child, as did the Swedish astronomer Hugo Gyldén and his wife, Theresa. Only when Kovalevskaia expected to be settled in one place for a long period of time did Fufa accompany her, as in the summer of 1889, when Sofia, Iulia, and others took a house in Sèvres, France, for four months. And even then a nursemaid was hired for the day-to-day chores entailed in Fufa's care.

In her lack of involvement in actual child care, Kovalevskaia was typical of women of her birth and upbringing. It was the custom for women of the continental European gentry and educated classes to leave the care of their children to nursemaids and other servants. Even village teachers in straitened financial circumstances often managed to set aside a small sum of money for a local girl to look after their offspring. And people of more affluent means would seldom spend time with their children, even if the mother had no other career but the management of her household and family.[16]

Kovalevskaia herself, for example, had rarely seen her mother as a child. In fact, she remembered her mother as a strange, exotic creature in a ball gown who would occasionally waft into the nursery to be kissed on the cheek by her children before going off to a reception.[17] Even when Kovalevskaia was older, the emotional distance between herself and her mother did not diminish. As a result, Kovalevskaia felt that her father cared more deeply about the fate of his offspring than did her mother. She was closer to her father and harbored some resentment toward her mother.

Partly because of her own childhood experiences, Kovalevskaia tried to make the time she spent with Fufa memorable for her daughter, even if the actual amount of time she devoted to her was relatively small. Fufa later recalled with great delight that although Kovalevskaia was often preoccupied and distracted when she was working on a problem or struggling through a personal drama, occasionally she would drop everything to devote the day to her little girl. Fufa said that no one could enter into the concerns of children with such single-minded enthusiasm and sincere, uncondescending sympathy as could her mother. Fufa's friends adored her mother; to them she seemed like a special comrade rather than a parent.[18]

One of the best memories Sofia Vladimirovna retained of her time with her mother was of the months in the winter of 1888 when Kovalevskaia was readying her book *Memories of Childhood* for publication. Kovalevskaia allowed her daughter to sit in on sessions during which she read portions of her memoirs to her women friends and had them translated into Swedish as she read them. Fufa was fascinated by the memoirs and delighted that her mother had wanted her to listen to them. Fufa later said that the memoirs "drew me very close to my mother; I started to think about her more as a person who had experienced much in her own right rather than as just my mother."[19]

Thus, although Kovalevskaia and Fufa spent much time apart, Fufa later recalled the periods with her mother as wonderful occasions, full of meaning for the child. It is not at all clear whether Fufa saw her mother less often than did the majority of children of her time and class. In fact, Kovalevskaia spent more time with Fufa—or at least better-quality time—than she herself had spent with her own mother.

Unfortunately, Kovalevskaia died of pneumonia in early 1891, when Fufa was just twelve years old. After her mother's death, Fufa lived with the Gyldéns until she finished the gymnasium (secondary school) in Stockholm, then returned to live with Iulia Lermontova in Russia. She completed the women's medical courses in Moscow and worked as a doctor for the Russian Red Cross in the Soviet Union and abroad until her retirement. After her retirement, she became a medical librarian and translator, helped edit Kovalevskaia's literary papers for publication, and seems to have become obsessed with her mother's memory. She died childless in the Soviet Union in 1953, at the age of seventy-four.

Any discussion of Kovalevskaia's career and familial involvements would be incomplete without mention of her love affair with Russian sociologist and jurist Maksim Kovalevskii (1853–1916). Kovalevskii, a distant relative of Kovalevskaia's deceased husband, visited Stockholm in the winter of 1887–88 to deliver a series of lectures. He and Kovalevskaia were immediately drawn to each other. They were both homesick for Russia and, like many Russians abroad, tended to consider most Western Europeans shallow and superficial in their commitments.

Maksim admired Kovalevskaia intensely, although he seems to have occasionally felt some resentment at her fame and her ability to immerse herself completely in her mathematics for days on end. The pair traveled through Europe together on Kovalevskaia's vacations, and in 1890 they began to live together whenever their respective commitments would allow them to so arrange their schedules. There is some evidence that they were intending to marry in the spring of 1891.[20]

In contrast to the myths one usually finds in the secondary literature on Kovalevskaia,[21] Sofia and Maksim's relationship did not disrupt her career. In fact, Kovalevskaia produced some of her best literary and mathematical work during her friendship with Maksim. She wrote her classic paper on the "Kovalevskaia top" (for which she was awarded the Prix Bordin of the French Academy of Sciences in 1888) and published her acclaimed *Memories of Childhood,* several essays, and a novella as well.

WHEN ONE views Kovalevskaia's life as a whole, it seems clear that for the most part she did not allow her career and home life to interfere with each other. Instead, she concentrated her energies on whatever needed her attention at the moment. The only period in which she was unsuccessful in blending her mathematics with her personal life was from 1874 to 1880. These were her years of conventional marriage, her time of subordinating herself to her husband, her family, and the traditional, patriarchal society of tsarist Russia.

Obviously, conventionality was not the answer for Kovalevskaia. She looked back on the years 1874 to 1880 with regret and a certain amount of self-disgust; she often said that were it not for the birth of her daughter, those years would have been a complete waste. For Kovalevskaia at least, her personality and the social conditions of Russia at the time combined to ensure that she could not pursue her scientific interests and a traditional life-style at the same time.

Kovalevskaia's nihilist philosophy made it easier for her to try alternative life-styles and to defy convention when necessary. She could ignore those who deplored her decision to live apart from her husband because her nihilism taught that breaking away from traditional family ties was not always a bad thing. She could dismiss the advice of those who attacked her as an inadequate mother and told her never to separate herself from her daughter because her own childhood experiences indicated that a small amount of time spent in loving contact with Fufa was better for both of them than Kovalevskaia's constant but purely nominal presence would have been. And her nihilism told her that a mother with professional and intellectual responsibilities outside of the home would have a more wholesome, broadening effect on her child than one who was interested in nothing more than the set of her ball gown and the latest theatrical presentation.

Throughout her life, Kovalevskaia was inspired by and drew comfort from the thought that she was opening up a new path for women; that she was proving by her example that women could succeed in science, that they could create. Moreover, she was showing that science could go hand in hand with a radical perspective on the world; that nihilism and mathematics and the struggle for women's rights were compatible with one another.

Kovalevskaia seems to have succeeded best when she departed most from traditional ways of doing things. Her Heidelberg "women's commune" was a support to several pioneers of higher education for women—Lermontova and Evreinova, as well as Kovalevskaia. And in her later years in Stockholm, Paris, and Berlin she managed to bring up her daughter while playing a central role in the mathematical community of her time. Finally, her innovative application of abstract Weierstrassian analysis to problems in mechanics and mathematical physics showed that breaks with tradition could be useful in her mathematical work.

For most of her life, Sofia Kovalevskaia had good fortune in her career and personal involvements: her father allowed her to study mathematics with a tutor, her women friends were sympathetic and supportive, her mathematical colleagues actively helped to integrate her into their professional community. Moreover, she was born at a time when the nihilist philosophy gave impetus to those women who wished to enter the sciences and provided an almost familial environment for those who dared to defy convention.

Of course, in a short paper it is impossible to chronicle all aspects of Kovalevskaia's career and familial involvements. Inevitably, some oversimplification of the picture has had to be employed. The dramatic nature of Kovalevskaia's relationship with her husband has barely been touched upon, and I have had to gloss over the details of Kovalevskaia's rich and varied life. Moreover, in tailoring this essay to the concerns of the present volume, I have presented an interpretation of Kovalevskaia's life that in some respects goes against the established folklore about her. But the evidence justifies a reinterpretation of her life. Kovalevskaia's "sad story" has been used for too long to "prove" that "love and mathematics make a dangerous mixture for a woman" and that women's "emotional frailties" bar them from successful careers.[22] In fact, Kovalevskaia's story illustrates precisely the opposite. Her experience shows that a satisfactory, if not ideal, merging of career and home life is indeed possible and practicable; her life can be a source of inspiration and encouragement to women scientists and women science students today.

10

HELENA M. PYCIOR

Marie Curie's "Anti-natural Path": Time Only for Science and Family

BIOGRAPHY seeks not merely to recount the details of a subject's life, but also to grasp the thematic patterns that governed that life. Although by no means a complete biography of Marie Sklodowska Curie,[1] this essay seeks to analyze the thematic patterns most closely associated with her successful combination of the roles of scientist, wife, and mother, and then widow and single parent. The essay is divided into three main chronological parts. The first, covering Curie's early years, emphasizes the importance of her parents as role models, as well as the development of a close relationship, built on shared dreams and mutual support, with her sister Bronia. The next two sections—the core of the essay—are devoted to the scientific collaboration and family life of Marie and Pierre Curie, and finally Marie Curie's twenty-eight years as widow, single parent, and independent woman scientist. The overriding theme of these later periods is the Curies' "anti-natural path"—a simple way of life that allocated the couple time only for science and family. The essay also shows that Curie's marital status and family arrangements were key elements of the sociocultural matrix in which she practiced science.

Marie Curie was born on November 7, 1867, in Warsaw, Poland, which was then under the control of the emperor of Russia. She was the last of five children of Bronislawa (née Boguska) and Wladyslaw Sklodowski, members of the Polish urban intelligentsia with roots in the small landed gentry.

Her father taught physics at a gymnasium (secondary school) for boys in Warsaw; and until a few months after Curie's birth, her mother was principal of a private girls' boarding school. Curie was actually born in the boarding school, at the time not only her mother's workplace, but also home for the Sklodowski family.

Despite her auspicious birth to loving and intellectual parents, Curie experienced neither a carefree childhood nor an easy young adulthood. In 1876 she lost a sister to typhus and, at the age of nine, her mother to tuberculosis. "This [latter] catastrophe," Curie wrote late in life, " . . . threw me into a profound depression."[2] Added to her personal loss, the young girl lived in an atmosphere of political intimidation and oppression. Russian domination of central Poland restricted the professional life and income of Wladyslaw Sklodowski and tainted his children's lives with distrust, hypocrisy, and hatred. These strong emotions intruded even into the private Polish schools of the children's early years and the Russian gymnasia of their adolescence. Since the private schools routinely defied the Russian proscriptions against teaching in the Polish language and about Polish culture, periodic visits by school inspectors became harrowing occasions for conspiracy between teachers and students. While in private school, for example, the precocious Curie was sometimes called upon to mask her feelings and discuss Russian culture in the Russian language with an inspector. Later she enrolled in an imperial gymnasium, the only path to a recognized diploma. Although she graduated first in her class, Curie as an adult remembered most of the gymnasium's teachers as "Russian professors, who, being hostile to the Polish nation, treated their pupils as enemies."[3]

Even with her nationalist objections to the gymnasium, the young Curie wrote a friend: "In spite of everything I like the school . . . and even love it."[4] This emotion was a testament to her strong academic bent and also a tribute to her parents' intellectual interests and egalitarian views. Her parents took their daughters' educational needs as seriously as their son's. Following his wife's death, her father "devoted himself entirely to his work and to the care of . . . [his children's] education."[5] He infected all his children with his love of science and literature.

The love of learning instilled by his parents led Curie's brother, the leading student of his class at an imperial gymnasium, to the study of medicine at the University of Warsaw. Despite first rankings in their respective

---

Opposite: Marie, Bronia, and Helena Sklodowska (*left to right*) with their father in 1890. Courtesy Archives of the Curie Laboratory, Paris.

classes, Curie and her sister Bronia were unable to follow their brother to the university. The university was closed to women; higher educational opportunities for Polish women were to be found only in foreign universities. But foreign study involved expenses Wladyslaw Sklodowski was unable to afford, despite his sincere support of his daughters' scholarly aspirations.

Abandoning hope of any immediate formal higher education, Curie found employment as a private tutor in Warsaw. This was an important, formative period in her life, during which she and Bronia experimented with Polish positivism and cemented their close relationship. From contact with the positivists, Curie developed an abiding belief that one "cannot hope to build a better world without improving the individuals. To that end each of us must work for his own improvement, and at the same time share a general responsibility for all humanity."[6] This ideology of education as the chief means of social progress cut across gender and class lines; Curie's own voluntary activities of the period included reading to dressmakers and collecting a library of Polish books for other women workers.

But how were the Sklodowska sisters to effect their "own improvement," while Curie was contributing to the family income through her private lessons and Bronia was running the family household? Bronia confided to her sister her ambition to study medicine at the Sorbonne; Curie, in turn, expressed a hope for advanced work in science, sociology, or literature—disciplines given special recognition by the Polish positivist movement. The sisters shared their frustrations as the passing months seemed to bring them no closer to these goals. Finally, Curie evolved a plan: she would enter service as a governess, the only barely lucrative position open to a woman of her social and educational background, and use her earnings to support Bronia's five years of study at the Sorbonne; upon receipt of her medical degree, Bronia would help finance Curie's work at the same university.

From 1885 through 1890, then, Curie worked as a governess for three different families, devoting over three years to the Zorawski family, who lived about 60 miles north of Warsaw. The normal conditions of the life of a governess weighed heavily on Curie. She taught the family's children from seven to nine hours a day. Moreover, as a governess she was obliged to offer social companionship at the Zorawskis' convenience. Thus she often interrupted her evenings of study to make a fourth player at the family card table, play checkers, attend dances, and converse with the family and their friends. Despite such efforts at maintaining the proper decorum of the governess, Curie committed a serious transgression of the position when she fell in love with and agreed to marry the family's eldest son, Casimir.

Somewhat naively, Curie had failed to grasp the vast social gap his parents envisioned between them and their governess. Thus she seems to have been surprised by the Zorawskis' violent opposition to the proposed marriage, as well as by Casimir's ready acquiescence to his parents' wishes.

Although she had been rejected as a potential daughter-in-law, Curie's obligation to Bronia kept her with the Zorawskis for two more years. Throughout this awkward period, she hid her sagging spirits behind a brave face. "As for me," she confessed in 1888, "I am very gay—and often I hide my deep lack of gaiety under laughter. This is something I learned to do when I found out that creatures who feel as keenly as I do, and are unable to change this characteristic of their nature, have to dissimulate it at least as much as possible."[7] Other letters reveal that during this period Curie's "lack of gaiety" ran deeply indeed. She was isolated from her family in Warsaw, barely able to afford the postage stamps by which she maintained minimal contact with them, abandoned by her first love, and pushed to the limits of her endurance by the demands of her position. She was led even to question the value of her own life: "My plans for the future? I have none, or rather they are so commonplace and simple that they are not worth talking about. I mean to get through as well as I can, and when I can do no more, say farewell to this base world. The loss will be small, and regret for me will be short—as short as for so many others."[8]

Yet the position as governess did not destroy completely Curie's self-confidence and intellectual drive. Throughout her stay with the Zorawskis, she continued to prepare for formal higher studies, no matter how elusive they came to seem. Even the letter that contained the dark hints of the worthlessness of her life offered a list of textbooks she was studying, with the note: "When I find myself quite unable to read with profit, I work problems of algebra or trigonometry, which allow no lapses of attention and get me back into the right road."[9]

By 1888 Bronia no longer required Curie's financial support. Intent on assisting his daughters' higher studies, Wladyslaw Sklodowski retired with a pension from the gymnasium and took a difficult position as director of a reform school. Now supported by her father, Bronia urged her younger sister to begin to save for her own higher studies. By March 1890 the two sisters were seriously discussing the possibility of Curie joining Bronia at the Sorbonne. Here Bronia, in one of the many crucial roles she played in her sister's life, took the initiative: "You must make something of your life sometime. . . . You must take this decision; you have been waiting too long."[10] But Curie hesitated, claiming that, if not stupid, she was unlucky;

that she had long ago abandoned the dream of Paris; and finally that familial obligations to her father and another sister, Helena, required that she remain in Warsaw.[11]

During the next year and a half, however, events intervened that cleared the way for Curie's move. Living at home with her father, she rejoined Warsaw's underground "floating university," which now offered her use of a meager laboratory at the Museum of Industry and Commerce. Here, carrying out experiments described in physics and chemistry textbooks, she developed her "taste for experimental research,"[12] a taste she knew could be satisfied only abroad. Also, two obstacles to her escape to Paris were removed: seeing Casimir Zorawski for one last time, she abandoned any remaining hopes of their marriage; and she accumulated the sum required for the trip to Paris and a year's tuition at the Sorbonne.

Fall of 1891, the beginning of the second major stage of Curie's life, found her, at the age of twenty-four, enrolled at the Sorbonne. Initially, the supportive Bronia gave her sister room and board in the small apartment she shared with her husband and new baby. Then Curie, wanting freedom from all distractions, moved into a single room of her own. She could barely afford to pay the monthly rent, had little money or time for food, and sometimes went without heat to conserve her meager resources. Yet she thrived on her solitary life as a student in Paris, and passed the *licence* in physics (comparable to receiving a bachelor of science degree) in 1893 as the first in her class.

Infected by Parisian science, Curie determined to pursue a degree in mathematics. Again there were obstacles to overcome: concern for her father, whom she thought needed her companionship, and lack of money. A careful weighing of her own needs against those of her father resolved the first issue. Her father seemed happy and healthy living near her brother; and as she at this point wrote of her own needs: "It is my whole life that is at stake. It seemed to me, therefore, that I could stay on here [Paris] without having remorse on my conscience."[13] Curie's financial problems were temporarily resolved as a Polish woman mathematician helped her secure an Alexandrovitch scholarship for the next academic year, 1893–94.

The year 1894 was among the most eventful of Curie's life. Not only did she pass the *licence* in mathematics, ranking second among all candidates, but she also met Pierre Curie. At the time of their meeting, Pierre Curie

---

Opposite: Newlyweds Pierre and Marie Curie in Sceaux (home of Pierre's parents). Courtesy Archives of the Curie Laboratory, Paris.

was thirty-four years old and professor at the École de Physique et Chimie in Paris. The son of a physician who had turned his apartment into a hospital for the wounded of the Paris Commune of 1871, Pierre Curie had been raised as a freethinker and educated at home through about the age of sixteen. Following receipt of the *licence* in physical sciences from the Sorbonne in 1877, he had built a solid reputation as one of France's leading young physicists. This early reputation was based on research into the physics of crystals, which was conducted in collaboration with his brother, Jacques, in the late 1870s and early 1880s and resulted in the discovery of piezoelectricity; and independent work on crystallography and magnetism, initiated in 1883 when Jacques left Paris for a professorship at Montpellier.[14]

By 1894, when Marie first approached Pierre (with a request for space in his laboratory so she could conduct the magnetic research that the Society for the Encouragement of National Industry had hired her to do), she and Pierre seem to have decided upon single lives for themselves. Marie planned to return to Warsaw, comfort her father in his advanced years, and fulfill a positivistic mission of teaching science to further the Polish cause. Pierre Curie had concluded that a conventional marriage was not for him. Through 1894, in fact, Pierre seems to have clung to his earlier intense collaboration with Jacques—when the brothers had "lived entirely together," sharing scientific research and recreation and even "arriv[ing] . . . at the same opinions about all things, with the result that it was no longer necessary . . . to speak in order to understand each other"[15]—as the ideal of a life devoted to science. As Pierre's surviving notes indicate, he was not hopeful of finding a wife who would tolerate his complete absorption in science, let alone participate in it. Thus, prior to meeting Marie, he had written that "women of genius are rare," and speculated that marriage to an ordinary woman would place limits on his anti-natural path of almost complete devotion to science.[16]

After a courtship of over a year, Marie and Pierre Curie married and jointly embraced the "anti-natural path," which permitted the couple time only for science and for their extended family.[17] Traditional to the extent that Marie assumed responsibility for the domestic scene,[18] their marriage was unconventional in many other respects. Marie Curie, who never fussed about her wardrobe, "wore no unusual dress" for their simple civil wedding[19]—no matter to the groom, who later in life amused himself at a banquet by calculating the number of scientific laboratories that could be financed through sale of the jewels adorning the female guests.[20] The newly married couple even refused some furniture offered by the Curie family, as so many more objects to dust. "I am arranging my flat little by little," Marie wrote her brother in 1895, "but I intend to keep it to a style which will give me no worries and will not require attention, as I have very little help: a woman who comes for an hour a day to wash the dishes and do the heavy work."[21]

This decision to concentrate on the essentials of life—with sparse wardrobe and furniture, almost no social ties beyond their extended families and select Parisian scientists, and occasional bike rides into the countryside—fit the couple's budget and Pierre Curie's simple, eccentric ways. It also took into account Marie Curie's perfectionist streak. What Marie Curie did, she did well: in her early days of marriage, when the couple could afford only limited domestic help, she painstakingly learned to cook;[22] certainly in her laboratory,[23] and probably in her home as well, she enforced high standards of tidiness. Thus a traditional marriage could easily have drained all her energies; instead the Curies calculatedly pared their family life down to the essentials, thus freeing Marie Curie for a scientific career.

The arrival of their first daughter, Irène, in 1897 tested the couple's commitment to the anti-natural path. Around the turn of the century, if marriage did not end a woman's career, motherhood almost certainly did. But a combination of fortuitous conditions, including her husband's moral support, again saved Marie Curie for science. "It became a serious problem how to take care of our little Irene and of our home without giving up my scientific work. Such a renunciation would have been very painful to me, and my husband would not even think of it; he used to say that he had got a wife made expressly for him to share all his preoccupations. Neither of us would contemplate abandoning what was so precious to both."[24]

Equally fortunate was the availability of child care services. After trying unsuccessfully to nurse Irène, Curie embraced the French custom of turning care of her infant over to a nurse. From that point on, the couple hired a series of nurses and domestic servants to assist with the household chores and to watch Irène and eventually her younger sister, Eve. But even hired help was not the major factor facilitating the Curies' successful (and guiltless) combination of parenthood and first-rate scientific careers. Marie Curie's own explanation emphasized, rather, the willingness of Dr. Eugène Curie, Pierre's father, to assume responsibility for the two Curie girls. Following his wife's death (a few weeks after Irène's birth), Dr. Curie joined his son's household. "While I was in the laboratory," his daughter-in-law later admitted, "she [Irène] was in the care of her grandfather, who loved her tenderly and whose own life was made brighter by her. So the close union of our family enabled me to meet my obligations."[25]

Curie was now free to devote weekdays and some evenings and weekends to study and research. The director of the École de Physique et Chimie, where Pierre Curie was a professor, permitted her to work in the school's laboratory. There, after a year of research on magnetization, she turned to radioactivity, which became not only the subject of her doctoral thesis, but the focus of her research for the rest of her life. Her selection in late 1897 of radioactivity as a research topic—apparently made independently of Pierre—was possibly a stroke of intuitive genius and certainly a brave gamble. Radioactivity had been discovered only the year before, when Henri Becquerel noted that a compound of uranium, put on a photographic plate wrapped in black paper, left an image on the plate. The excitement surrounding Becquerel's discovery aside, "few scientists [of the late nineteenth century] thought radioactivity had much of a future."[26]

Curie soon formulated her own distinctive approach to radioactivity. First she developed a technique for measuring the intensity of radiation (actually measuring the conductivity of the affected air with an apparatus developed earlier by Pierre and Jacques Curie), which she applied to many different compounds. At this early stage, she discovered that thorium was also radioactive,[27] and formulated the hypothesis that radioactivity was an atomic property. She now latched onto this hypothesis—and relentlessly pursued its chemical consequences.

As an atomic property, she reasoned, radioactivity was probably not restricted to two elements. Borrowing materials from various French scientists, she measured the radiation emitted by all standard metals and nonmetals, rare elements, and an assortment of rocks and minerals. These measurements produced an anomaly: while only compounds containing uranium or thorium proved radioactive, some of the uranium compounds proved more radioactive than pure uranium. For example, the intensity of the radiation from pitchblende (an ore of uranium oxide) was four times that from metallic uranium. Here Curie made a bold hypothesis: there was a new (undiscovered) element in the pitchblende.[28]

At this exciting point, Pierre Curie joined his wife in the study of radioactivity, a multidisciplinary phenomenon that the couple soon realized involved chemistry, physics, and medicine. Theirs was a fruitful collaboration, in which Pierre served primarily as the physicist, and Marie, although trained especially in physics and mathematics, as the chemist. Together they discovered polonium (named by Marie Curie in honor of her homeland), which proved to be one of three new elements contained in pitchblende. Working with Gustave Bémont, they next discovered radium, while a coworker, the chemist André Debierne, found actinium.

Having established the existence of radium and polonium by somewhat indirect means, the couple then turned to preparation of radium as a pure salt. This was an exhausting task, involving physical and chemical techniques, whose practical execution seems to have fallen primarily to Marie Curie. As a first step, radium-bearing barium was extracted from pitchblende residue. A ton of residue gave 10 to 20 kilograms of crude barium sulfate. The sulfates were then purified and converted into chlorides. Next there followed fractional crystallization, a complex chemical process designed to eliminate some of the barium from the chlorides. Performed several thousand times, fractional crystallization finally left pure radium chloride. As a result, by 1902 the Curies had established on solid chemical grounds the existence of radium as an element by preparing it in a pure form and calculating its atomic weight.[29]

In this early period of radioactivity research, the Curies did far more than add new elements to the periodic table. Their work "was instrumental in opening up most fruitful avenues of research in a previously untrodden field."[30] Marie Curie herself coined the field's name, "radioactivity." Also, the Curies' astonishing results with radium helped legitimate the field, especially in the eyes of chemists. The honors showered upon the couple, including the Nobel Prize for physics, shared with Becquerel in 1903, were so many victories for the new field. They were also victories for the cause of women in science. In fact, both Pierre and Marie Curie took seemingly calculated steps to ensure that the male-dominated scientific community of the period did not ignore Marie's part in their research. Here her husband's experience as a scientific collaborator, as well as his honest and modest nature, which had been fostered by his early liberal education, proved important. But perhaps more crucial was Curie's self-confidence, which spurred her to publish independently the results of work for which she alone deserved credit and, in her accounts of collaborative efforts, to record carefully those experimental results and ideas that belonged primarily to her. Thus in 1898 she alone published a note containing the announcement of the discovery of the radioactivity of thorium, as well as the bold hypothesis that pitchblende contained a new element.[31] Marie Curie's subsequent accounts of radioactivity never failed to mention these independent contributions, just as they did not fail to note separately the research on radioactivity her husband conducted alone or with collaborators other than herself.[32]

Although concerned about establishing Marie Curie's scientific reputation, the couple did not blatantly challenge the many early twentieth-century scientific and academic mores that favored Pierre over Marie Curie. Thus in 1903 the Royal Institution invited Pierre Curie to lecture on radium; Marie Curie accompanied her husband to London, but merely sat in the audience as he spoke on their joint project. Furthermore, even after receipt of the Nobel Prize, the couple neither expected nor asked that Marie Curie receive such traditional fruits of scientific success as a major professorship and her own laboratory. Instead, the couple concerned themselves with Pierre Curie's academic career. The same honest and modest personality, combined with disdain for convention, that permitted Pierre Curie to collaborate with his wife as an equal also blocked his easy advancement in the French academic world. His professorship at the École de Physique et Chimie afforded him neither prestige, for which he did not care, nor a well-equipped laboratory, which both he and his wife ardently desired. An offer of a professorship to Pierre Curie, and a position in his laboratory for Marie, made in 1900 by the University of Geneva, resulted in new French opportunities for both Curies: for Pierre, an assistant professorship at the Sorbonne; and for Marie, a lectureship at the École Normale Supérieure (for women) at Sèvres. Finally, in 1906, as Marie Curie later emphasized, "because of the awarding of the Nobel prize and the general public recognition, a new chair of physics was created in the Sorbonne, and . . . [Pierre] was named as its occupant."[33] Marie Curie, in turn, was to become director (*chef de travaux*) of the laboratory the university promised to create for her husband. Pierre Curie's tenure as a full professor, however, lasted just a few months: in April of the year of his appointment, a horse-drawn cart killed him as he tried to cross a Parisian street.

Pierre Curie's death affected Marie Curie both personally and professionally. In a matter of seconds she went from wife to widow and single parent. The sorrow of Curie was profound. But she chose—as in her earlier school recitations for Russian inspectors and her governess years—to mask her feelings once again. She now assumed an impassive facade, hoping to shield her daughters from her "inner grief, which they were too young to realize."[34]

With her husband's death, Curie also went from female scientific collaborator to the perhaps more difficult position of independent woman scientist. Men scientists could no longer comfortably view her as a scientific wife sharing her husband's laboratory, research, and honors. She was now a single woman scientist, whose Nobel Prize and research entitled her to independent roles in the French academic community and the international community of science. Curie herself, as well as French academics and male scientists, had to adjust to her new position.

The Sorbonne was the first institution forced to deal with the anomalous widow. What was to be her university position in the absence of her husband? This question was one of scientific life or death for Curie, since continuation of her research required a laboratory and therefore some sort of academic association. The French government, however, initially viewed her situation from a different perspective. Treating her as the widow of a great man rather than a scientist in her own right, they offered her a widow's pension. Declining the pension, Curie stated her desire to continue to work. Then staunch supporters—including Jacques Curie, Paul Appell (once her professor and now dean of the Faculty of Science at the Sorbonne), and Georges Gouy (a close friend of Pierre)—put Marie Curie forward as the French scientist best able to continue her husband's research. Within a few weeks of Pierre's death, the Sorbonne appointed Marie Curie its first woman professor. Years later, attributing her appointment to the emotion surrounding her husband's death, Curie candidly acknowledged: "The University by doing this . . . gave me opportunity to pursue the researches which otherwise might have had to be abandoned."[35]

She immediately threw herself back into scientific research and teaching, emphasizing all the time that she was carrying out the dream shared with her husband of a life devoted to science. Her major research following Pierre Curie's death—much of it done in collaboration with André Debierne[36]—centered on completing the proofs that radium and polonium were true chemical elements. She turned to this work because of Lord Kelvin's doubts about the elemental nature of radium, published in a letter of 1906 to the London *Times,* and her own earlier doubts concerning polonium, the salt of which (unlike radium) had never been isolated. She again performed the complex and tedious extraction processes on pitchblende. She produced "perfectly pure" radium salt and, with Debierne, polonium salt as well. Then she and Debierne went a step further. In 1910, in a paper written jointly for the *Comptes rendus,* a weekly publication of the Académie des Sciences, they announced that they had clinched the proof of radium's elemental nature by preparing radium as a pure metal.[37]

The accomplishments of her early widowhood were crowned by a Nobel Prize in chemistry, awarded in 1911 to Curie alone for her longstanding work on radium and polonium. This unprecedented second Nobel Prize was a testament not only to her scientific genius, but also to her success at beginning to carve her own niche as an independent woman scientist in the male-dominated scientific community of her period. Of course, even in her early widowhood, Curie enjoyed advantages unknown to any other contemporary woman scientist. At the time of Pierre's death she was already a Nobel laureate. Moreover, besides permitting her to continue her research, her Sorbonne professorship established her as France's official expert in the field of radioactivity. As a result, the scientific community could not easily ignore her.

Still, there were undercurrents of tension and paternalism as male scientists of international stature were compelled to deal directly with Marie Curie without the intermediary of a husband or other male collaborators. For example, prior to the International Congress of Radiology and Electricity of 1910, at which Curie and Debierne were France's major representatives, Ernest Rutherford (1871–1937) and Otto Hahn (1879–1968) "wondered what Madame Curie would request, and they prepared, with their friends, to soften her by a little flattery, in that they all agreed that whatever new unit of radioactivity should be set up it would be called a 'Curie.'"[38] But flattery did not prevent Curie from insisting on a definition of the new unit (named in honor of her late husband) that differed from that established at an early session of the congress. Feeling ill, Curie left the session before its end and, later that evening, in the privacy of her hotel room, prepared a statement on the issue. The statement, delivered to Rutherford, declared that she "desired" a change in the propositions adopted at the session and, in particular, acceptance of her definition of the "curie."[39] "She had her way," Robert Reid has noted, "her bald dictatorial statement was accepted. But she created her critics, some of whom were becoming less tolerant of her mistressly attitudes."[40]

The claim of Curie's "mistressly" behavior seems to rest primarily on her failure to use normal conference procedures to argue her position on the new unit. The decisions of the radiology congress, however, were not generally reached through public debate. Furthermore, the nature of Curie's participation in the congress differed from that of her male peers even before her retreat to her hotel room. As David Wilson's biography of Rutherford shows, "The Congress was called to confirm what the inner circle of experts had already decided,"[41] and this circle—consisting at least of Rutherford, Hahn, Bertram Boltwood (1870–1927), and Stefan Meyer (1872–1949)—had decided to handle Curie with flattery rather than frank and open discussion. Denied an equal voice in the inner circle and shunning public debate, Curie simply defined her own mode of expression at the congress—direct communication by letter with the inner circle.

The strained relationship with Curie was symptomatic of the general inability of male scientists to adjust quickly to independent women scientists. For example, in 1913 Boltwood raised a fuss over the arrival of a woman—Ellen Gleditsch, one of the first of many women scientists trained by Marie Curie[42]—to work in his laboratory at Yale. Writing that he had tried unsuccessfully to dissuade Gleditsch from coming to Yale, Boltwood added in a letter to Rutherford: "Tell Mrs. Rutherford that a silver fruit dish will make a very nice wedding present!!!" Mary Rutherford's response was equal to the situation: "Are you engaged to the charmer yet, I forget who she was?"[43]

Even men generally regarded as supportive of women scientists—including Curie's peers, Rutherford[44] and Hahn—related differently to female and male colleagues. Hahn collaborated on radioactivity with Lise Meitner (1878–1968) for over thirty years. They published results jointly, and Hahn sometimes took additional, informal steps to assure that Meitner's contributions were not overlooked. In 1912, for example, Hahn wrote to a reluctant "Rutherford begging that due credit should be given to his collaborators von Baeyer and Lise Meitner."[45] Yet Meitner did not participate fully in all the special occasions of Hahn's laboratory. For example, Rutherford's visit of 1908 to the laboratory started inauspiciously for Meitner. As Meitner recalled years later: "When he [Rutherford] saw me, he said in great astonishment: 'Oh, I thought you were a man!' "[46] Her gender exposed, Meitner then fell at least partially into the role of hostess for Mary Rutherford. "While Mrs Rutherford did her Christmas shopping, sometimes accompanied by Lise Meitner," Hahn recounted, "Rutherford and I had long talks."[47] Following the visit, Rutherford wrote to Boltwood about Meitner—but he misspelled her name and, describing her as "a young lady but not beautiful," noted that she was no threat to Hahn's bachelorhood.[48]

Given these patterns of gender differentiation evidenced by key male scientists of the radiology congress, it is not surprising that tension between Curie and the inner circle persisted at least through 1912. In his account of the congress, Rutherford stated that Curie had agreed to prepare the international radium standard.[49] But after Stefan Meyer and his Austrian colleagues prepared a radium standard by late 1911, Rutherford, Boltwood, and Meyer plotted the best strategy for assuring that Meyer's standard was accepted along with Curie's. In a letter to Boltwood, Rutherford explained that Curie was being "very obstinate" about her standard and predicted that Meyer's standard would make Curie "far more reasonable." Rutherford then proposed an informal meeting to compare the two standards. But making an issue of Curie's gender, he warned Boltwood: "It is going to be a ticklish business to get the matter arranged satisfactorily, as Mme. Curie is rather a difficult person to deal with. She has the advantages and at the same time the disadvantages of being a woman."[50] The potentially awkward confrontation between Curie and the inner circle of radioactivity, to which this correspondence seems to have been leading, did not materialize. The inner circle arranged to have the two standards compared in April 1912 while Curie was recovering from surgery. Debierne represented Curie in the affair, after which Rutherford noted: "I think we perhaps got through matters very much quicker without Mme. Curie, for you know that she is inclined to raise difficulties."[51]

Actually, the international radium standard was one of the least of Curie's worries during the academic year 1911–12. She was now already beginning to suffer serious side effects from her fourteen years of research with highly radioactive material, and she had personal problems. With the passing years, responsibilities as a single parent weighed more heavily on her. Honoring the wishes of Eugène Curie (who remained with her following his son's death), Curie had moved her extended family to the Parisian suburbs, and she spent at least an hour a day commuting back and forth to her laboratory. Concern for her daughters' education also led her to design a special elementary school, run for two years by herself and other Sorbonne professors. Then, at the beginning of 1910, she lost her father-in-law. His death was painful and costly in time as she "passed all her free moments at the bedside of . . . [the] sick man who was both difficult and impatient."[52] The death left Curie without a special caretaker for her two daughters, then aged twelve and five; although employing Polish governesses (the first of whom had been a distant relative) for the girls, she now sometimes had to take time away from her outside work to meet their needs. As late as 1919, for example, she admitted to Rutherford that a serious illness afflicting Eve had kept her away from her laboratory and caused her delay in responding to one of his earlier requests.[53]

In addition to the death of her father-in-law, Curie endured a scandal concerning her relationship with Paul Langevin (1872–1946). Langevin was a leading French physicist who had studied under Pierre Curie and, as one of his many scientific projects, continued Curie's work on magnetism.[54] He and Marie Curie had taught together at Sèvres, and both before and after Pierre's death he was a select member of the small social group with which Marie Curie's anti-natural path permitted her contact. Many factors promoted intellectual sympathy, possibly leading to physical intimacy, between the two scientists. Langevin, however, was a married man with four young children—and a wife who supplied Parisian newspapers with letters supposedly exchanged between her husband and Curie. The letters, whose authenticity seems never to have been completely verified, implied that the two scientists shared an apartment and showed that Curie had urged the unhappily married Langevin to force his wife to accept a divorce.[55]

The Curie-Langevin scandal, as developed in the international press, impugned not only Curie's personal morals, but her scientific reputation as well. Writing in *World Today,* Henry Smith Williams used the Curie-Langevin relationship to resurrect the question of whether or not a woman was capable of independent, creative scientific research. He noted that "discriminating critics" had always believed major credit for the Curies' work on radioactivity was due Pierre, but that these same critics had been confused in 1910 when the widowed Marie Curie announced the isolation of the metal radium. Thus it had seemed for a while, Williams admitted, that Curie's "work could not be impugned as shining in the light of any man's reflected glory." But, he continued, the Curie-Langevin correspondence—with "passages that seem pretty clearly to suggest an intellectual dominance of the man over his woman colleague, such as scholars of the first rank are not supposed to accept"—reopened the question of the scientific talents of Marie Curie and women in general. Although declining to take a definite stand on the question, Williams related the opinions of "one of the most famous of European men of science" that Marie Curie was "an average plodder" and that, in general, a woman could excel in science only while "working under guidance and inspiration of a profoundly imaginative man . . . [and while] she was in love with that man."[56] The article thus pointed to the conclusion that Curie's scientific accomplishments were to be attributed to the love and inspiration of Pierre Curie, and later Langevin.

A much more subtly constructed account of the scandal, leading by innuendo to a similar conclusion, appeared in the *New York Times*. Ignoring the fact that the widowed Curie had collaborated primarily with Debierne, the *Times* article described Langevin as Curie's "co-worker in scientific research" and referred to their "constant close association in their scientific researches." The article used extracts from the Curie-Langevin correspondence that had appeared in Parisian newspapers not only to elaborate on such personal details as the couple's apartment, but also to explore the nature of their intellectual sympathy. Here Curie was subtly portrayed as Langevin's subordinate, even in the area of radioactivity:

> The bond of interest Prof. Langevin and Mme. Curie found in their scientific work is shown in another letter cited. "I have been thinking of your letter," writes Prof. Langevin, "and will tell you what to say on the consequences of the discovery of radium." . . . Mme. Marie Sklowdowska [*sic*] Curie is the most prominent woman to-day in the scientific world. To her has always been given the joint credit with Prof. Curie of the great discovery of thirteen years ago.[57]

Curie's self-confidence and talent for masking her feelings, combined with the assistance of two key women, helped her survive the scandal. Just a month after news of the correspondence with Langevin broke, she claimed her Nobel Prize in chemistry. At her side was her faithful sister Bronia, who periodically returned from Poland (where she had lived since 1898) to support her younger sister through difficult times, including Eve's birth, Pierre's death, and now the scandal. The Nobel Prize and Marie Curie's acceptance address—in which she continued her practice of delineating her independent contributions to the study of radioactivity[58]—at least partially neutralized the adverse professional effects of the scandal. But still the scandal exacted a heavy personal toll. It seems to have assured that for the rest of her life Curie had no close personal relationship with any man and kept an impassive face turned toward the outside world. Combined with serious medical problems, the scandal kept her away from research for over a year and made her more dependent than ever on other women. During this period, besides accompanying her to Stockholm, Bronia saw her through an operation on her kidneys; and Hertha Ayrton, the English mathematician and physicist, gave her privacy and care in a house on the Hampshire coast.[59]

Curie resumed scientific work only in late 1912, and even then not the driven, prolific research of her early widowhood. She now took time away from research to travel abroad for honors, and at home to oversee construction of the new laboratory promised Pierre in 1905. The foreign honors and conferences, as well as her new laboratory, seemed to help her regain her self-esteem in the wake of the Langevin scandal and to solidify her position in the international community of scientists. Rutherford had first planned to bring Curie to England in early 1912 for receipt of an honorary degree as one of the "little things [that] . . . help to smooth matters over."[60] Curie's illness, of course, intervened. But Rutherford extended a similar invitation for 1913, and Curie traveled to Birmingham, where she received an honorary degree from the local university and participated in the annual meeting of the British Association for the Advancement of Science, along with Rutherford, Niels Bohr, and J.W.S. Rayleigh. At Birmingham there were hints of the diplomatic skills Curie was honing for survival in the male scientific community. In late 1911 she had written to Rutherford to thank him for the kindnesses he had shown her during the first Solvay Congress, and now at Birmingham, perhaps deliberately repaying Rutherford's flattery in kind, she described him to reporters as "the one man living who promises to confer some inestimable boon on mankind as a result of the discovery of radium."[61] Following completion of her laboratory, moreover,

Curie wrote to Rutherford on equal terms. Thus, in 1919, armed with increased confidence and the official stationery of the laboratory, she (unsuccessfully) suggested to Rutherford that the two of them simply decide on names for the emanations of polonium, radium, and actinium and impose the names on the field.[62]

By 1918, as World War I ended and the Laboratoire Curie finally opened, Marie Curie was over fifty years old. Serious illness, scandal, depression, the distractions of building a new laboratory, and an international war had kept her away from intensive research for about seven years. In the postwar period she resumed her scientific research, while throwing her major energies into fashioning the Laboratoire Curie into one of the world's leading research institutes. Work for the laboratory was professional, but it also had a personal element, for there were early indications that Irène Curie would follow her parents into the field of radioactivity. Whereas the younger daughter, Eve, had an artistic rather than scientific bent, Irène participated in Marie Curie's efforts to bring X-ray equipment and technicians to the French battlefields and later in 1918 joined her mother's laboratory as a *préparateur*. Indeed, until Irène's marriage in 1926 to Frédéric Joliot, she and her mother traveled the anti-natural path together, sometimes to Eve's dismay.

Neither the financially strapped postwar French government nor the French industrial and private sources that Curie resourcefully tapped were able to provide all the radioactive materials and scientific equipment desired by Curie. Quite remarkably, however, Curie's familial status now helped attract additional funds for her laboratory. In this later stage of her fund-raising activities—as in so many other crucial moments of her life—Curie was assisted by a woman, this time the American journalist Marie Mattingly Meloney. The two women first met during an interview in 1920. As Curie described the deficiencies of her laboratory, Meloney quickly formulated an American fund-raising campaign for the double Nobel laureate. Curie was promised a gram of radium, worth about $100,000; and in return Curie agreed to visit the United States as a model for American women. As her published articles and surviving correspondence indicate, Meloney objected to the direction the first feminist movement had taken in the United States and especially to the willingness of some of the feminist pioneers to forego marriage and motherhood for careers. She determined to bring Curie to the United States as the perfect role model of a woman who had successfully combined career and family.[63]

When Meloney's private appeal to her friends failed to raise enough money for a gram of radium, she established the Marie Curie Radium

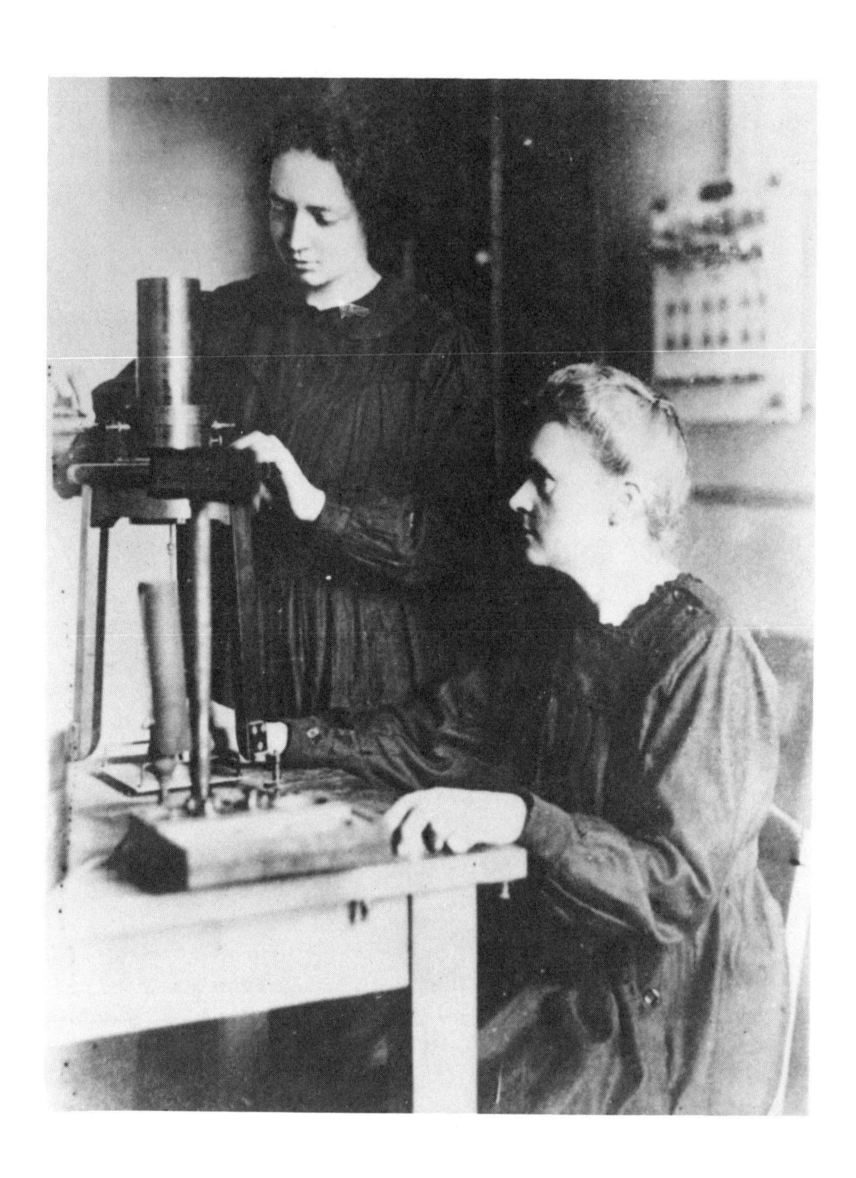

Fund, which was charged with collecting the required sum through an appeal to all American women. At this point American women—as diverse as academic women, wives of male philanthropists, Polish-American women, women afflicted with cancer (who had a special interest in the therapeutic potential of radium), and Girl Scouts—became Curie's benefactors. The fund raised over $150,000, and Curie made a triumphant tour of the United States during which she was hailed variously as a working mother, a female scholar and scientist, and the healer of cancer.

The warm reception accorded Curie by American women contrasted with the behind-the-scenes objections raised at Harvard and Yale to proposals for honorary degrees for her. Although Curie's visit conveyed a generally optimistic view of women in science,[64] these objections told a different story of at least a few male scientists and scholars at major universities who would deny an honorary degree to a woman scientist even though she was a double Nobel laureate. Thus, in December 1920 Charles Eliot, then president emeritus of Harvard, informed Meloney that the university's Physics Department was "not in favor of conferring such a degree" on Curie. The department included William Duane, who had worked in Curie's laboratory for six years on his way to a professorship at Harvard in 1913. According to Eliot, Duane favored awarding the degree, telling his Harvard colleagues that Marie Curie "had a large share in the details of the researches" on radium. Yet, Eliot added, she had done no work of great importance following her husband's death.[65]

These remarks infuriated Meloney. Responding to the insinuation that Pierre Curie's contributions to radioactivity were greater than his wife's, Meloney reminded Eliot that Curie's two Nobel prizes had already settled this issue in her favor. Then Meloney criticized Eliot's claim of the insignificance of Curie's activities during her widowhood.[66] This claim affected Meloney to the quick, for it overlooked not only Curie's independent scientific work, but also her responsibilities as a single mother. Meloney noted the irony of the situation: she was bringing Curie to the United States as an example of a working mother, but Eliot, who had written a series of articles extolling motherhood and large families as the highest "social and political service" women could contribute to mankind,[67] refused to give Curie any credit for the time and energy she had devoted to her family. Harvard's refusal of the honorary degree stood, and it was announced in June 1921 that the touring Curie, if forced by ill health to choose between

---

Opposite: Marie and Irène Curie in the Laboratoire Curie in 1923, midway between Marie's second Nobel Prize (1911) and her eldest daughter's Nobel (1935). Courtesy Archives of the Curie Laboratory, Paris.

acceptance of an honorary degree at Wellesley's commencement exercises and attendance at a special Harvard reception in her honor, would go to Wellesley.[68]

Unlike Harvard, Yale gave Curie an honorary degree—apparently at the instigation of medical doctors, however, and without prior consultation with the university's scientists. When informed of Yale's intentions, Boltwood objected to the degree. He also initially refused to serve on a special committee of the American Chemical Society set up to welcome Curie, and to open his laboratory for her inspection, even though, as he reported to Rutherford, "the Madame . . . [has] expressed a particular desire to visit New Haven and call on me." Later, upon discovering that his "action was likely to be misunderstood and to cause some hard feeling," he joined the committee and honored Curie's request to see his laboratory.[69]

These slights, which Curie may not have known about, mattered little as the touring scientist met thousands of admiring Americans at scientific and medical banquets, on major college and university campuses, and in small and large cities around the nation. Her self-confidence seems to have blossomed with the almost uniformly favorable public reception. Even the visit to Boltwood's laboratory went off well, with the Yale scientist admitting that he "was quite pleasantly surprised to find that she was quite keen about scientific matters and in an unusually amiable mood."[70] Boltwood's admission of surprise at finding a double Nobel laureate "keen about scientific matters" seems strange. But it is possible that the tour of 1921 was a unique occasion that forced Boltwood to deal with an especially confident Curie on equal terms. Significant reinforcement of Curie's positive self-image was, in fact, one of the lasting legacies of her American tour. As Eve Curie later explained, the tour convinced her mother that "at fifty-five [she] was something other than a student or a research worker: Marie was responsible for a new science and a new system of therapeutics. The prestige of her name was such that by a simple gesture, by the mere act of being present, she could assure the success of some project of general interest that was dear to her."[71]

Even as Curie now lent her name and time to scientific conferences, committee work, and medical causes, her life continued to center around the Laboratorie Curie. When she drained French sources, she used annual income from the American Radium Fund and her new American connections to acquire additional radioactive materials and the latest scientific equipment, essential for survival in the highly competitive field of atomic and nuclear physics of the 1920s and 1930s. Her international reputation so attracted research students and colleagues that seventeen different nations were represented at the laboratory in one year of her directorship. The combination of good apparatus, strong radioactive sources, and talented personnel resulted in a continuous string of publications from the Laboratoire Curie and at least two major discoveries of the late 1920s and early 1930s: the spectral analysis of alpha rays by Salomon Rosenblum in 1929 and, even more important, the discovery of artificial radioactivity by Irène and Frédéric Joliot-Curie in mid-January 1934, just half a year before Marie Curie's death from the effects of years of close work with highly radioactive materials.

The Joliot-Curies' discovery was a fitting culmination of Marie Curie's life, in which science and family were the most important elements. "Believe me," Curie reflected late in life, "family solidarity is after all the only good thing."[72] Curie's appreciation of family was a product of the supportive Sklodowski environment, which fostered her dreams and talents from childhood through young adulthood. Still, she waited until the age of twenty-eight before finding in the eccentric and modest Pierre Curie a husband with whom she could blend science and family. Even then her successful combination of the two necessitated sacrifice—in particular, a life that embraced the anti-natural path.

As the demands of science conditioned the Curies' family life, so Marie Curie's marital and familial status affected her scientific career. Her marriage to Pierre Curie assured her laboratory space and contact with France's leading male scientists. It also set in motion a scientific collaboration that produced a joint Nobel Prize. Her husband's death in 1906, followed by that of her father-in-law a few years later, meant additional domestic responsibilities for Curie and somewhat less time for science. Moreover, Pierre's death altered her professional status: no longer a scientific wife sharing her husband's laboratory, research, and honors, she was now forced to carve out a niche as an independent woman scientist. In a period still tainted by gender distinctions, the change in Curie's marital status from wife to widow thus forced adjustments not only in her small family unit, but also in her own professional life—at the Sorbonne and within the inner circle of radioactivity. As the Langevin scandal of 1911 showed, there was special pressure on her as a widow to prove that she was a creative scientist in her own right. Finally, the Laboratoire Curie flourished at least partially because of Curie's successful blending of science and family—which, after all, formed the basis of her invitation to the United States, where she found supplementary funding to keep the laboratory on the cutting edge of atomic and nuclear physics and thereby facilitate work such as that completed by the Joliot-Curies.

Thus, Marie Curie's very practice of science—including her early scientific opportunities, collaborative style, relationships with professional colleagues, scientific reputation, and effectiveness as director of the Laboratoire Curie—was fundamentally conditioned by her marital and familial status. In short, while the thematic patterns associated with Curie's combination of the roles of scientist, wife, mother, and then widow and single parent are not the only interesting patterns in her life, they may be the dominant ones.

Opposite: Marie Curie and other scientists at the 1927 Solvay Congress in Brussels. Also pictured are Einstein, Langevin, Bohr, Heisenberg, and Planck. Courtesy of the Institut International de Physique Solvay and the Niels Bohr Library of the American Institute of Physics.

INSTITUT INTERNATIONAL DE PHYSIQUE SOLVAY

CINQUIÈME CONSEIL DE PHYSIQUE — BRUXELLES, 1927

11

PEGGY A. KIDWELL

Cecilia Payne-Gaposchkin: Astronomy in the Family

> One wanted to believe that the price was not impossible for these accomplished women, that there were fathers, husbands, babies successfully flourishing beside the beautiful work. For there so rarely were.
>
> MARY GORDON, *Men and Angels*

THIS ACCOUNT of the family life of the astrophysicist Cecilia Payne-Gaposchkin (1900–1979) is marred by oversimplification, individual biases, and limited information. Payne-Gaposchkin was an intensely private person, and her carefully worded reminiscences and correspondence leave unanswered numerous questions about her family relations. Her children and several close acquaintances survive, and their comments offer valuable insight. They also show how complicated relations among family members can be and how different one woman can appear to several people. In correspondence and interviews one also encounters firmly entrenched, often unconscious assumptions about what families are and should be. Any scholar interpreting such evidence introduces his or her own bias. Historians routinely examine limited and conflicting records and produce accounts colored by individual judgments and prejudices. An essay that focuses on the family of Cecilia Payne-Gaposchkin suffers from the additional danger of understating the central emphasis of her life, the practice of astronomy. From her student days, Payne-Gaposchkin's strongest hopes and energies were devoted to the stars. She was intensely loyal to her family, but most interested in her research. Writing about a male astronomer, this might be assumed; in the case of a woman it must be stated outright.

Limited records and strong biases undoubtedly mar this attempt to suggest the complex family life of Cecilia Payne-Gaposchkin, just as focusing on her family shifts attention from the work she considered most important. Nonetheless, Payne-Gaposchkin was one of the first women to combine a professional career as a research scientist with marriage and child raising. Her experiences, though imperfectly recorded, deserve attention.

Family ties shaped the work of women astronomers long before Cecilia Helena Payne was born. In the eighteenth and much of the nineteenth centuries, when women were barred from advanced education and from most paid astronomical work, relatives offered one of few ways for them to study the stars. At the end of the eighteenth century, German-born Caroline Herschel assisted her brother William in his observations of variable stars, planets, and nebulae; she also discovered several comets on her own. Similarly, the nineteenth-century Englishwoman Margaret Huggins assisted her husband, William, in studies of the solar spectrum, while her compatriot Annarella Warrington Smythe worked with her husband, the amateur astronomer W. H. Smythe. Generally these women took up astronomy as a result of the work of their relatives. A few, most notably the Scotswoman Mary Somerville, were far more interested in the physical sciences than their relatives were. Helpful kin provided moral and financial resources, as well as access to libraries and scientific societies closed to women.

Some of the first women paid for astronomical work had been trained in the family. Maria Mitchell, the first professor of astronomy at Vassar College in New York, learned to observe the stars from her father, William. He not only provided the telescope with which she discovered a comet, but arranged for her to do calculations for the U.S. Coast and Geodetic Survey. She was well prepared when Matthew Vassar sought a woman to teach astronomy at his college.

Family ties not only led women to become college professors, but to be nautical instrument makers, authors of books on astronomy, and observatory assistants.[1] Such work was not to interfere with responsibilities like running a household, bearing children, entertaining as befit family position, and tending the young, the sick, and the elderly. For those wealthy enough to hire servants, some tedium might be avoided. Nonetheless, the physical strain of bearing several children must have weighed on both Mary Somerville and Annarella Smythe. Maria Mitchell devoted considerable energy to taking care of her elderly father, who came to live with her in the Vassar observatory. Low-paid observatory assistants who raised children or cared for elderly relatives faced harder tasks. Williamina Fleming single-handedly brought up her son on her salary as Curator of Astronomical Photographs of the Harvard Observatory. In 1900 she was one of several Harvard employees asked to keep a diary of her activities for several months. With six working days a week, she had to handle her laundry and other household chores on Sunday morning. In her journal she commented tartly that her Sunday mornings must seem matter-of-fact and different, compared with those of the officers of the university.[2]

Although women who worked in astronomy had diverse family responsibilities, few of those with regularly paying jobs were married. Computing was so tedious and poorly paid that it offered few incentives to those with other sources of income. Observatories were autocratic institutions in which women had little opportunity for promotion. At women's colleges, faculty not only received low wages, but often were expected to live and eat with students, with board and room a significant fraction of their pay. Female faculty did come to have more off-campus housing by the end of the nineteenth century, but restrictions against hiring married women to teach there remained in force.[3] While marriage was deemed an asset in male candidates for teaching and observatory positions, it was a detriment for women. Female computers and professors were expected to resign their posts if they married, the better to assist their husbands. For example, Annie Scott Dill Russell, one of the first women hired by the Greenwich Observatory, resigned after her 1895 marriage to astronomer E. W. Maunder. E. W. Maunder remained on the Greenwich staff, and his wife assisted, without pay, in his study of sunspots. Similarly, when Priscilla Fairfield, a faculty member at Smith College, married Bart J. Bok in 1929, she left Smith and became much interested in Bok's work on galactic structure. Although this match ended Priscilla Bok's career, her spouse was deemed an excellent choice for the Harvard University faculty. Restrictions against the hiring of married women could be lifted in emergencies. Annie Maunder actually was paid at Greenwich briefly during World War I and Priscilla Bok was considered an instructor at Radcliffe College during World War II.[4]

In her family life, as in her astronomy career, Cecilia Payne-Gaposchkin did not follow traditional patterns. By the beginning of the twentieth century family had become far less important to the training of women in astronomy. From her relatives, Cecilia gained a strong sense of the importance of artistic and intellectual pursuits, but no special knowledge of the sciences. It was as a student at Newnham College of Cambridge University that she became fascinated by the emerging discipline of astrophysics. From Newnham she went on to become one of the first graduate students in astronomy at the Harvard College Observatory. She obtained her Radcliffe doctorate in 1925, writing a brilliant dissertation on the temperature and composition of stellar atmospheres. At the same time she became part of the observatory family, participating not only in its work, but in picnics, parties, and astronomical lectures. Unable to find a research position in England, Payne remained at the Harvard Observatory, first as a National Research Council fellow and then as a staff member. In 1934 she married the Russian-born astronomer Sergei Gaposchkin. Their sometimes stormy union produced three children, as well as a book and several papers on the subject of variable stars. Both worked at the observatory. In 1956 Payne-Gaposchkin also became a professor on the faculty of Harvard University; she was the first woman promoted to this rank at Harvard. Her marriage as well as her association with the observatory endured until her death in 1979.[5]

Payne was born into England's intellectual circles. Her father, Edward Payne (1844–1904), earned his way through Magdelen College, Oxford, was elected to a fellowship, and studied successfully for the bar. While practicing law he also studied English exploration and colonial history, writing two volumes on the history of America.[6] In 1899, at the age of fifty-five, Edward Payne married Emma Pertz (1866–1941), the daughter of a Prussian army engineer. Emma Pertz's extensive intellectual lineage included an early member of the American Philosophical Society, a distinguished German librarian, and the Scottish geologist Leonard Horner. She herself was a painter. The Paynes soon had three children, Cecilia, Humfry, and Lenora. They lived in Wendover outside of London; Edward Payne continued his legal work. He had been suffering from heart trouble and dizziness and in December 1904 was discovered drowned in a canal near the family home. Payne admired her father's scholarship, his love of music, and his kindness; she would name her eldest son Edward. Even at the end of her life, she found the memory of his death too traumatic to discuss.[7]

Emma Payne was a woman of firm and conventional opinions. Her daughter described her as "prismatic and pruniferous"[8] (Charles Dickens had introduced the phrase to describe a character in his novel *Little Dorrit* who survived adverse circumstances, but lacked both humor and charity). Left an annual pension of 120 pounds, Emma Payne spent her resources carefully. Through what her daughter-in-law called "a miracle of courage and self-sacrifice," she not only met the physical needs of her children, but arranged for them to attend plays and concerts and to travel in Europe in the tradition of the English upper classes.[9] Payne did break with conventional practice when it came to her daughters' education. Although many in England still found the notion of a female undergraduate faintly absurd, all of Emma Payne's children prepared for professions. Cecilia Payne's decision to attend Newnham College was inspired by one of her mother's relatives, Dorothea Horner. Although Horner never told Payne about her own days at Newnham or her botanical research, her example inspired Cecilia to attend the college and study science.[10] Humfry soon was studying law and then Greek archeology at Oxford, and Lenora became an architect.

Cecilia Payne's only close living male relative was Humfry. She grew up under the guidance of her mother and aunts, influenced as well by women at girls' schools and women's college. This environment may have encouraged her, as it did other women, to disregard conventional notions about the subjects of study appropriate to women. It did not lead to mingling with boys. Social mores dictated that she not play with her brother's friends and the few dances she went to were an embarrassment. She found Cambridge University equally segregated by sex, with women set apart in classrooms. In the laboratory, men were disinclined to take women students seriously and were uninterested in Payne romantically. She did befriend the disabled New Zealand veteran Leslie J. Comrie and came to greatly admire her professor of astronomy, Arthur S. Eddington. Eddington, however, was greatly devoted to his mother, and both he and Comrie regarded Payne mainly as a promising astronomer. Indeed, in 1923 Comrie recommended Payne as one who "would devote her whole life to astronomy and . . . would not want to run away after a few years training to get married."[11] Neither Comrie nor Eddington was able to find a job for Payne in England, but they encouraged her to go to the Harvard Observatory.[12] When Payne arrived in Cambridge, Massachussetts, in 1923, she joined a "family" of researchers. Harlow Shapley, the observatory director, was a youthful father who was happy to advise her in both practical and professional matters. Other staff substituted for family she had left behind in England; they shared accommodations and entertained one another. Not only Payne but several of her colleagues found at the observatory the companionship and social status others achieved through kinship.[13]

Payne, living in a foreign country and devoting night and day to astronomy, found it natural to center her social life as well as her work around the observatory. The institution was physically and to a large extent financially separate from both Harvard and Radcliffe. When Payne arrived, only the director was a faculty member. As a Radcliffe graduate student, Payne did have a peripheral place in the Harvard University community. Once she obtained her Ph.D. in 1925, she discovered that women employees of the observatory had little status in this world. She noted in 1930 that in seven years at Harvard, her acquaintances there had been limited to observatory staff and one physics professor. Wives of Harvard faculty simply did not call on women of her rank. Payne may have overstated her isolation and, in any case, rarely would have been home to receive callers; nonetheless, she clearly did not feel a part of the larger university community.[14] She was elected an associate of the Harvard Faculty Club in 1931, but by then her orientation toward the observatory family was well established.

Harlow Shapley and his wife, Martha, fostered ties among the staff, friends, and benefactors of the observatory. Cambridge residents were invited to an annual series of Open Nights that featured a lecture on an astronomical topic followed by a glimpse of the heavens through telescopes set up on observatory grounds. A smaller group of astronomy enthusiasts attended the monthly lecture and annual picnic of the Bond Astronomical Club. Observatory staff also assembled regularly for both formal colloquia and informal reports on current research. Payne participated in all of these activities, as a speaker on Open Nights, an officer of the Bond Club, and a speaker and questioner at talks.[15]

Important though open nights, the Bond Club, colloquia, and less formal meetings were to the observatory community, social events did even more to create familial ties. The Shapley residence was on the observatory grounds and staff were regularly invited and expected to attend parties there. Entertainment included Ping-Pong, bridge, dancing, conversation, and music. There were parties to honor arrivals and departures, marriages, holidays, and scientific meetings. Sometimes preparations became quite elaborate. For example, when the American Astronomical Society met in Cambridge in 1929, observatory staff prepared their own version of the musical *H.M.S. Pinafore*. Payne played the leading role of Josephine, indulging a childhood love of theatricals. The performance proved sufficiently successful to be repeated for the Bond Club and was long remembered by Payne and other participants.[16]

Joining the observatory family had its price. When Payne was a graduate student and postdoctoral fellow, she was relatively free to choose which research problems she would study. As a staff member, she was expected to follow the research program of the observatory. Harvard did not have an advanced spectrograph, which would have made it possible for Payne to extend her early work on the temperature and composition of stellar atmospheres. Instead, by 1928 she was spending more and more time on a

The Harvard Observatory family in the 1920s. Cecilia Payne is sixth from right; Harlow Shapley is second from left. Courtesy of Katherine Haramundanis.

longstanding observatory program of measuring stellar magnitudes and distances. Payne found photometric work tedious and unproductive and sometimes protested against this shift. Shapley claimed that he had rescued her from an entirely too mathematical and physical subject by pointing out "the ultimate glories of domination in the field of standard photometry." Payne regretted her salvation and found the glories of photometry dim indeed.[17]

Before Shapley became director in 1920, women at the observatory rarely married. He changed this tradition, intentionally or otherwise, by beginning a graduate program in astronomy. A fellowship designed to encourage women to study at the observatory proved the starting point for the graduate program. The first student, Adelaide Ames, came in 1922. Cecilia Payne, Mary Howe, and Harvia Hastings Wilson arrived the following year. This influx of young women, combined in a few years with the presence of male graduate students, led to several weddings. Some, like Harvia Wilson and Helen Roper, married outside the astronomical community. Others, like Mary Howe, Bart J. Bok, Emma T. R. Williams, and Emily Hughes, married astronomically minded spouses who had no official connection with Harvard. Helen Sawyer married a fellow graduate student in astronomy, Frank S. Hogg. Eric Lindsay and Carl Seyfert married members of the observatory staff. By 1934 Harvard had so departed from its image as an institution devoid of romance that the astronomer J. H. Moore called it "a regular matrimonial bureau."[18]

Those women astronomers at Harvard who married in the late 1920s and early 1930s were part of a trend. Deborah J. Warner has pointed out a general increase in the number of women astronomers who married at this time.[19] Similarly, comparing entries in the 1921 and 1938 editions of *American Men of Science,* Margaret Rossiter found that the percentage of women listed who were married increased, both for astronomers and for women scientists generally. She also notes that married women had much higher unemployment rates, especially in 1938. This might reflect the antinepotism policies of the 1930s that barred a woman from paid work at the same institution where her husband was employed. In any event, Radcliffe astronomy graduate students who married rarely advanced far in their careers. Some who worked toward or obtained master's degrees, such as Mary Howe, Agnes Hoovens, Harvia Wilson, and Helen Roper, left paid astronomical work when they married. Those with doctorates persevered, finding tenuous positions as assistants or instructors where their husbands worked. For example, Helen Sawyer Hogg was at the Dominion Astrophysical Observatory and then the David Dunlop Observatory, Carol Anger Rieke at Johns Hopkins and then the University of Chicago, and Emma T. R. Williams Vyssotsky at the Leander McCormick Observatory of the University of Virginia.[20]

Payne's reaction to the romances blooming about her is difficult to determine. She participated in wedding showers and enjoyed the hospitality of both married and unmarried friends. She deeply admired several men, including her former teacher Eddington. One 1924 letter describes her joy when they chanced to view Niagara Falls together. Payne's fondness for Harlow Shapley was tinged with resentment when he drew her away from research on stellar spectroscopy and jealousy when he worked more closely with others.[21]

Eddington was a confirmed bachelor and Shapley a married man. At least one other man did make a bid for Payne's attention early in her Harvard years. After she completed her doctorate in 1925, she sailed home to England for the summer. The mathematician Norbert Wiener, then teaching at MIT, was on the same boat. Wiener, who had just broken off a romance with Margaret Engemann, planned to go climbing in the Alps and to visit German mathematicians. He met Payne on ship and soon reported to his parents that "she has a keen sense of humor and is good company but has gone through the one-sided development characteristic of the English scholarly woman, accentuated by a measure of poverty."[22] He wrote in a similar vein to his sister Bertha, commenting that Payne was jolly, shy, widely read, and of a cultivated, artistic taste.[23] He soon confided to another sister that he would be pleased to have a wife who continued her scientific career.[24]

Payne and Wiener corresponded occasionally during the summer. He came to London in late August and was delighted by how rest and a haircut had improved her appearance. A few visits to the theater, attendance at a meeting of the British Association for the Advancement of Science (BAAS), and dinner with Mrs. Payne did nothing to dampen Wiener's enthusiasm. He warned his sister Constance: "I don't promise not to be engaged by the time I return home."[25]

Payne's reaction to Wiener is less clear. She did not mention Wiener's attentions in a letter she wrote that summer to her good friend Margaret Harwood. Neither Payne nor Wiener referred to their meeting in autobiographical accounts. They took the same ship back to the United States, but went their separate ways. Payne was not interested in playing the role of a scientist's spouse; she saw herself as the equal of other astronomers. Wiener could reasonably expect to find a wife who would run his household and give a highest priority to his career. He settled his differences with Margaret Engemann and they married in the spring of 1926. Payne set out to study the spectra of stars of high luminosity.[26]

Payne had spoken at the 1925 meeting of the BAAS from what she called "motives of pure advertisement": she hoped to find a job in England.[27] When no opportunity arose, she returned to the Harvard Observatory, first as a National Research Council fellow and then as a staff member. According to her autobiography, three deaths in the early 1930s turned her attention from strictly astronomical matters. Adelaide Ames, whom Payne had known at the observatory from their student days, drowned accidentally in the summer of 1932. Payne admired her generous nature and resolved to be more like her, embracing life and doing her part as a human being. The following May, a dear friend from Payne's college days died tragically. And the astronomer William Waterfield, who had spent some years at the observatory and at its Boyden Station in South Africa, died in a motorcycle accident. Payne renewed her intention to be more open-hearted and promptly fell in love for the first time in her life. When this infatuation was not reciprocated, Priscilla and Bart J. Bok encouraged her to take a long trip abroad.[28] This idea dovetailed neatly with Payne's research plans. In the spring of 1933 she had coauthored papers on the spectra of novae and nebulae. She also hoped to make a more thorough study of all kinds of variable stars, developing the last section of her second book, *Stars on High Luminosity*. Harvard photographic plates were a rich source of information about stars that vary in brightness, and Payne planned to write a monograph on the subject with the Russian astronomer Boris Gerasimovic. Gerasimovic had spent 1926 to 1929 at Harvard, and visited there again when the International Astronomical Union met in Cambridge in 1932.[29] Eager to present her spectroscopic results, ready to consult with Gerasimovic on their book, and hopeful that a change of scenery might improve her spirits, Payne planned her travels for the summer of 1933.

Payne's itinerary included stops in England, Holland, Denmark, Sweden, Finland, Estonia, and the Soviet Union, climaxing with the meeting of the Astronomisches Gesellschaft in Göttingen. In addition to talk of spectra, photometry, and variable stars, Payne encountered political turmoil disturbing to anyone who embraced life. By the time she reached Estonia, she had met several astronomers who wanted to visit Harvard.[30] In Russia she not only talked to someone who wanted to do variable star work at Harvard, but to a woman who simply wanted to leave the country. Payne was appalled by the poverty and poor working conditions of Russian astronomers, whose instruments might be out of service for years, who lacked food and firewood, and who feared to talk freely.[31]

The impoverished state of Russian astronomy was much on Payne's mind as she traveled on to Germany. Then, at the astronomical meetings in Göttingen, she met the Russian emigré Sergei Gaposchkin. According to his own account, Gaposchkin (1898–1984) was one of nine children of a Russian day laborer and his wife. In 1915 he was forced to leave full-time school, although he continued to study sporadically while working as a fisherman, a factory worker, a soldier, and a policeman. In 1920, not long after both his parents had died, Gaposchkin's boat was caught in a storm on the Black Sea; he finally came ashore in Turkey. After diverse further adventures, he managed to continue his education in Bulgaria and then in Berlin. He completed a Ph.D. in literature at a Russian university in Berlin in 1928 and earned a doctorate in astronomy at Berlin University in 1932. P. Guthnick, director of the Babelsberg Observatory of Berlin University, thought well enough of Gaposchkin to ask him to continue there as an assistant.[32]

According to Guthnick, Gaposchkin had enemies within the Babelsberg Observatory staff. After the National Socialists came to power in 1933, they put sufficient pressure on Guthnick for employing the Russian that he feared for his own position. Guthnick appealed for help to a distinguished Danish astronomer, who in turn wrote Shapley. Shapley replied that unless some miracle occurred, he could not possibly do anything for Gaposchkin. The economic depression had forced him to lay off competent staff members, and they had first rights to any jobs that might open up.[33] When Guthnick heard of Shapley's reaction, he wrote directly to explain that Gaposchkin had not been a direct target of the National Socialists, but the victim of a personal intrigue that could only have succeeded in such a time of unrest. Shapley mentioned this curious note in a letter to Payne in early August.[34]

Gaposchkin refused to leave his fate in the hands of observatory directors. He bicycled for four days from Berlin to Göttingen to attend the Astronomisches Gesellschaft meeting and appealed directly to Payne. Payne herself had overcome considerable barriers to become a professional astronomer and sympathized with those who did likewise. She had resolved to be more open to the needs of other people. Moreover, she was much attracted to Gaposchkin, finding him small but strong and of a remarkably happy disposition. After hearing his story, she spent a sleepless night and then, with her usual thoroughness, set out to find a place for him at Harvard.

Payne's campaign began with a lengthy reply to Shapley's offhand comment about Gaposchkin. She briefly summarized his life and mentioned several eminent astronomers who recommended him. She herself deemed him a "good but not brilliant" astronomer. As a former member of the White Army, Gaposchkin would have no chance for a position in his homeland. Having stated the facts as dispassionately as she could, Payne then argued that Harvard had a "national and personal responsibility" to aid someone with Gaposchkin's dedication to astronomy. Although she promised him nothing, she began inquiries in London about securing an international passport for him and trusted that Shapley could handle U.S. emigration authorities. Payne offered to try to raise a salary for Gaposchkin herself and suggested that he have a fellowship to do photometric work. Previously she had had an assistant paid by Radcliffe College and had also donated some money to the observatory, but she hoped that Shapley would do the fund raising. The German government apparently would pay deportation expenses. In conclusion, Payne restated her deep conviction that the Harvard Observatory could and should come to the aid of Gaposchkin. She promised to raise the matter again when she returned to Cambridge in a few weeks.[35]

Two days later, Sergei Gaposchkin himself wrote Shapley a long letter in German pleading for a job. He included an English version, in Payne's handwriting, of the account of his life published in his thesis. Shapley replied that he didn't know whether he could offer Gaposchkin a job. Once Payne returned to Harvard, she applied the full force of her persuasive powers. In October Shapley wrote to offer Gaposchkin the post of a research assistant for one year. He promptly accepted and arrived in Cambridge in late November. The miracle required to find him a job had occurred.[36]

Payne continued her research in photometry, with forays into spectroscopy. By January of 1934 she was planning another trip to Europe to continue her work with Gerasimovic. Although a distinguished Norwegian astronomer warned against the trip, in late February Shapley was seeking funds for it.[37] Other plans intervened, however. Payne's initial attraction to Sergei Gaposchkin had deepened. She found him a forthright, congenial companion who shared her devotion to astronomy. He would require her financial assistance if they married and hence would allow her to continue astronomical research. Sergei had benefited greatly from Payne's efforts and depended on her for his research work. He also did not have a permanent appointment at Harvard and may well have considered the benefits of alliance with an American citizen (Payne had received her citizenship in 1931).[38] Both astronomers were ready to settle down. Without making any prolonged analysis of the consequences, they decided to marry quietly in New York and informed Shapley. Ever one to enliven an occasion, the director arranged for friends in New York to chaperone the couple and sent roses, a telegram of congratulations, and a letter of advice. Payne and Gaposchkin married in New York's City Hall, and Shapley's friends provided a champagne-and-cavier luncheon.[39]

The marriage of Payne and Gaposchkin astonished their acquaintances. To the observatory family, the union of a thirty-five-year-old Englishwoman who had spent over a decade away from home with a Russian whose parents had been dead for years was an "elopement." When the honeymooners returned, a reception and shower awaited them.[40] Emma Payne was disturbed by the unexpected marriage, but Shapley sent assurances that her daughter had found a congenial mate.[41] The reaction of astronomers is best summarized by Henry Norris Russell's comment "I sincerely hope that it turns out splendidly, but I keep wondering how it happened."[42] The new Mrs. Gaposchkin dropped word of her marriage casually into professional correspondence. She wrote to tell the astronomer R. S. Dugan that her roommate and associate Frances Wright had been unable to work on a project for Dugan because she was moving. Wright was moving, she continued, so that the newlyweds might have the apartment. She then hastened to add that she planned to continue her career as as astronomer.[43] To aid those who wished to look up her scientific papers, she adopted the last name Payne-Gaposchkin. Otherwise, she later explained, her writings might be listed under two different names, and she feared she would suffer a "split personality."[44] While Payne-Gaposchkin made such adjustments, Shapley found a way to keep Sergei Gaposchkin on the observatory payroll.

Both at Harvard and in the astronomical community generally, Cecilia Payne-Gaposchkin was, as Shapley put it, "*the* Gaposchkin." Her salary was more than twice as large as that of Sergei Gaposchkin, she taught courses for graduate students in astronomy and eventually for undergraduates, she supervised projects, edited observatory publications, and eventually became a faculty member. Payne-Gaposchkin also served on several commissions of the International Astronomical Union, was elected to the American Philosophical Society in 1936, and joined the American Academy of Arts and Sciences when it admitted women in 1943.[45]

Although Sergei Gaposchkin did not receive this recognition, he was hardworking and devoted to the study of eclipsing variable stars. Having married a Russian emigré, Payne-Gaposchkin no longer felt free to visit Russia, broke off all correspondence with Gerasimovic, and began working with Sergei on a survey of observations of variable stars.[46] It was her least successful book. After it was finished in 1938, the Gaposchkins and several assistants embarked on a more general study of the variable stars photographed on Harvard plates; these results continued to be published into the 1950s.[47]

In their marriage, the Gaposchkins found companionship and mutual assistance. Cecilia took a girlish delight in Sergei's attentions. She enjoyed the pride he took in her accomplishments and relied on him to do many routine household tasks. He depended on her not only for collaboration and assistance in dealing with astronomical associates, but in mundane matters such as writing letters and papers in English.[48]

(Overleaf)

Edward, Cecilia, Katherine, and Sergei Gaposchkin, about 1940. Courtesy of Katherine Haramundanis.

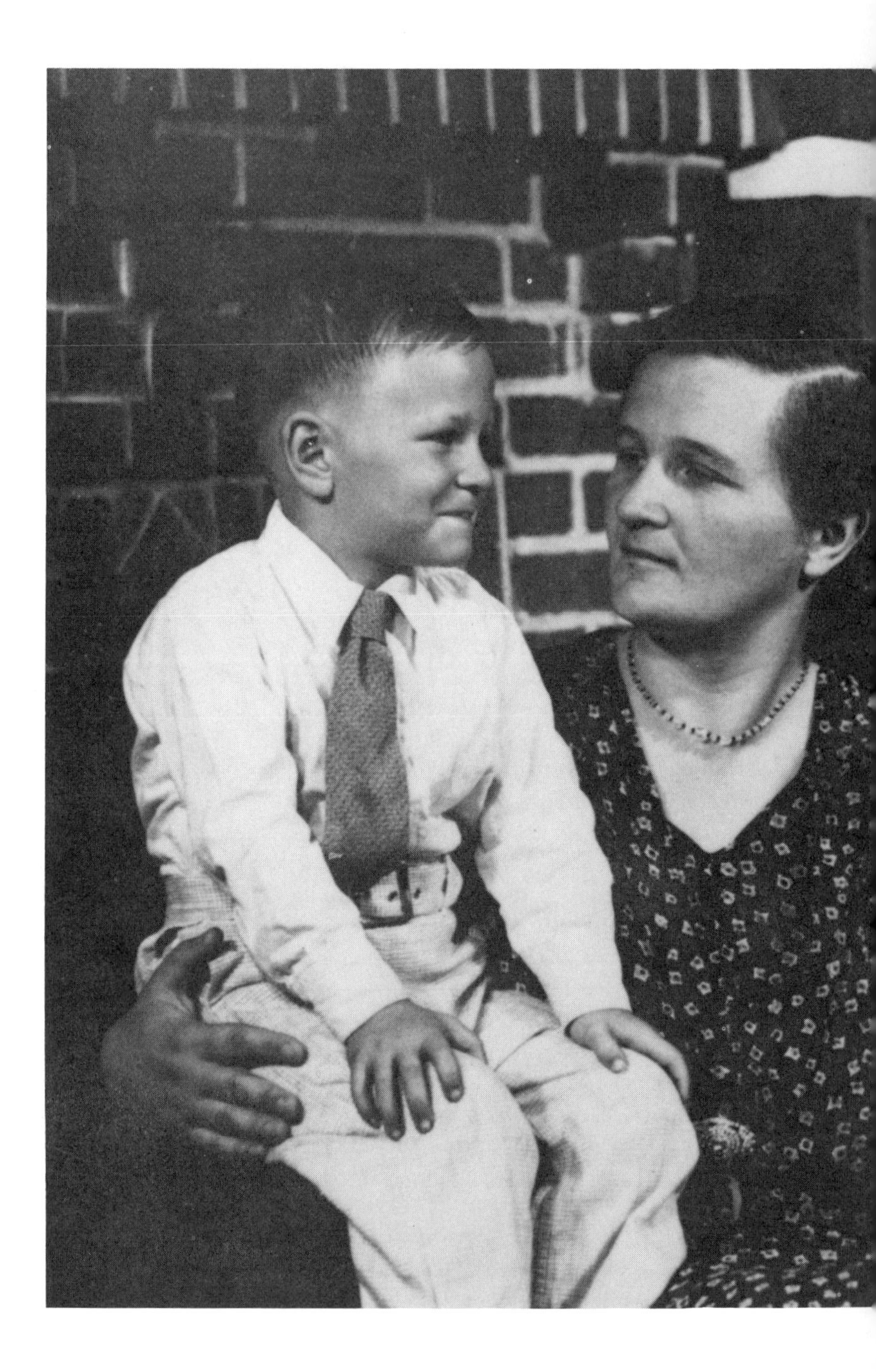

In choosing to marry and remain an active, paid astronomer, Payne-Gaposchkin strained conventional notions about women's roles. In choosing to marry Sergei Gaposchkin, she strained her relations with the observatory family and the astronomical community in other ways as well. Sergei was rude to many who might have helped him, and undiplomatic with those who gave him opportunities.[49] He was quick to take offense, yet spoke his own opinions with little regard for what others might think. He openly stated his admiration for female assistants at the Harvard Observatory and expected them to be as impressed by his muscular physique as he himself was. In short, Sergei was a misfit in the astronomical community, to the extent that an astronomer might comment when he seemed to be acting normal.[50]

Through these years of collaboration, the Gaposchkin family grew steadily. Sergei and Cecilia's first child, Edward Michael, was born in May 1935. Katherine Lenora followed in early 1937. Finally, another boy, Peter, was born in 1940. The family moved to a house in Lexington, Massachusetts, and added rooms as needed.[51]

At the time of her pregnancies, Payne-Gaposchkin withdrew from observatory activities for as long as convention dictated. But she saw no need to curtail her research simply because she had children. Thus she presented a paper to the American Astronomical Society five months before Edward's birth and directed research on photometry and variable stars at the Harvard Summer School in Astronomy the following summer. Similarly, the summer after Katherine was born, Payne-Gaposchkin spent six weeks lecturing at the Yerkes Observatory in Wisconsin. During her third pregnancy, she overstepped accepted bounds and agreed to speak at Pembroke College of Brown University. When Shapley heard of this plan, he warned; "The Observatory administration would be open to severe censure if you went to Providence in your present condition." Payne-Gaposchkin canceled her talk and did not lecture in Providence until after Peter was born.[52]

During the 1930s the Gaposchkins had been able to hire someone to tend their children, just as Emma Payne had employed others to care for Cecilia, Humfry, and Lenora. With the outbreak of World War II, servants could find better-paying jobs and the Gaposchkins took to bringing their children to work with them. Peter Gaposchkin did not do well under this regime, and according to the wisdom of the time, his mother was responsible. When Peter had not learned to talk by his third birthday, she became so worried she feared she might have a nervous breakdown. After many consultations, she arranged to have Mrs. Dean B. MacLaughlin, the wife of an astronomer doing war work in Cambridge, tend Peter. Mrs. MacLaughlin's "casual, cheerful kindness" was just what he needed. To the immense relief of the entire family, he soon was talking.[53] Child care remained a problem, and the Gaposchkin children spent much of their early years roaming the observatory. Sergei's office was easy for them to enter, while Payne-Gaposchkin worked in an isolated nook that could only be reached by wandering through several other offices. Observatory staff were expected to tolerate the children.[54] Shapley was well aware that this was not the usual practice at observatories and specifically warned Payne-Gaposchkin not to take her children to work when the family visited the Mount Wilson Observatory in 1948.[55] In the 1950s the Harvard Observatory Council formally voted to warn one of the children not to disturb those reading in the observatory library. Payne-Gaposchkin was out of town when the vote was taken and took the action as a personal insult.[56]

Spending six days and often a few evenings each week at the observatory, Payne-Gaposchkin had little time for details of child raising or housekeeping. Home improvements, gardening, cleaning, and much of the cooking were left to Sergei, the children, and whatever help was available. Payne-Gaposchkin did cook at times and enjoyed needlework. She encouraged her children to think, took delight in their achievements, and relished their companionship on her travels.[57]

Although Payne-Gaposchkin spent relatively little time tending to the physical needs of her family, she strove to establish them in academic careers. In addition to finding a job for Sergei and collaborating with him, she arranged observing time for him and defended his articles from attacks by other astronomers.[58] Edward, her eldest son, worked part-time for his parents at the observatory in 1955, while he was finishing his undergraduate work. Later, when he was studying at Cambridge University in England, his mother asked Princeton astronomer Martin Schwarzschild to suggest a thesis topic for him. Edward obtained a Harvard doctorate in geophysics in 1969. By then, he had spent ten years in the satellite geophysics group at the Smithsonian Astrophysical Observatory, which was housed on the observatory grounds. After his mother's death, Edward left the observatory and now does private consulting.[59] In part to pay for the education of her daughter, Katherine, Payne-Gaposchkin wrote the elementary textbook *Introduction to Astronomy*. Like Edward, Katherine worked at the observatory for a time. She also assisted in a later edition of the textbook and subsequently became a technical writer. Following in her mother's footsteps, she married a foreign-born man and had children.[60] Peter Gaposchkin had a more difficult time finding a place in life. He completed a doctorate in theoretical astrophysics at the University of California at Berkeley, writing his dissertation on the Friedman universe. But his personality proved unsuited to astronomical research, and he has worked at several other occupations.[61]

Nurturing the Gaposchkins took much of the time that Payne-Gaposchkin had devoted to her observatory family. Nonetheless, she not only incorporated her children into workdays, but continued to attend colloquia, open nights, informal seminars, and Shapley parties. In 1949 she was one of the founders of the Observatory Philharmonic Orchestra, heading the violin section and conducting occasionally.[62] She also became more closely associated with the running of the observatory as a member of the Harvard Observatory Council from the mid-1940s.[63] During World War II the Gaposchkins felt that it was their duty as educated people to arrange a forum on international affairs. At these meetings, experts on specific problems clarified some of the issues that had led to war and suggested postwar solutions. Naturally enough, the forum met at the observatory.[64]

Payne-Gaposchkin remained loyal to the observatory throughout the period of rapid change that followed Shapley's retirement in the early 1950s. Shapley's successor, Donald Menzel, had the director's house torn down to make way for other buildings. Feuding over Menzel's appointment as director, his personal style, and the growing size of the observatory combined to decrease the familial atmosphere. Payne-Gaposchkin's sphere grew to include a position on the Harvard faculty and then affiliation with the Smithsonian Astrophysical Observatory. She retired from these posts in turn, however, retaining up to the end of her life only her position at the Harvard Observatory.

Payne-Gaposchkin shifted her research to conform to the needs of the Harvard family, changed collaborators to further her husband's career, and refused to have Sergei's flamboyant personality destroy their marriage. Such loyalty came at a high price, particularly when prospects of professional advancement arose. In November 1938 Otto Struve suggested that she be nominated as secretary of the American Astronomical Society. H. N. Russell vetoed the idea, suggesting that Payne-Gaposchkin's "professional and domestic obligations" already taxed her to the limit.[65] In 1944, when Struve was asked to rank astronomers who might be compared to Fred Whipple as candidates for a professorship at Harvard, he mentioned five astronomers who might be considered, including Payne-Gaposchkin. Struve noted that Payne-Gaposchkin's "domestic situation" might be a drawback.[66] In 1947, when Russell was asked to recommend Payne-Gaposchkin for a possible Radcliffe professorship, he took care to provide an answer for anyone who might wonder what credit should be assigned for the papers the Gaposchkins had written jointly. Comparing the two astronomers, he commented "she is his equal in industry (this is high praise) and his superior in knowledge and judgment."[67] Despite Russell's efforts, the appointment went elsewhere.

Payne-Gaposchkin's family also influenced her opportunities outside astronomy. For example, in March 1941 I.R.S. Bourghton, a classics professor at Bryn Mawr College, wrote Shapley to ask whether Payne-Gaposchkin would make a good president of that school. He knew of Humfry Payne and wanted to know more about Payne-Gaposchkin and her family. Shapley was well aware of the complexities of administration and also was in no mood to lose a distinguished and inexpensive teacher and staff member. Replying to Broughton, he praised Payne-Gaposchkin's accomplishments as an astronomer and noted her wider interests in music, literature, and the activities of the American Association of University Women (AAUW). At the same time, he cited three reasons he could not recommend Payne-Gaposchkin for the presidency. First, she was an excellent research astronomer and should not be distracted by administrative duties. Second, she had what Shapley called a "genius temperament" and might not get along with professors of a conventional sort. Finally, Shapley noted that while Payne's father and brother had been distinguished scholars and Humfry had proved an able administrator, the Gaposchkin family was not so attractive. Without going into any details, he said that Sergei's personality might cause serious problems.[68]

Payne-Gaposchkin mentioned her combination of the roles of wife, mother, and astronomer in speeches she gave for the AAUW. In a 1937 talk she suggested that the psychological effects of discrimination impede women's achievements in the sciences even more than lack of opportunity. As she put it, "The woman scholar today is expected to live the life of a recluse, as was required of men scholars a hundred years ago. And it is more difficult for women than men to do without a full and happy life." Payne-Gaposchkin hastened to say that she didn't mean to imply that a woman must be married to be happy. A newspaper account of the talk noted that she was married and the mother of two children.[69]

Payne-Gaposchkin returned to this topic at the Denver national convention of the AAUW, held in June 1939. She was asked to speak as one of the most successful scholars who had received an AAUW fellowship as a graduate student; her topic was "Problems of the Woman Scholar." She suggested that those who became scholars should be willing to learn every necessary tool, possess good health and great courage, and think that nothing else would satisfy them—as nothing else was what they would receive. She also warned that even the best of men would, in unguarded moments, cast aspersions on a woman's ability. Indeed, Payne-Gaposchkin herself was troubled by women's lack of professional "staying power" and stressed, as she had before, that women scholars should be able to lead normal lives, with marriage and children if they desired.[70]

The Gaposchkin family also figured in a 1940 report of a talk Payne-Gaposchkin gave to a sectional AAUW meeting in New York City. A *New York Tribune* reporter asked about her children and she said: "They seem to be just normal little chaps to me—and we are trying to let them live normal, healthy little lives."[71] Similarly, an Associated Press feature published in California's *Oakland Tribune* the next year mentioned that Payne-Gaposchkin was not only an astronomer, but "the mother of three children, a housewife, a lucid philosopher and a writer of as yet unpublished detective stories."[72]

In all of the public accounts mentioned thus far, Payne-Gaposchkin's domestic duties were only a small part of the story presented and certainly did not occupy the leading paragraphs of articles. Similarly, a March 1946 article in the *Boston Globe* included a picture of the four oldest members of the Gaposchkin family, but was primarily an account of Payne-Gaposchkin's work on novae.[73] A November 1949 article in the *Boston Traveller* suggests the glorification of housework as a role for women that occurred in postwar America. Payne-Gaposchkin had been awarded the Radcliffe Alumnae Association's medal and was cited for managing to "combine scholarly work of rare excellence with the efficient management of a household and the rearing of three children." Reporting on her achievements, the *Traveller* headlined "Finds Housework Real Tough Job: Mrs. Gaposchkin Regards Mother's Task with Awe." Payne-Gaposchkin's comments on the difficulty of getting all parts of a meal on the table at the right time are discussed first, her work on variable stars second.[74]

Perhaps in reaction to this glorification of housework, and perhaps also out of sensitivity to her family, P ayne-Gaposchkin was not eager to publicize her private life in later years. For example, in June 1954 F. G. Strauss of the AAUW wrote the observatory to ask if someone might interview her for an article on combining scholarly work with raising a family. Payne-Gaposchkin was in England for the summer and wrote that she was disinclined to expose her personal life to the AAUW.[75] Indeed, a 1957 article from the *Journal of the American Association of University Women* that announced Payne-Gaposchkin's receipt of the AAUW Achievement Award said nothing about the children and mentioned Sergei only in passing.[76]

During their later years, the Gaposchkins continued to do research, to travel together and apart, and to pursue interests in literature and the arts.

Thus Payne-Gaposchkin went to England and Europe with Katherine in 1949 and spent the summer there by herself in 1954. Astronomical research took Sergei to Mount Wilson in 1952 and on a year-long journey to Australia in 1956. On this second journey, Cecilia joined him for several months. In the late 1950s and early 1960s, she spent considerable time in California and England, including a sabbatical at Mount Wilson on which Sergei accompanied her. Katherine, Edward, and their families were near at hand in Massachusetts. Payne-Gaposchkin also spent several weeks of 1966 in Berkeley with Peter.[77]

As her own family matured, Payne-Gaposchkin took a new interest in her ancestors. She had inherited correspondence of her great-grandmother Julia Garnett Pertz and devoted considerable time to deciphering and annotating it. In 1979 she arranged to have *The Garnett Letters* printed privately.[78] She also used the letters as a basis of a *Harvard Library Bulletin* article on an antebellum plan for removing the evils of slavery.[79] Payne-Gaposchkin went on to consider her own life. By 1979 she had completed her autobiography, a carefully stated account of her origins and astronomical career. Here she discusses her father, diverse female relatives, the Harvard Observatory under Shapley, and her first meeting with Sergei, but says remarkably little about her mother, her siblings, and her marriage. Both the title and the closing poem deserve special mention. The poem, by Payne-Gaposchkin herself, is titled "Research." It begins with the lines

O Universe, o Lover
I gave myself to thee
Not for gold
Not for glory
But for love.[80]

Looking over Payne-Gaposchkin's life, her first and last love was astronomical research. From the time she heard Eddington as an undergraduate until shortly before her death, studying the stars was her first priority. She might rescue Sergei from political turmoil, she might make household arrangements, she might marvel at her babies' play, she might try to better the position of her children as they grew, she might comment occasionally on women and the family, but she was most interested in knowing and understanding the stars.

Payne-Gaposchkin titled her autobiographical account "The Dyer's Hand," suggesting the influence of both the Harvard Observatory and her marriage on her research. The phrase comes froms Shakespeare's Sonnet

111, an apology by one without great fortune who had to resort to public means and public manners to survive. Payne-Gaposchkin picked up the lines

> Thence comes it that my name receives a brand,
> and almost thence my nature is subdued
> to what it works in, like the dyer's hand.

She called the third part of her autobiography, which concerns her meeting with Sergei and the years that followed, "The Dyer's Hand Subdued." Her outstanding beginning as an astronomer was indeed subdued by both the public demands of Shapley's research program and private matters.

Shakespeare closed his sonnet with a sentiment that sounds quite unlike anything Payne-Gaposchkin would say publicly: a call for sympathy. The couplet reads

> Pity me then, dear friend, and I assure ye
> Even that your pity is enough to cure me.

In these pages, I have outlined the course of one woman scientist's family life, from its beginnings among scholarly Englishmen and women through the unusual family of the Harvard Observatory under Harlow Shapley to complex relations with husband and children. Payne-Gaposchkin honored her families, cherished the companionship they provided, and relished activities with them. At the same time, in a culture where women were not expected to combine marriage and professional life, and with a thoroughly unconventional spouse, Payne-Gaposchkin's family life restricted her choices, strained her loyalties, and often left her exhausted. Payne-Gaposchkin's life, like that of other scientists, can only be fully understood if one considers her family, the strengths it provided, and the constraints it imposed.

12

PNINA G. ABIR-AM

Synergy or Clash: Disciplinary and Marital Strategies in the Career of Mathematical Biologist Dorothy Wrinch

THE DICTUM that "no [wo]man can serve two masters" ruled firm among the first generations of college-educated women, who often felt compelled to choose between a career, usually teaching, and marriage. Whether understood in an absolute manner or merely as a pragmatic injunction ("but I have only had time for one"),[1] this dictum continues to face academic women, albeit in attenuated forms, since problems of the quality and timing of career and marriage have come to replace the very possibility of their combined existence. Some of the boldest solutions to this dilemma were attempted in Britain under the stimulus of the post–World War I climate of social liberation, epitomized by such progressive legislation as the granting of the vote to women (and the immediate emergence of the first generation of women parliamentarians), and the Sexual Disqualification [Removal] Act, which regulated academic degrees for women.[2]

Dorothy Maud Wrinch (1894–1976) was a scientist who insisted on achieving success in combining an academic career with marriage, motherhood, and personal freedom, and who accomplished this feat in two

(Overleaf)

Dorothy Wrinch giving a talk at the fourth meeting of the Biotheoretical Gathering, 1933, on topological approaches to cell division. Photo by C. H. Waddington. Courtesy of Gary Werskey.

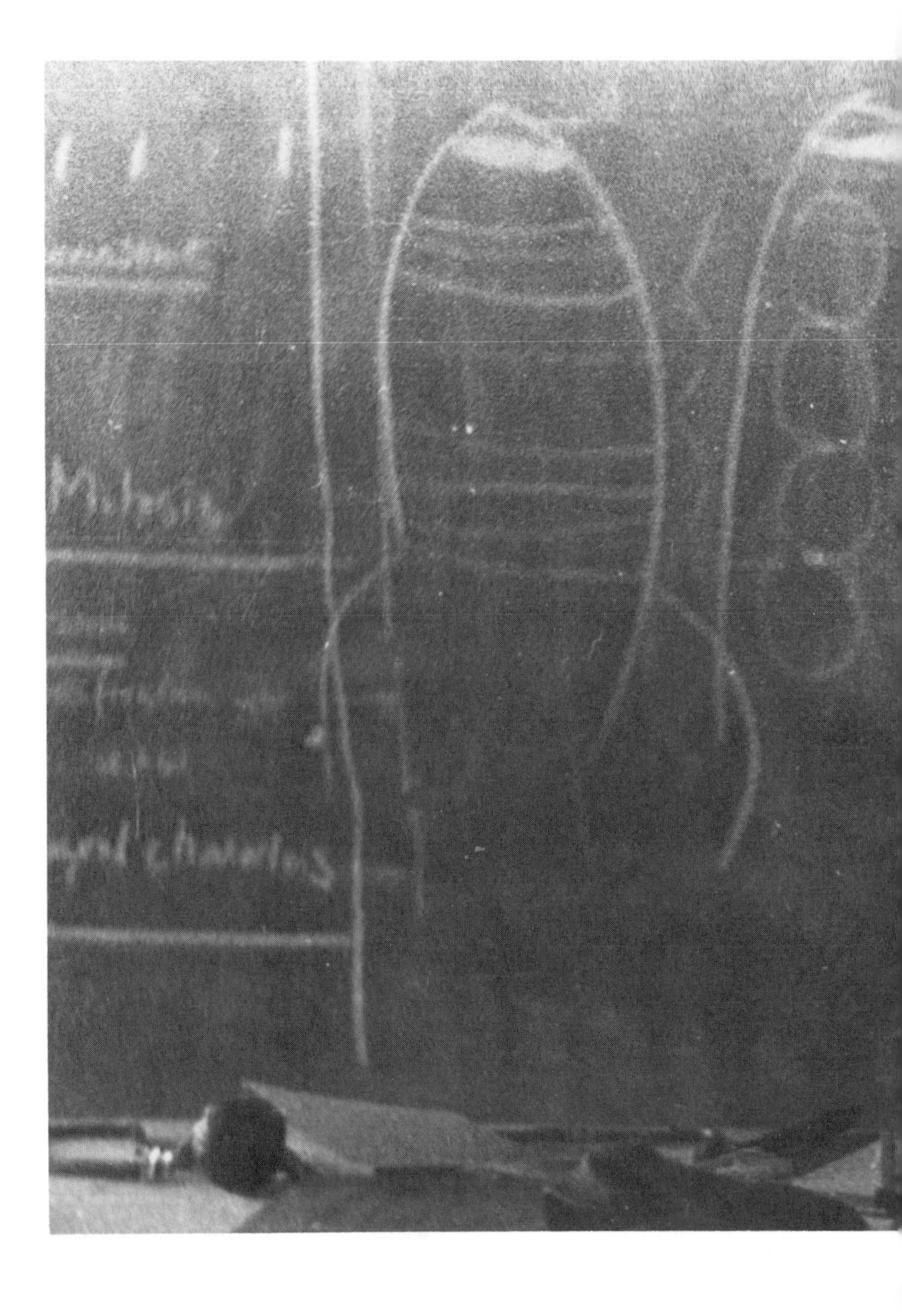

different disciplines and countries. She was a Victorian-born polymath who moved from Britain to America in 1939, when she was forty-five. Wrinch is best known for her work on protein structure in the late 1930s, when she was the first to propose a theory covering many then accepted facts. In later days her theory, like others proposed in the 1930s, did not become part of the scientific consensus; but hers in particular came to be regarded by both scientists and some historians as "notorious." This scapegoatism, grounded in a variety of disciplinary, philosophical, social, and gender-related biases, was further reinforced by Wrinch's own lifelong obsessive defense of her theory and refusal to follow the shifting scientific frontier.[3]

Wrinch's nonconventional stance in science bears on the most important episode in the transition from classical to molecular biology, namely the evolution of solutions to the problem of protein structure. That stance cannot be understood without a proper exploration of her career, especially prior to her involvement with proteins, but also afterward. Since that career was the product of uniquely intersecting disciplinary and marital trajectories, we are confronted not only with a biographical problem, but also with one pertaining to the major problem of social theory, namely the connection between individual action and social conventions. Moreover, since the social conventions in this case focus on both science and gender, Wrinch's story has direct relevance to history of science and women's studies.

Born in Rosario, Argentina, to British subjects Ada Souter and Hugh Edward Hart Wrinch, a member of the Institute of Mechanical Engineers of London, Dorothy Maud Wrinch grew up in Surbiton, greater London, where her father was employed at the local waterworks. She attended the Surbiton High School, a public day school drawing its pupils from the immediate neighborhood.[4] As gifted students often followed the college choice of their headmistress, following her school's preparation for the Oxford and Cambridge Joint Board Examination Wrinch entered Girton College, Cambridge, with a scholarship in 1913.[5] The choice of mathematics as her subject was influenced by her father, who would coauthor mathematical papers with her in the 1920s.[6]

At Girton, Wrinch acquired a reputation as a "glutton for work," showing both the ability and the determination that would position her in 1916 as the only Girton woman Wrangler (highest ranking among those graduating with honors by grades in the final examinations) in the Cambridge Mathematical Tripos.[7] She was also moderately active in college affairs. Thus, speaking in a college debate, she proposed a motion that "the social and intellectual achievements of America are not worthy of praise." Though the correspondent of the *Girton Review* said the proposer made a few "good points," the motion was lost despite its prophetic description of Wrinch's experience once she moved to America, where she was to spend the last thirty-seven years of her life. In addition, Wrinch was one of the "Second VI" tennis team, was very interested in musical activities (although she didn't perform), and served as secretary and, in her third year, as president of the Mathematical Club.[8]

Wrinch's life took its first unusual turn when, despite her great achievement upon graduation, she did not start advanced training in mathematical research, a necessary path for those who wanted to become research mathematicians.[9] Instead, she took Part II of the Moral Sciences Tripos (now the Philosophy Tripos) while ranking Class II (the middle third of the grades in the final examinations) on the Honours List of 1917. She was compelled to take the Moral Sciences Tripos as a condition for getting a Girton scholarship for her fourth year, even though her only purpose was to study symbolic logic with Bertrand Russell (1872–1970). She had attended Russell's lectures at Trinity College on "Our Knowledge of an External World" in her first year, but was unable to attend his lectures on symbolic logic because of the "exigencies of the Mathematical Tripos." Once the tripos were behind her, however, she soon approached Russell with her renewed interest. Following his advice, she managed to obtain a scholarship from Girton, another from the Surrey Education Committee (for residents of the county), as well as some article writing and translation work on Bolzano from Russell's friend mathematician Philip Jourdain (1879–1919), then editor of the *Monist.* All these financial resources were needed to allow her to attend Russell's projected lectures in the fall of 1916, though she was to stay an additional year in Cambridge because of the condition attached to the Girton scholarship.[10]

In the summer of 1916 Wrinch began to make contacts with other logicians on Russell's behalf, both for scholarly purposes, such as a projected translation by Jean Nicod (1893–1930), and social ones, such as the long walk over the Surrey Downs during which she introduced her Girton friend Dora Black (later Mrs. Russell), a recent graduate in modern languages with great interest in social activism, to Russell and Nicod.[11]

During 1917–18 Wrinch stayed on at Girton as a research scholar. She continued to serve as a go-between for Russell, then in London, and his Cambridge contacts, such as the organizer of his projected lectures on mathematical logic, to whom she passed along Russell's syllabus. She also sent him books, including her copy of the *Principles [of Mathematics]* and papers by Nicod and Norbert Wiener (1894–1964) that he had requested. Russell also asked her for Alfred N. Whitehead's (1861–1947) books on descriptive and projective geometry and requested that she verify the quotations and references and make an index for the second edition of his *Philosophical Essays*. She gladly agreed to do these tasks, and then she asked for his help with the "difficult section on convergents in *Principia Mathematica*," which he agreed to give.[12]

Throughout 1918 Wrinch continued to supply Russell, then in prison for his antiwar position, with books. His interest at that time shifted to behaviorism, psychoanalysis, and abnormal psychology. It was during the summer of 1918 that her letters on books and logic came to include occasional references to women and love. Thus, while Wrinch opened her letter of May 10, 1918, following a report on sending books, with "I hope the Sabine women will please," she replied the following August to his letter of July 31, concluding a three-page disquisition on her difficulties with atomic and molecular propositions in logic: "Does it not seem that it is the I-functions not the E-functions which are important in love?"[13]

Logic remained, however, the official link or excuse for an increasingly charged emotional relationship. Thus, while reporting to the jailed Russell on a philosophical weekend with the Mind Association, she told him how pleased she was to hear enthusiastic opinions of him, delivered to her because of her image as a disciple: "As I sat at your feet, everyone confided their private opinions of you to me." Indeed, she seems to have become not only a scholarly disciple, but also an adoring and confiding one, as her letter concluded: "With regard to my message a few weeks ago. Do you remember a conversation we had at Frascati's one Sunday night? I am very much disappointed and sick of life . . . you [said], it is apparently no fun unless the relation is symmetric—and unfortunately I am not in love! It's merely distressing, I find. We all miss you very much. DMW"[14]

Further revealing exchanges took place later in August 1918, when Wrinch reported to Russell on various tasks she undertook on his behalf, such as dealing with his publishers and correcting the typescript of his *Introduction to Mathematical Philosophy*. She also told him of her long discussion with George Santayana, whom she found overly talkative and dogmatic (in comparison to the more logical and flexible Russell), adding interesting details on her efforts at the time to find a job through ads. Eventually she obtained a teaching job at University College, London, where she was to get 100 pounds per term (the equivalent of $60 a week) for eight lectures a week in algebra, trigonometry, calculus, and solid geometry to first- and second-year mathematical honors students. She considered that opportunity magnificent, though she remained annoyed at the apparent need to get permission from the Girton mistress to accept the job: "I am [loathe] to get that woman K. Jex-Blake to let me take it—I wish all these people didn't have to be consulted about everything."[15]

Wrinch seemed to experience additional difficulties with Cambridge authorities, collegial and mathematical. She reported to Russell that the Cambridge mathematician and Trinity Fellow G. H. Hardy (1877–1947) had encouraged her to work out her disagreement with one of Russell's conclusions in volume three of *Principia Mathematica* for publication in the *Proceedings of the Cambridge Philosophical Society,* something that would please the Cambridge Research Department. However, Hardy's "heart failed him," she said, because he sent her paper to Whitehead for refereeing and he "kept it for the last ten weeks."[16]

While Russell remained in jail, Wrinch maintained contact with other colleagues of his; she told him of her visit with Jourdain, his preoccupation with the "axiom of choice," her belief that Jourdain's "proof" was invalid, and her observation that the Jourdains were pathetic.[17] She also met again with Santayana, but concluded that she wasn't attracted by what his type of philosopher had to say because those philosophers were not keen on epistemology, "which seems to me now to be the most thrilling part of philosophy (logic being considered as a separate science)."[18]

By 1918 Russell was providing the twenty-four-year-old Wrinch with both intellectual and social excitement. The social contact provided a great experience for her, including as it did introductions to social figures such as Lady Ottoline Morrell over formal dinners. Wrinch, who saw herself as a poor scholar, had to borrow plumes and clothes, thus prompting the joke that she might be wearing the clothes of Colette, the actress with whom Russell had been involved during World War I. Wrinch seemed to enjoy her sudden access to people who led a life of material comfort:

> I am glad that you enjoyed the joke about my wearing Collette's clothes! I am in borrowed plumes here too! They fit me remarkably well—and it will be trouble having to give them up. Lady O[ttoline Morrell] was much entertained—I do ache for fine clothing and splendour! You would not believe me when I said that I coveted the rings of the lady at the next table when we dined at Frascati's but I did. It is difficult to be the poor scholar and ache for material glories at the same time. Each seems to get in the way of the others. Lady R[ussell] is kind to say that I look slightly less like Miss Jex-Blake than before.[19]

Her Girton image, symbolized by the dour mistress Katharine Jex-Blake, and her modest social and economic background impeded Wrinch from joining the ranks of those who, like Russell, had a privileged heritage. She looked with excitement at the coming winter, when she expected to take a flat of her own and entertain Russell as a guest with a view to discussing not just logic, but also the analysis (rather than the psychology) of emotions. Yet she remained concerned with the greater difficulty experienced by women to combine work and pleasure. In a letter to Russell in which she urged him to use her first name—after three years of friendship—she wrote:

> I now have your letter of the 21st. . . . I was very glad to get a note on the 22nd. I am very much cheered up by what you say about the winter. It will be lovely if you will help me sometimes. And there are great plans on foot as to a flat for me! You will come and I will cook you a nice dinner, won't you. . . . I drink in all you say about logic . . . but long to discuss it with you. . . . *Very* glad to have jolly letter of the 21st. Of course, or at least I suppose of course—one only considers other pleasures if they can be worked in with one's work—and not if they exclude it. But these things are much more difficult to fit in, if one's a woman. It looks to me as if they might prove so diverting that there would not be sufficient time for work.[20]

Though the rest of this letter, written over four days, includes mostly details on Wrinch's interests in the Russellian topics of mathematical logic and psychology (especially on her projected paper on "mediate cardinals," her view of Jourdain's "MULTAX" or "axiom of choice" as occupying a "most peculiar position in logic," and of psychoanalysis as "extremely interesting but . . . not quite scientific enough"), it also reflects her expectation that logic would provide help with sorting dilemmas from everyday life, especially on the topic of pleasure, which she often discussed with Russell. The question for her was how could one decide between "an improbable great pleasure and a more probable little one" or "the analysis of a thing being worth the risk"?[21]

That Wrinch liked to express personal questions in symbolic and logical terms is obvious from her comments, a week later, on Russell's accepting the request to call her by her first name, to the effect that she did not mean "the relation to be symmetric." But she gladly consented to his mentioning of "Xian names" and switched from the way she had addressed him since 1914 (Mr. Russell) to "Bertie." Restating her interest to talk philosophy with him soon, she also promised to do, if needed, typescript corrections for his *Introduction to Mathematical Philosophy* on additional copies, while also soliciting a report from him, to be sent to Hardy, to facilitate the publication of her paper on serial types.[22]

Similar expectations for future joint discussions of philosophical problems and reports on her work in progress continued into the summer of 1919, when Wrinch expressed her excitement at the arrival of Wittgenstein's book and her dream that he had landed in England "last night" and "we were all rushing to sit at his feet . . . this does not include you. You were on a throne or a dais, W on his dais and the rest of us on the floor!!"[23] Also at that time, Wrinch was invited to stay at a farm in Lulworth, where Russell entertained guests at "reading parties" and she was very grateful for the time she had there: "I had enjoyed myself at Newlands most frightfully! Thank you so much for all the lovely times we had. You really are the most delightful person in the world to stay with! September has been one long pleasure and I shall never forget it. Ever yours affectionately."[24]

This crescendo of intimacy in the period 1917–19 came to an end shortly afterward when the forty-seven-year-old Russell, who had long been separated from his wife, became involved with Dora Black. Black was then leading a bohemian life in London and apparently had fewer reservations than Wrinch about putting pleasure before work. Russell and Black were also drawn together by their common interest in various socialist and feminist causes, their libertine views of marriage, a desire to have children, and their bolder attitudes toward social conventions. Following trips to the Soviet Union and China, Russell and Black were married in 1921, shortly before the birth of their first child, John Conrad. Eventually, they began joint work in social and, especially, educational reform, including, the famous progressive school in Beacon Hill, with Russell abandoning his academic and logical priorities in favor of free-lance activities.[25]

Under these new circumstances, Wrinch found herself not only deprived of Russell's attention and connections, but also stuck with his legacy of research interests that did not suit the then prevailing disciplinary divisions by academic expertise. In effect, Wrinch's mathematical interest in logic and epistemology, derivative of Russell's formative influence, would position her as doubly marginal to the communities of mathematicians and philosophers that she would address in the 1920s with equal commitment. Mathematical logic was, and remained, too mathematical for philosophers and too philosophical for mathematicians, while most philosophers at that time, as Wrinch observed, were not interested in epistemology. Moreover, with Jourdain's death in 1919, she did not have any other mentors to help her find a job and thus she returned to Girton on Yarrow Scientific Fellowships for the period 1920–23.

A play written by Wrinch on the theme of betrayal and sections of a novel on parenthood she wrote in the late 1920s capture some of her sense of loss, frustration, and embarrassment when the friendly trio of Russell,

Black, and Wrinch suddenly dissolved, with her being left out despite her longer emotional involvement with both Russell and Black and despite her "rights" as the one who had introduced them to each other.[26] A year later, in 1922, after carefully investigating various alternatives, the twenty-eight-year-old Wrinch quickly and quietly married John William Nicholson (1881–1955), who had recently been appointed Fellow and director of studies in mathematics and physics at Balliol College, Oxford.[27]

Like Wrinch, Nicholson had excelled in mathematics at Cambridge, ranking slightly below A. S. Eddington (1882–1944), the most prominent British astrophysicist in the interwar period, knighted in 1905. Unlike Russell, he was of modest origins, his father having been clerk of an ironworks in Yorkshire. Following his graduation from Trinity College, Nicholson remained at Cambridge's Cavendish Laboratory (whose director, J. J. Thomson [1856–1940], was later Master of Trinity) as a research student (Isaac Newton Student in 1906 and Smith Prizeman in 1907). Eventually he became a lecturer and was one of the mathematical "coaches" for Girton students in 1913, the year Wrinch entered the college; he left shortly afterward for a professorship at King's College, London.

Before moving to London, Nicholson was active in and a president of (in 1906) the ∇^2V Club, for mathematical physicists, which gave him the opportunity to meet with Niels Bohr (1885–1962) in 1911. At the second meeting, Nicholson read "an interesting paper on electrons in metals which was followed by a discussion in which Dr. Bohr took a prominent part."[28] The atomic theory developed by Nicholson was eclipsed by Bohr's, and Nicholson never got proper recognition for his pioneering work in this domain.[29]

Prior to leaving London for Oxford in 1921, Nicholson was active in London-based scientific societies as vice-president of the Physical Society, council member of the Royal Society (he was elected Fellow of the Royal Society in 1917), the Royal Astronomical Society, the London Mathematical Society, and chairman of the British Association Committee on Mathematical Tables. Once in Oxford, he managed to secure students for Wrinch at the women's college, Lady Margaret Hall (hereafter LMH), so she could move from Cambridge, where she had continued to hold a Yarrow Scientific Research Fellowship for the year after their marriage in the summer of 1922.

Initially, Wrinch was determined to move to Oxford only if she could find a suitable position, and she so informed the LMH principal who asked her to give tutorials following the death of a male tutor. At Wrinch's suggestion, Nicholson assumed responsibility for tutoring LMH women students until Wrinch was able to move and take them. But when her conditions for relocating were not met during 1922–23, and she gave no indication that she planned to come, LMH principal Linda Grier and her council decided to appoint a tutor from Balliol, to whom Nicholson delegated the responsibility for the LMH students:

> We felt that it must have been extremely difficult for you with the mass of work which you have, to deal with our very immature students. We understood, of course, that you were doing this with a view to Mrs. Nicholson taking the work later. But as I understand that there is no certainty of her coming to Oxford at the present we feel clear that we could not trespass on your kindness any longer, and as Mr. Newboult appeared to be doing the work efficiently we are asking him to continue it.[30]

A slightly alarmed Nicholson replied to the LMH principal that his wife definitely would be coming to Oxford next autumn. His letter also intimated that their conditions for her relocation had become more flexible by then and amounted to the appearance of a position, rather than a real one. The availability of a few pupils thus tipped the balance in favor of transforming their marriage from a commuting one to one of sharing residence. It enabled Wrinch to save face, so she need not admit she was following a husband without work of her own:

> There is no longer any doubt that my wife is fully able to take your people next year. In fact, since first she became responsible for them last October, either directly or through me, she has been endeavoring to arrange to deal with them herself from October next at the latest, and now it is certain that she can. In fact, they have formed an important part of the arrangements whereby she is now able to come. We are very sorry, that owing to my accident, neither of us has written specifically to you about the arrangements for next term.[31]

Nicholson not only used his position as director of studies at Balliol to negotiate with the LMH principal on his wife's behalf while sparing her the embarrassment of publicly admitting that she had to give up her strong aspiration to combine a suitable position with her marriage, but also to inform his junior delegate tutor that his services would no longer be needed. Nicholson further allayed the LMH principal's concern at having to reverse her college's decision and deprive a good tutor of work by tipping her with information to which he was privy by virtue of his position, namely that his delegate tutor had been elected to a lectureship at St. John's College and presumably would be too busy to continue with his work at LMH: "I am very glad that Mr. Newboult has been able to see me through, both at Balliol and Lady Margaret, so well as he has this term. I have told him that my wife is coming. He has, of course, now been elected to the St. John's Lectureship and will have a considerable amount of work."[32]

The principal, however, stood firmly by the rules and college priorities of securing tutors on schedule rather than responding to the needs of spouses in a commuting marriage. This initial clash between the principal's primary concern with the organization of tutorials and with proper rules of appointment, and Wrinch's priorities of research and family resurfaced, creating recurring embarrassments of the type evident in the principal's reply: "I am much concerned to learn from Professor Nicholson's letter that you had been expecting to take our students. I had received no intimation from you that there was any possibility of your coming to Oxford next term. In fact, I had understood from what you said and from what others had told me that there was no likelihood of your doing so unless you had an assurance of more work, and even of a definite post, in some of the other women's colleges."[33]

Eventually, a compromise was found when the principal split the students between the former tutor and Wrinch, with her receiving the new students only. This arrangement was more meager than the one Wrinch had rejected a year earlier (when she had choice of all students, i.e., four instead of two), but the principal assured Wrinch LMH had just begun to regularize the appointment of a mathematical tutor, implying that Wrinch would have better prospects in the years to come.[34]

Following her first year at Lady Margaret Hall (1923–24), Wrinch was invited by its council to be lecturer for the following year as well, an arrangement that prevailed throughout the 1920s and 1930s, though later on long-term appointments of three and seven years at a time were made. However, because of the smaller size of Oxford women's colleges, as well as the relatively small number of women mathematical students, Wrinch's position at LMH was far from a full-time job. This meant that she had to seek additional tutorials at the other four women's colleges, none of which could support a full-time mathematical fellow.[35]

In addition, Wrinch's status as a married woman lecturer put her in an awkward or anomalous position in the Oxford social system, because she did not fit into the two prevailing categories—resident dons in the women's colleges and faculty wives (who were usually nonacademic or inactive lecturers). This social problem was further compounded by the fact that both Wrinch and her husband were recent arrivals on the Oxford scene.

The mathematics and science communities at Oxford were smaller than those to which they were previously accustomed in Cambridge or London and the inbred customs were stronger.[36]

Eventually, through the good offices of Lady Margaret Hall's principal, the principals of the other women's colleges (St. Hilda, St. Hugh's, Society for Home Students, and Somerville) agreed to sponsor and pay for Wrinch's lectures to their students on a per-term basis. This arrangement involved a great deal of coordination and uncertainty, since each college had different policies and even different pay scales. Thus, she had to cancel lectures for a term in 1924 for which no college sponsor could be found in time, but in 1926 four colleges wanted Wrinch on their lecture list, though she was scheduled to give only three lectures, one each term. Each college paid 10 pounds for its share in the scheme, the total of which still amounted to only one-quarter of the salary of a university lecturer (200 pounds).[37]

Wrinch obtained her first long-term appointment in 1927, when the Council of Lady Margaret Hall appointed her a lecturer in mathematics for three years.[38] In that same year Wrinch was one of the first two women to appear on the list of electors to the board of the Faculty of Physical Sciences, the only woman mathematician and the only one married to another elector (her husband, as director in mathematics and physics at Balliol, would also be on that list).[39] She was thus the first woman to qualify for a university lectureship in mathematics at Oxford, which meant that her lectures, once approved by the mathematical subfaculty, were also open to men students.[40]

Wrinch's still tenuous position improved further in the spring of 1929, when the Finance Committee of Lady Margaret Hall agreed to join in a guarantee of a minimum salary of 200 pounds a year, plus 10 percent pension premiums[41] (the equivalent of a university lectureship held by men without, however, the additional 300 pounds male lecturers had as college fellows), depending on a similar decision by the other women's colleges.[42] Other marks of stabilization of her position, though all small, came in 1929: membership in the Education and Library committees, a special retaining fee to cover one course of lectures initiated by the Finance Committee, extension of her duties to supervision of physics students, and intimation that the following year, when her three-year appointment was due to expire, she would be reelected for seven years, as was customary for university lecturers. Indeed, early in 1930 the council reappointed her for seven years, effective October 1930.[43]

Wrinch managed to show an impressive research productivity in these years. By 1929 fifteen of her forty-two publications were submitted and approved as the basis for a D.Sc. degree, the first such degree given by Oxford University to a woman.[44] The degree also enabled her to call herself Dr. Wrinch rather than Mrs. Wrinch Nicholson, an appellation she preferred to Mrs. Nicholson, used by her principal. She published all her work, including collaborative work with her husband, under her maiden name. This practice was unusual among married academics at that time, who would either adopt their husband's name or hyphenate both into one name. Her insistence on her own professional identity stemmed both from her Girton education, which promoted female scholarly independence, and from the socialist and feminist beliefs she had absorbed in Russell's circles.[45]

By 1929 Wrinch accumulated scientific credit of her own, having been sole author of 75 percent of her published work. Her record, though impressive in its versatility and quantity, was scattered among several disciplines. Out of the forty-two papers published by the time she was awarded the D.Sc. in 1929, only three count as work in pure mathematics by both topic (real and complex variable analysis, Cantorian set theory, transfinite arithmetic) and the journal's focus. Moreover, on the few occasions when she addressed an audience with a prominent representation of pure mathematicians—e.g., a 1924 paper in the *Proceedings of the London Mathematical Society* and communications at the International Congress of Mathematics in 1928 (Bologna) and in 1932 (Zurich)—her work was in mathematical physics or applied mathematics.[46] Indeed, Wrinch seemed to confuse her social facility in the company of mostly male pure mathematicians at these congresses and related professional occasions with their recognition of her work. At the 1928 congress held in Bologna Wrinch must have impressed her colleagues with yet another social accomplishment by receiving daily cables from her husband on the welfare of their then one-year-old baby daughter, Pamela, who had remained in England with him.[47]

Mathematical physics (especially applications of potential theory in electrostatics, electrodynamics, vibrations, elasticity, aerodynamics, and seismology) was another major interest; she wrote seventeen papers in the field, including two coauthored with her husband in 1925 and 1927. Only one of their two collaborative papers took advantage of his FRS status to be published in the *Proceedings of the Royal Society*.[48] Wrinch was drawn to mathematical physics as a result of her access to her husband's work, an established mathematical physicist who was actively collaborating at that time with former Cambridge colleagues.[49]

In addition, about half of Wrinch's publications, or a total of twenty papers in the 1920s, were in philosophy of science.[50] This work reflected her long-suppressed desire to escape the world of mathematics to which she had been confined by her education, her desire to be associated with a prestigious descipline, and marriage, as well as her capacity for interdisciplinary, unconfined thinking. In due course, she was to explore the relationship between science and philosophy, with special interest in scientific theories, first in physics and later in biology.

In the 1910s Russell's logicism had deflated for Wrinch the ultimate status of pure mathematics as irreducible and led her to distancing herself from pure mathematics. By the late 1920s Russell's legacy of logicism and his later concern with the geometrical and logical foundations of physical theories pointed Wrinch toward a new avenue of direct involvement with scientific theories.

In the early 1920s she addressed herself to the mathematical and philosophical status of the theory of relativity at a symposium of the Aristotelian Society, where she, the only woman participant, shared the podium with distinguished philosophically minded scientists, including Alfred N. Whitehead and Lord Haldane.[51] In the early 1920s she also wrote general papers on the scientific method, especially on inference, nineteenth-century inductive logic, and philosophy of probability, often in collaboration with H. Jeffreys (1891–).[52]

In the late 1920s Wrinch addressed herself to the philosophical principles reflected by a range of scientific theories, including the electron theory, quantum theory, and theory of relativity (which she continued to refer to as a "kind of achievement which stands alone in Scientific Theory"),[53] eventually venturing into "embryonic sciences" such as physiology, genetics, psychology, and sociology.[54] She became convinced that further progress in scientific theories depended on closer cooperation of scientists with mathematicians and logicians, since geometry and logic had proved so capable, in the theory of general relativity, to unlock the secrets of the universe. Like other mathematically oriented philosophers of science at the time, Wrinch came to believe that physical reality was not only expressed but actually determined by the (arbitrary) postulates of mathematics and logics. This belief in the ultimate capacity for scientific reductionism of these last bastions of certainty was to both enable and constrain Wrinch's strategy as a theoretical biologist in the 1930s and later.[55]

In 1930 Wrinch's husband suffered a nervous breakdown caused by prolonged excessive drinking, and his office as university lecturer at Oxford was terminated as of December 31, 1930.[56] It seems that Nicholson's heavy drinking, long protected by the cultural tolerance of this habit, had already begun to affect his health and capacity to carry his duties by 1926, when he resigned as examiner from Balliol. His condition remained out of control in the late 1920s, with students seeking him in vain for tutorials,[57] and in the summer of 1930 Wrinch sought and obtained a separation deed. As she put it in a letter marked "very confidential please" to her "Dearest Bertie," her miseries came to an end with her becoming, once again, a "free woman." The tragedy, as she saw it, was John Nicholson's intellectual waste as a "good mathematician," rather than the dissolution of their marriage or the emotional impact it had on her own "manic depressive personality":

> At last my miseries are to be, to some extent, at an end! It has been a fearful time, and I could not come and see you when I was overcome with anxieties. I like to be happy when I see my friends. John and I have now parted with a nice little separation deed duly executed. I can scarcely believe that it is through at last. Now I have gone one step towards being again a free woman. . . . I get wildly happy when I make an epsilon of progress. But this is only my manic depressive personality, I expect! . . . With regard to J [husband, John], all is tragedy: I can hardly bear to think of it, but there was no other way. But it is fearful when one thinks of his work. It is an awful grief to me to think of a good mathematician going so utterly to pieces.[58]

This posture of emphasizing intellectual over emotional matters was further displayed in Wrinch's negotiations with Balliol's home bursar about Nicholson's books. They were eventually split between his college and his family late in 1932 when the official solicitor gave permission to dispose of them, and any other belongings, in view of Nicholson's permanent confinement to an asylum.[59]

In seeking possession of fifty to one hundred books from her husband's study in Balliol, Wrinch did not mention familial rights, but rather her needs as a scholar to have access to books they used to share, "having both the profession of mathematics and physics." Others, she added, could be of future use to their daughter because "the range of scientific interests of the patient was very wide."[60]

Nicholson's breakdown prompted Wrinch to seek to branch out into other fields and attempt to escape from Oxford, temporarily if not permanently. Her first attempt to explore a new field involved sociology, especially the study of professions, a topic to which she was drawn as a result of a book she published in 1930, under a pseudonym, on the difficulties of professional people, and women in particular, to combine careers with parenthood.[61]

There she described a series of cases, obviously taken from among her acquaintances, in which the choice of parenthood and the marriage it entailed destroyed not only the professional future of talented women, but also their domestic life as they lost their former charm and conformed to matronly social roles. In contrast to this category of frustration, she adduced the opinion of professional women who had independence but were forced to give up parenthood, since it was then deemed incompatible with a career. In the manner of the then prevailing positive eugenics, she deplored the "decline of breeding among professional workers," while offering some suggestions to arrest this decline, most notably the establishment of child-rearing institutes. Backed by the resources of scientific research through auxiliary services set up by the Consumer Cooperative Movement, the ultimate goal of such institutes was to secure "the right to breed . . . for every man and woman in the professional world."[62]

Wrinch's main concern, obviously induced by her experience at that time as a married lecturer with a two-year-old daughter, was to articulate the need of professional women for freedom from both domestic and financial worries, a need amply demonstrated by women dons who had to pay the price of childlessness for that sort of ideal working environment: "All women capable of professional activities should have freedom from trivial domestic and financial anxieties and a sympathetic environment to enable them to function as well as they possibly can . . . the barren women have, indeed, done all women a service in showing that women, like men, require these things if their development is to reach its highest pitch."[63]

Bertrand Russell, the author of several books on children's education and social reform,[64] was one of the three friends and the only man to whom the book was dedicated (by initials only) "with gratitude and affection." He commented favorably on Wrinch's analysis of the conflict evoked by marriage and motherhood for professional couples. But he also noted her utopian proposals for solving those conflicts, most notably the idea of child-rearing institutes. Having been a practitioner rather than imaginer of social reform, Russell was much more aware than Wrinch of the profound social impediments to innovation, especially the distinction between utopian ideas and their social acceptance. Prior to attempting (unsuccessfully) to interest his American publisher (Norton & Co.) in her book, he inquired how carefully she wished to preserve the anonymity of her authorship. She had chosen a pseudonym, presumably to protect her scientific reputation by appearing to subscribe to the norms of science as an exclusive vocation.

> I have now finished your book on "The Retreat from Parenthood," and I think it a quite extraordinarily good book. The first part of it is written with a restrained

> and intellectualized passion that makes it very good reading. Your proposals in the second part are very constructive, and I wish I could suppose that they would be carried out. . . . My own belief is that the sort of thing that you and I consider desirable in regard to the rearing of children is scarcely possible without a social revolution . . . are you very anxious to preserve anonymity strictly, or may the authorship of the book be mentioned privately?[65]

Sybil Moholy-Nagy, the wife of the Hungarian-born abstractionist Bauhaus painter Laszlo Moholy-Nagy (1895–1946), then living in London and to whom the book was also dedicated, congratulated Wrinch on her perceptive analysis of parenthood, but was also quick to point out that her solutions might suit exceptional women only.[66] Unfortunately, there are no surviving comments of the third person to whom the book was dedicated: Margery Fry (1874–1958), principal of Somerville College from 1926 to 1930; sister of Bloomsbury group member Roger Fry, a painter and a biographee of Virginia Woolf; social reformer; and a friend who helped Wrinch with personal crises in the 1930s.[67]

Wrinch's desire to study the problems of conflict between parental and professional commitments from a comparative sociological perspective formed the basis of applications for U.S.-bound traveling fellowships from the Rhodes and Rockefeller trustees. The LMH principal proved most sympathetic to Wrinch's request for nomination and further approached on her behalf the only Oxford-based Rhodes trustee, the warden of New College, to personally recommend Wrinch as a "most exceptional person" and inquire privately into her chances. By the time the principal obtained an informal answer to the effect that Rhodes Fellowships were given to men's colleges only,[68] Wrinch had already submitted her application, which included impressive testimonials. She further decided not to withdraw it, possibly hoping that her unusual merits would induce a change in the Rhodes trustees' sexist policy. Indeed, the LMH principal stressed her wide interests and accomplishments in science, art, and literature, her remarkable determination to carry out her diverse work, and her skill at making friends through her impressive conversation.[69]

More specific testimonials came from the Oxford professors of physics and mathematics. A.E.H. Love, FRS, Sedleian Professor of Natural Philosophy, stressed her versatility in various mathematical fields (especially mathematical logic, analysis, and applied mathematics), singling out her papers "Some Boundary Problems of Mathematical Physics" (1924) and "On the Asymptotic Evaluation of Functions Defined by Contour Integrals"

(1928) as of special interest. He further stressed that her work had been accomplished under dual scholastic and domestic pressures, concluding that she was a unique and outstanding investigator and teacher among Englishwomen.[70]

Similar words came from a younger mathematician colleague who had recently been elected to the Rouse Ball professorship of mathematics at Oxford, E. A. Milne, FRS:

> As a fellow student with her at Cambridge and as a colleague at Oxford I have had ample opportunities of forming a judgement on her original work in mathematics. . . . I know of no mathematician who has made significant contributions to so many different branches of mathematics, from symbolic logic through pure analysis to diverse questions of mathematical and cosmical physics. I would especially like to give testimony to her deep-seated enthusiasm for scientific investigation, to her grasp of scientific method and her zeal for all that science stands for.[71]

Milne further stressed Wrinch's ability as an inspiring teacher, her activity in university affairs, and the force of her personality as important assets to mathematics and liberal education at Oxford.

In addition, T. Percy Nunn, professor of education at the University of London, who knew Wrinch from her Girton days and later from philosophical discussions at the Aristotelian Society, stressed her broad interests, "at once scientific and humane in temper," and her originality on problems of scientific method.[72]

While Wrinch's being denied a Rhodes Fellowship was a clear case of sexism in academic policy, her parallel denial of the Rockefeller Fellowships was more of a result of her strained circumstances at the time of application, especially her desperate need to get away. Accordingly, Wrinch applied for two Rockefeller Fellowships, one in sociology and one in mathematics. Wrinch's hastiness turned out to be self-defeating, because the application for a fellowship in sociology was interpreted as evidence that her interest in mathematics was no longer serious, and the application in mathematics made it clear that she was a novice in sociology. Both applications were turned down, and her principal could only apologize to the Rockefeller representative, saying that if she had known about it she would have dissuaded Wrinch from putting in two simultaneous applications. However, she excused Wrinch's hasty applications as due to personal distraction and anxiety, prompted by "circumstances, which it is impossible for me to describe" and which "made it intensely important that she should, if possible, get away next year."[73]

Wrinch's attempt to solicit the cooperation of the soon-to-be British authority on professions, Professor (later Sir Alex Morris) Carr-Saunders (1886–1966), also failed. He saw no way to put his material for the book *Professions* (Oxford, 1933) at the disposal of an amateur, however enthusiastic, able, and determined, who seemed to have no idea of the steps involved in such a cooperative arrangement. Subsequently, Wrinch also failed to receive a fellowship from the American Federation of University Women, and the sociological option faded from her plans for drastic professional change.[74]

In the spring of 1931, however, Wrinch received a small research allowance from her college of affiliation, the Suzette Taylor Fellowship, which together with another small allowance from Girton, the Hertha Ayrton Fellowship, and the leave of absence she was granted, enabled her to spend the period between August 1931 and April 1932 in Vienna accompanied by her five-year-old daughter. She mentioned Professors Pauli, Przibram, and Thirring as her main scholarly contacts in Vienna, but it was the tense political situation in Vienna that made an impact on her there: "We feel very much in the middle of things here, with the little Fascist revolution the other day and so many exciting happenings but I do wish I could see what is going to happen. Don't you think that this feverish economising is the straight road to a final break up of the present system and where will we then find a new system which is neither Fascism nor Bolshevismus????"[75]

Though her request for renewal was granted, she decided to postpone the tenure of the second Suzette Taylor Fellowship to 1933–34. Her reasons included the drop in the value of the British pound on the Continent, the failure to get the fellowship from the American Federation of University Women, as well as examination duties at a difficult time for her subfaculty. But she also revealed that she was using the fellowship in order to extend her mathematical applications from physics to biology, adding that the experiments in cell division needed to provide data for testing the relevance of classical potential theory, her major area of work in applied mathematics, were under way and were likely to take more time.[76]

Indeed, in the summer of 1932 Wrinch became a founding member of the Biotheoretical Gathering, a group of philosophically and ideologically minded scientists keen on developing a theoretical biology (they are still known in the secondary literature as the Theoretical Biology Club, a term popularized in 1936 when *Order and Life,* a book dedicated to the group's members by its author, Joseph Needham [1900–], offered a public summary of the group's ideas). The founding members included, besides Wrinch and Needham, who was a reader in biochemistry at Cambridge:

Joseph Henri Woodger (1894–1981), reader in biology in London; C. H. Waddington (1905–1975), lecturer in zoology at Cambridge; and John Desmond Bernal (1901–1971), lecturer in structural crystallography at Cambridge. Later meetings of this group included occasional members drawn from both science and philosophy, most notably Max Black (1909–), Lancelot Law Whyte (1896–1972), Karl Popper (1902–), and two women scientists, Dorothy Moyle Needham (1896–), the biochemist wife of Joseph Needham and also a former Girtonian, and Dorothy Crowfoot Hodgkin (1910–), a graduate student of Bernal. She spent her entire career at Oxford as student and fellow of Somerville College, university lecturer, and professor of chemical crystallography. Her career culminated in the Nobel Prize for chemistry in 1964. The group's founders and members were talented, eccentric individuals who found themselves at the margins of their scientific fields. All were attempting to redress their marginality through joint efforts at exploring scientific utopias, such as new combinations of scientific fields or theoretical perspectives for strongly empiricist fields. They shared Wrinch's expectations that philosophy and mathematics could revolutionize biology along the lines of their previous success in physics, especially general relativity. Therefore, they tried to pool their disparate resources in science and philosophy through meetings that had the additional advantage of being social vacations. They all hoped, perhaps unrealistically, to forge bridges between separate disciplines through personal rapprochements among themselves.[77]

By 1934 Wrinch had become totally absorbed with the biological applications of her mathematical field of potential theory; she used the second year of her leave of absence, 1933–34, and its associated fellowships to "gather data to test my theories as to the applicability of potential theory to chromosome mechanics." This data gathering involved stays in biological laboratories, including those of Maurice Caullery in Paris, J.B.S. Haldane at the John Innes Institute of Horticulture Research, W. Wilson at Bedford College, London, and E. Freundlich of Berlin, then temporarily at University College, London. Indeed, the final report she submitted in February 1935 on her two years of tenure of the Suzette Taylor Fellowship captures her

(Overleaf)

Wrinch and colleagues at a meeting of the Biotheoretical Gathering, 1935. First row (*left to right*): J. D. Bernal, B. P. Wiesner, and C. H. Waddington; second row (*left to right*): D. E. Woodger, J. H. Woodger, D. M. Wrinch, and J. Needham. Courtesy of Gary Werskey.

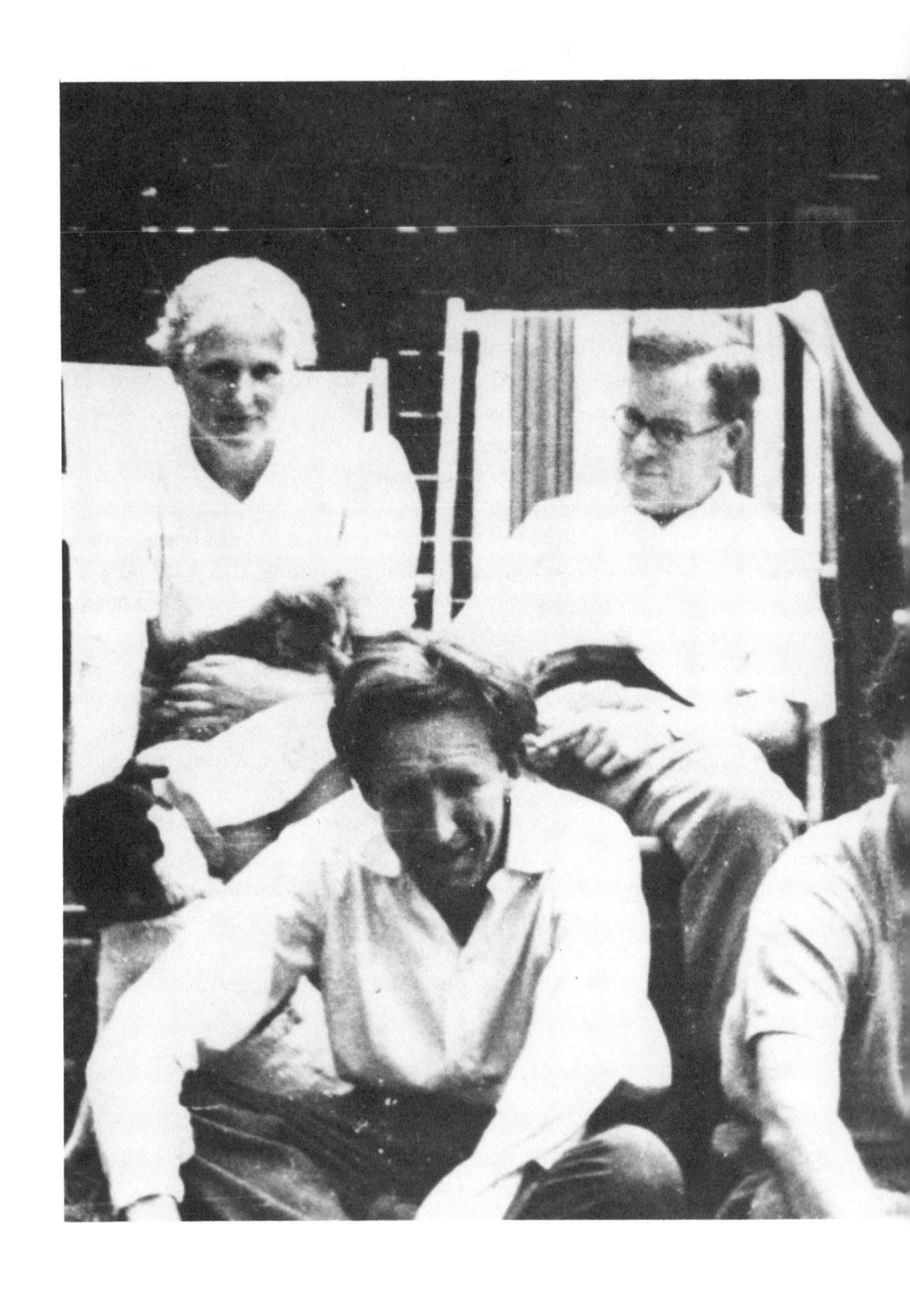

transition in the period 1932–34 from applied mathematics or mathematics applied to problems of physics to that applied to biology, especially to cell division and chromosome structure.[78]

Meanwhile, Wrinch had also been applying for a research grant from the Rockefeller Foundation, after meeting its European officers during her stay in Paris in 1934. Her application was part of a group proposal submitted by the founding members of the Biotheoretical Gathering for a research institute affiliated with Cambridge University. In the summer of 1935, following their visits to Cambridge, the Rockefeller Foundation's officers decided not to consider immediately the large and problematic proposal for a research institute, but rather to focus on personal grants to the partners in the proposal until various administrative hurdles pertaining to their projected investment in a larger scheme would clear.[79] On October 1, 1935, Wrinch was awarded a research grant to pursue her work on mathematical applications to biology, once the Rockefeller Foundation obtained assurance that her affiliation with the Oxford Mathematical Institute would continue.[80]

Wrinch was fortunate to receive the Rockefeller grant, not only because she had failed to obtain other grants and had already used up the smaller grants given by both her colleges of affiliation (Girton and LMH), but especially because this generous grant of $12,500 was for five years. Thus she was relieved from dependence on the meticulous annual negotiations for tutorials that she had been compelled to conduct for the preceding dozen years. In addition, her personal circumstances—suddenly becoming a single parent, with the social, economic, and cultural anomaly of being married but unaccompanied and unsupported (the marriage would be dissolved by a decree of the Church of England in 1938)—required that she get away from Oxford, the scene of so much personal and professional misery, as much as possible. Though the Rockefeller Foundation's grant stipulated that she remain affiliated with Oxford (its policy required that grants be administered via academic institutions as guarantees against charges of interference), the grant did release her from most of the social control to which she was subjected by providing the means and justification for numerous absences. It also made her financially independent of the teaching fees; within a month after her award, she indicated to the principals that a junior substitute would be required and she recruited from London a woman doctoral candidate to take ten to twelve hours of her teaching per week.[81] The grant propelled her, at least informally, into the position of director of studies, rather than mere tutor, thus bypassing the formal reasons, personal as well as professional, that had so far prevented her from converting her 1929 D.Sc. into a real position at Oxford or elsewhere.

The grant further satisfied her repressed tendencies to branch out into new, transdisciplinary domains, especially into theoretical biology, which she had been exploring as part of the Biotheoretical Gathering since 1932. Indeed, in her euphoria at this turn of fortune, Wrinch drafted a memorandum to the principals informing them of her need for extensive travel for her research and seeking their advice, or rather consent, for prolonged but vaguely defined projected absences.[82]

The period between 1936 and 1939 marked the peak of Wrinch's career and of her most important contribution to science, namely the first theory of protein structure, or the cyclol theory.[83] Her theory, like all other theories and interpretations proposed in the 1930s to account for the paradoxical ability of proteins to combine enormous functional or biological diversity with striking structural or molecular unity, is not part of the positive scientific knowledge of today. Yet, at the time, it had a major impact on consolidating transdisciplinary research from half a dozen disciplines around the problem of protein structure. This problem, as I have argued elsewhere, was the first conceptual focus for a new type of biology, molecular biology. Coined in 1938 at a time when proteins' duality as both proper molecules (or rather macromolecules) and as biologically active agents came to justify a transdisciplinary discourse beyond the formerly stiff boundaries between classical biology and physics, "molecular biology" found its first raison d'être in the problem of protein structure. The ancient controversy between mechanism and vitalism was then believed to be resolved by decoding the structure of proteins. Their structure held the secret to their capacity for a wide range of biological functions, including, at that time, genetic duplication. Hence, the problem of protein structure was then called "the secret of life."[84]

Wrinch's theory required a two-dimensional rather than the linear link between the amino acid monomers in proteins, to the effect that fabrics rather than chains would be formed. Those fabrics were characterized by special symmetries because they were composed of hexagonal arrays. Thus, the fabric fitted the trigonal symmetry associated with insulin crystals, then perhaps the best-known crystallographically due to the pioneering work of Wrinch's friends in the Biotheoretical Gathering, Bernal and especially Crowfoot. Trigonal symmetry turned out to be rare in other protein crystals, however. By assuming that the side chains of the protein were all on one side of the fabric while the other side was smooth, Wrinch's theory could account for many facts of protein chemistry. For example, the immunological properties of proteins could derive from the surface with side chains, while film properties could be explained in terms of the smooth surface of the cyclol fabric. Other key "facts" she presented as compatible with her theory were ultracentrifuge studies by Theodor Svedberg indicating the breakdown of protein into subunits, the molecular weight of which was a submultiple of the native protein from which they split; and organic-analytic studies by Max Bergmann according to which certain amino acids occurred in periodic distributions.[85] Svedberg and Bergmann were world-renowned authorities on the physical and organic chemistry of proteins, respectively, though, as it turned out, the regularities of structure that they had discussed and on which Wrinch relied as crucial corroborating evidence for her model were only artifacts.

In 1937, following a successful tour of American centers of protein studies late in 1936, Wrinch elevated her proposal from the status of a "mere working hypothesis" to that of a "theory," and even secured prestigious publication of a comprehensive version of it through the offices of the president of the Royal Society.[86] Her suddenly released energy and long-repressed desire to reach the peaks of professional success propelled Wrinch into becoming a celebrity, an uncrowned queen of protein studies, and she was invited to expound on her theory at many academic centers.

A full account of the scientific commotion around her theory in the late 1930s is beyond the scope of this paper; I treat it elsewhere as a problem in the historical sociology of scientific knowledge.[87] Some mention, however, must be given here to the controversy that erupted in 1939, because it affected Wrinch's second phase of synergy between her disciplinary and marital trajectories.

Wrinch's model initially stirred great interest among protein workers, since it was the first coherent attempt to account for increasing proof of the molecularity of proteins at a time protein studies were recovering from three decades of the "dark age of biocolloidology."[88] During that time, protein studies were dominated by the colloid theory, propagated by leading physical chemists, who established hegemony over the proteins' "world of neglected dimension" (i.e., the domain between the small molecules of chemistry and cell biology) and, despite a certain level of dissent, perpetuated the view that proteins were amorphous colloids and thus had no definite molecular structure. The colloid viewpoint began to lose its popularity in the late 1920s and the early 1930s. Yet, no alternative model was available

---

Opposite: Wrinch and her model of protein structure, said to be built by her twelve-year-old daughter; photo appeared as part of a 1941 newspaper feature on women in science. Courtesy of the Associated Press, the *New York Post,* and the Rockefeller Archive Center.

WOMEN IN SCIENCE

Woman Einstein—Dr. Wrinch

Solves Biological Problems Mathematically

By SIGRID ARNE
AP Feature Service Writer

BALTIMORE, June 3.—Dr. Dorothy Wrinch was a curly-haired tike of four when her engineer-father took her to a private school in Rosario in the Argentine. He announced to the head-mistress:

"This child is to be a mathematician."

Either he guessed uncannily right, or the tike was unique. Since then she has had a career of brilliant mathematical success—in Paris, Vienna, Oxford and London, where she lectured to advanced mathematical students from many countries. At Cambridge, England, she was the first woman invited to lecture.

Now she is a curly-haired woman of forty with a ready laugh and a perfectly feminine attention to such details as afternoon tea. What's important, she is the fulcrum of an exciting battle in science.

Mathematical Biology

Dr. Wrinch believes she has taken the first steps toward discovering the secret of the structure of protein molecules. Other great scientists agree with her, among them Irving Langmuir of the General Electric laboratories, and Nils Bohr of Denmark. Others "pooh-pooh."

If she has, green-eyed Dorothy Wrinch will be remembered long after today's dictators have been reduced to a chapter in history, because protein is the most important living matter. To understand it fully would be to know what life is. Some proteins are the viruses that give us such man-killers as infantile paralysis, and perhaps cancer. Again, to understand proteins might mean the defeat of those ills.

Usually such solutions come from chemists and biologists who work with the materials in test tubes. But chemists, who have worked with proteins, never have been sure that they had the whole story because proteins change their structures so swiftly.

The Mathematical Molecule

The novel point about Dr. Wrinch's work is this: She studied the available knowledge on proteins. Then she took a pad, a pencil and her own mathematical brilliance and built the facts into a consistent picture.

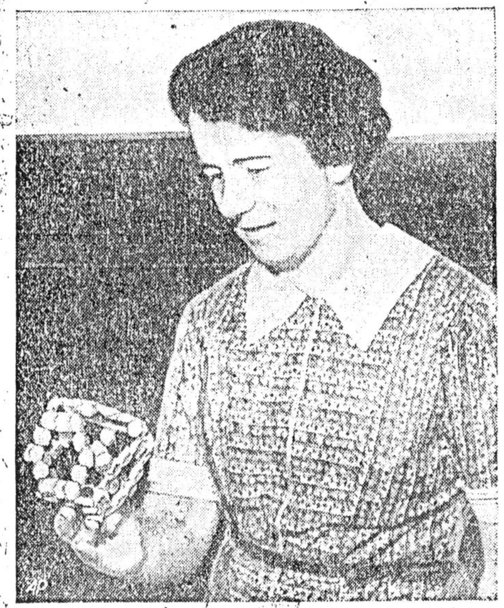

DR. WRINCH with her daughter's protein molecule model.

molecule consists of two tiny four-sided pyramids, set base to base. She says the molecule is hollow and the walls are a rigid net of rings of atoms.

It was the relation of the atom that led her to the discovery that a protein molecule can take only one shape. Back of that deduction lies mathematics too complex to indulge in here.

There is physical corroboration for Dr. Wrinch's theory. It lies in X-ray pictures other scientists have made of protein material. The pictures are muddled, but with Dr. Langmuir, Dr. Wrinch has shown the relation between the pictures and her predictions.

The driving force behind her studies has been a strong conviction that patterns in nature are extremely important, and that mathematics are necessary to studying patterns.

Her curiosity led her far afield in studies. She finished in mathematics at Cambridge and Oxford. Then she explored cytology, organic, physical and colloidal chemistry, physiology and biochemistry. Friends chuckled. She countered with: "She is fitting together what she learned in those separate sciences."

A Scientist's Witticism

She has worked in Baltimore intermittently since 1936. Since 1939 she has been at Johns Hopkins. She lives two miles from her laboratory with her daughter, Pamela, 14, who is so important to her doctor-mother that her pictures are filed in the office "with all the other protein material," as her mother puts it. (That's a scientist's witticism.)

Pamela evidently took protein theories with her alphabet. She recently startled her mother by presenting her with a set of protein molecule models. The child had made them as a surprise. They were correct.

Dr. Wrinch says her work "seldom drops her for more than a few hours." Then she plays the piano, indulges in a set of tennis, or goes for a stiff walk with Pamela through Baltimore parks.

—

This is the second in a series of five articles on the accomplishments of women in science.

before Wrinch's to order the often isolated facts emerging at the borderline of physics, chemistry, and biology in favor of proteins' molecularity as well as to explain proteins' parallel emergence as the most basic level of biological organization.

Thus, a number of special, transdisciplinary meetings were organized in the late 1930s by the British Association (1937), the Royal Society (1938), Cold Spring Harbor Symposia (1938), and the Rockefeller Foundation (1938), among others, to coordinate the rapidly expanding frontier of protein studies and their emergence as the raison d'être for a new sort of biology, molecular biology. At all these meetings, Wrinch's model was featured prominently, epitomizing as it did not only the unique status of proteins as defying classical disciplinary classification, but also changing views of the relationship between theory and facts in science. It even brought coherence to the Rockefeller Foundation's new policy of assisting biological progress by providing a theoretical rationale for its transfer of technologies from the physical to the biological sciences. Hence, the foundation not only gave Wrinch a long-term grant, but also undertook the coordination and funding of her numerous trips. Especially decisive for shifting her epistemological posture was her first trip, late in 1936, when she mistook her American reception, marred by curiosity of her persona as an attractive female theoretician, for scientific confidence in her model.

Indeed, shortly after she returned from America with an inflated sense of self-importance, she abandoned her formerly moderate posture of presenting her model as a "working hypothesis." Instead, she began claiming that her model was not merely in accordance with facts from various fields, but was a proven theory with predictive power. She so argued even while subjecting herself to a telling paradox: according to her own philosophy of science, a great theory should be so abstract that no empirical facts could possibly undermine it or, for reasons of symmetry, prove it. Though her model seemed to be just such an abstract construction, she was trapped, as a result of shifting her frame of reference from mathematicians and logicians to natural scientists, into claiming that her model was proven by empirical facts. Yet she remained unwilling to accept the possibility that the model could also be disproved by them.[89]

Wrinch's claim that her theory was supported by the X-ray data, which were and remained the most decisive source of evidence for any molecular structure including proteins, coupled with her lack of direct experience with experimental science, was perceived as an affront to the protein X-ray crystallographers, since it implied that they could not interpret their own findings. This was particularly annoying to Bernal, who not only pioneered the

X-ray protein crystallography, but was mathematically proficient and also had ambitions to propose his own theory of protein structure.[90] Moreover, as a veteran protein worker, he had been helpful to Wrinch in the early stages of her work, only to find that she used his students' and associates' data to upstage him by trying to monopolize priority for "solving" protein structure.[91] He felt compelled to deflate Wrinch's claims in *Nature*, an endeavor in which he was joined by other X-ray protein crystallographers, all of whom were trained by him directly or indirectly. These exchanges, in which both Wrinch and her opponents claimed more for or against her theory than the data warranted (Sir William H. Bragg, the doyen of X-ray crystallography, admitted that protein X-ray data at that time, being preliminary, were not in a position to corroborate or refute any theory), discredited her among British crystallographers. Moreover, her relationships with her former friends Bernal and Crowfoot became so strained that they would become a factor in her later decision to stay in America.[92]

In 1939 Wrinch obtained the Lady Carlisle Research Fellowship (250 pounds per year), to be held at Somerville College for five years, following warm recommendation letters from her friends D'Arcy Thompson and D. Jordan Lloyd and testimonials from the physicists Niels Bohr, F. Lindemann, and Irving Langmuir.[93] She resigned from Lady Margaret Hall, where her teaching duties only encroached on her time and where, despite the principal's benevolence, she continued to face minor embarrassments, such as a reminder that she had no permission to dine on a regular basis (not being a fellow but only a senior member), a denial of office space during her leave of absence from teaching, and the need to beg for her old office during her visit there in the summer of 1939. Her visit allowed her to escape from the heat of the even nastier controversy she then faced in America.[94]

The outbreak of World War II and the defense minister's decline of her offer to join the war effort like her male colleagues of comparable standing compelled Wrinch to return to America. There she had been, since the spring of 1939, a visiting scholar in the chemistry department at the Johns Hopkins University. She spent the remainder of her Rockefeller Fellowship (which was extended to August 1941, to enable her to find a position in America) engaging in a hopeless controversy with Linus Pauling (1901–).

Like Bernal, Pauling was a top contender for the position of protein theoretician, having published on this topic in 1936. He claimed that Wrinch's rival theory was refuted on thermodynamic grounds, because the cyclol bond, which was pivotal to her theory, was too unstable to exist in nature or in the laboratory. Wrinch, aided by the chemical physicist and Nobel Prize winner Irving Langmuir (1882–1957), labored hard to refute Pauling's claims (which indeed turned out to be wrong when the cyclol bond was discovered both in nature and in the laboratory in the 1950s), but fought her battle both in absolutist or conclusive terms derivative of her formative background in logicism, and on enemy ground where even the refereeing process favored Pauling.[95] Moreover, since her theory ran against a major tenet of accepted wisdom in chemistry (i.e., proteins were made of linearly linked amino acids by the peptide bond), chemists had no interest in supporting it, nor were they receptive to her esoteric and abstract arguments about symmetry and deduction imported from geometry and logic and held by her to be more basic than chemistry's facts. Pauling's determination to strike at his daring but vulnerable rival was matched only by his desire to take credit for solving the problem of protein structure all by himself, something he would accomplish only ten years later while exhibiting his model of the alpha helix to the popular press as no less than the solution to the much debated "secret of life."[96] Meanwhile, however, he was still far away from the cherished goal of solving protein structure and struck with great force at Wrinch, who, by proposing the first coherent theory of protein structure, associated herself with the definition of the problem and captured world attention. This confrontation wore heavily on Wrinch, who in the naivete of an enthusiastic newcomer might not have suspected that she attracted such heavy fire precisely because she had stepped into a domain Pauling considered his own:

> This new Pauling business gets me down. He is a most dangerous fellow. . . . Even decent people hesitate to stand up to LP [Linus Pauling]. He is bright and quick and mercyless in repartee when he likes and I think people just are afraid of him. It takes poor delta [her favored signature nickname] to point out where he is wrong: truly none of them would under any [circumstances]. The big paper on bond strengths and lengths has come back from JCPhys [*Journal of Chemical Physics*] with reports from six referees. They all seize upon my comments which apply to P and want them deleted. They are cowards.[97]

The net result of this controversy was devastating for Wrinch. Pauling's influence was translated not only into termination of her Rockefeller grant (and most grants for the rest of her career), but also hurt her prospects of finding a job. Despite wide scouting, the controversy, coupled with her celebrity status (which meant that only a respectable job could be offered), prejudice against women, and lack of patrons, prevented her from finding any job in 1940, a desperate situation for a single mother in a foreign country on the verge of war.

Her despair at that time was evident from her letters to her intimate friend British mathematician Eric H. Neville (1890–1961), who had remained, since their encounter in the early 1930s, a loyal friend, adoring lover, and staunch supporter in her professional battles. He would not leave his miserable marriage in the 1930s, when she most needed him, possibly because he and his wife had lost their only child. In the 1950s, when both were widowed and he proposed most ardently, she was no longer interested in marriage (it would have been her third) and recalled bitterly her unfulfilled need for his company in the 1930s. However, she willingly accepted his renewed adoration and help on the books she was then keen on producing as her scientific testament.[98]

Replying to one of Neville's dense and numerous letters, Wrinch confessed to a series of serious worries, both professional and personal. After commenting on the difficult situation of her friends Dora and Bertrand Russell, who had been divorced since 1935, Wrinch detailed troubles as a single mother without job security: "Darling Pam is growing so sweetly but she has the awful craving for comfort which is being satisfied partially at the moment by passionate devotion to the Church of all things, but I welcome it as she needs more comfort than I can give her. . . . I am extremely lonely, except for Pam who is adorable in her adolescence."[99] She expressed her severe worries about the future of her work because of the impending war situation, her newcomer status in America, and the conservative stance of many institutions there, especially toward women: "I fear greatly for the future of my work and also for everyone else's work, though females are evidently going to be frozen out first and soon. You talk about future possibilities here but I see none except for those already entrenched. I think the Paulings may weather the whole things—just—also Langmuirs and others in industry and commerce. But the private univ. like Harvard etc are feeling very rocky and state univs. are curtailing everything."[100]

Wrinch also explored job opportunities in Canada though her friend mathematician H.S.M. Coxeter of the University of Toronto, only to find that there was nothing for her. Her depression led her to reflect on her life, work, and friends during the past decade with further bitterness and despair:

> I feel so badly because of this change in the views of my whole lifetime. If you are able to hang on to what you thought in the last war you are lucky so far as your inner Geist is concerned. . . . One misses the comfort of common feelings with others for which I always long. . . . But it is awful to feel that one [I] have taken as my optimum the people who have proved dead wrong. It may be ok

> for the chosen few, but they are right out of touch with the ordinary people and the world as it is. . . . You say maybe one should have combatted snobbery among our friends, but I feel that these Haldanes and Darlingtons and JDBs were and are just unimportant.[101]

She further dwelt on their own relationship, which, despite his ongoing help and love, could not erase her sense of loneliness and social misery in the 1930s:

> I feel mildly jealous that you were able to go down to Cornwall with them, cos [because] when I came back from US and you met me at Southampton you could not even go to Oxford with me, tho[ugh] I critically needed your help. It is odd how these things, like your visit to that horrible seaside place near Crowfoots home on my birthday in 1935 when you left me in such a state of restlessness that it took me days to recover, stick in one's mind. Oh dear, it has been a sad history 1930–1939, hasn't it. . . . Of course, you saved my life in other ways so often, sums and cyclols and I am forever grateful for this.[102]

Wrinch concluded her letter to Neville with some insights into the reasons for her current difficulties with colleagues, difficulties related to her dual legacy of social militancy and intellectual interdisciplinarity, which led her to flaunt disciplinary conventions on scientific authority. Thus, she neither anticipated nor understood the resistance she encountered once she ventured to clash with such traditions or to claim priority for treatment of topics others had begun. This tendency stemmed both from her peripherality, socially as well as intellectually, to the several experimental fields into which she parachuted herself with her theory (and on whose ongoing data she depended for corroboration or refutation) and from an uncontrollable drive to strike, or at least project, success and fame. The prolonged repression of suitable opportunities to express her special capabilities in the preceding decades made that drive particularly acute. She also hit on an additional reason for so much criticism being directed at her personally even though many of the most combative papers were written in collaboration with Nobel Prize winner Irving Langmuir: "Fan [I. Fankuchen (1904–64), Bernal's major collaborator in the late 1930s] is here again. He is nice and tenderhearted and certainly clever with his techniques. He says that what angered everyone most was our attitude that WE had started the proper treatment of S2s. Well I reckon we did. Also he says that most of the anger is against Irving. But they don't dare attack him, so I get it."[103]

Within a month, she was again detailing her difficulties with colleagues in

America (Fankuchen's remarks had alluded to problems with the protein X-ray crystallographers in Britain). These were traceable to her controversy with Pauling, her isolation in America, but also her vulnerability as a woman theoretician pretending to subsume the worlds of (mostly male) experimentalists in a number of fields, while capturing all to herself the ultimate glory of uncovering the "secret of life." Indeed, the identification of Wrinch with her cyclol model was so strong that scientists who were attracted to her as a charming woman would pretend an interest in her models, knowing that was the best way to secure her attention and sympathy. This confusion of the personal and the professional is not uncommon among both men and women professionals. Wrinch herself was particularly adept at cultivating friends as resources for her theory and other professional projects. She complained, however, when this strategy backfired at a time she most craved professional recognition: "I met a very old friend of the Bucklers'. . . . He appeared to be thrilled to death with cyclols. . . . It turned out that he is one of five Trustees of the Institute of Thought [Advanced] Studies in Princeton. . . . I went to a grand dinner with him several times and each time he was more anxious to help me. . . . It is the same old story and it makes me perfectly sick. I believe he has no interest in the cyclols but thinks it is fun to gossip with me."[104]

Also depressing was the news communicated to her by a woman friend who was also unhappy with "official science," that an "ambush" had been prepared for Wrinch at a conference the preceding August by a number of male colleagues who felt "very pricked" when she failed to show up. Her geographical and personal isolation and her responsibility as a single parent made such confrontations all the harder to cope with. She even thought of quitting science altogether:

> I just can't seem to bear it and everything else alone. Sometimes I take poor lovely Pam into my confidence and for days on end treat her as an adult friend but it is a strain on her and I then have to go back to be her mother and protector which leaves me quite alone to stand against the world. Really and truly my life has been a complete washout, without any happiness except Pam and proteins. . . . I am convinced that there is no one that can do what I should be doing and I care still for the proteins so much, but there is nothing else and I can't even make a home for Pam and me alone. . . . But should I make a change? Advise me quick. . . . I have always had ambitions in other fields, executive things, etc., in which I know I should be first class. But the world is so uncertain at the moment . . . it would be a poor solution to get into business and find communism arrived . . . political things, out of the question here of course. Musical things too late. More general educational things I expect too late here. . . . I really would like to

quit. I suppose if I did quit off[icial] science I couldn't carry on at all and I don't know if I could bear that. . . . Well dear what a letter. There seems nothing to be done. Hitler has ruined everything, tho[ugh] most was ruined before.[105]

Her thirteen-year-old daughter, too, sensed the strain of the controversy and wrote a letter to Pauling demanding that he sort out his disagreements with her mother by proofs rather than by frequent personal attacks that only increased the misery in the world.[106]

Professional and personal salvation came late in 1940 when Professor Otto Charles Glaser (1880–1951), a biologist at Amherst College, Massachusetts, who had been corresponding with Wrinch since 1938, most ingeniously devised a scheme to engineer an appointment for her as visiting research professor at three colleges in the Connecticut Valley. Those included his own, a men's college, where he was prepared to offer joint seminars in biology and chemistry with her, and the nearby women's colleges, Smith and Mount Holyoke. As chairman of the biology department and the college's vice-president, Glaser was in a superb position to contact the presidents of the other two women's colleges, as well as to affect decisions at his own college.

Glaser devised this scheme at a time when his and Wrinch's broadsides at yet another trustee of the Institute for Advanced Study in Princeton, as well as Wrinch's own contacts there with Albert Einstein and Oswald Veblen, failed to produce a position for her.[107] Asking whether she could picture herself in residence in his region, he suggested that she conduct seminars for advanced students at times and under conditions least detrimental to her work. It was most helpful to have a well-placed insider explore the feasibility of the scheme, but giving her the freedom to reject any offers that might come to her as a result of the projected cooperation among the three colleges. An additional offer for an honorarium and a lecture before Amherst's Science Club were also included in the letter, as well as favorable comments on her daughter's cookies. Glaser's profound interest in Wrinch's professional future reflected the perception that the road to her heart led through her work and professional position. Indeed, following the delineation of the job scheme, he invited her to a professional dinner and affectionately concluded: "If I am altogether too officious you may scold me. If you do, I shall nevertheless continue to be concerned until you and your work have been granted the right to live."[108]

Wrinch's reply acknowledged that Glaser's "most kind and most helpful letter" gave her pleasure, and she agreed to his suggestions of seminars, living accommodations for herself in a college residence, and choice of school for her daughter. She was also happy to hear that she would have an office in the chemistry department at Amherst, though she was careful to suggest that she attach herself to different departments in different colleges so she would not step on the toes of people who, as veteran protein workers, might not welcome her controversial claims:

> Yes, I think we can press ahead at Amherst and go slower, if necessary at S [Smith] and MH [Mount Holyoke]. . . . I am notoriously poor at intuitions about people, but I have a feeling that I should go carefully in regard to chemistry at MH. EC [Emma Carr] is a very remarkable woman; does as good a job in her department than has many more eminent people. I greatly wish NOT to tread on her toes. I thought perhaps to stress the chemistry department at Smith and the physiology dept. at MH, partly with this in mind. . . . I thought perhaps you would keep her as happy as possible about the scheme.[109]

Wrinch concluded with the hope that their scheme was an opportunity "to anchor the Wrinches and the molecules for some time, even perhaps permanently," thus anticipating both her own marriage to Glaser three months later and the legal change of her daughter's name in 1947 from Pamela Wrinch Nicholson to Pamela Nicholson Wrinch. She described her sudden turn of fate from prolonged misery and depression in one of the last letters exchanged with Neville, once her second, happy marriage made their former intimacy superfluous: "I appreciate so very much your good wishes. I think you are one of the few people who know how desperately miserable and lonely I have been all my life. It is a miracle-no-less to find even so late in life, a partner and companion whom I love. I feel that as a true friend you should know that I fell deeply and truly though so suddenly in love."[110]

Wrinch went on to say that though she and Glaser had corresponded intensely since the winter 1940–41, she had no idea until July 27 that he wished to marry her. He first talked about his intentions to Wrinch's close friend geneticist Katherine Brehme Warren, who briefed Wrinch before Glaser told her of his feelings later that day. Then, she said, for the first time he spoke on things other than "molecules and geometry." She added that she gave a great deal of thought to the idea, which took her by surprise, but added that "the wonderful thing happened" and within a few days, she "was quite clear about it, too." Once they decided, they consulted with the three college presidents whether the wedding should be postponed for the duration of her projected appointment or not (all three welcomed the event) and went to her daughter's camp in Maine to get her blessing. Pamela, who had been cautioning her mother about remarrying, pointing out that she had already made one awful marriage, realized the decision had been made when she saw her mother so happy and completely transformed. The marriage took place in the laboratory at Woods Hole on August 20 with no guests but their scientist friends Kitty Brehme Warren as bridesmaid and John Fulton as best man.[111] The wedding was, however, heavily photographed and reported in the local newspaper as a social-scientific event, thus suggesting that Wrinch's flair for publicity, which had gotten her in trouble during the protein controversy, was undiminished. (In the late 1930s the popular press had carried pictures of her and her models with captions such as "Woman Einstein—Dr. Wrinch solves Biological Problems Mathematically.")[112]

The marriage to Glaser proved most rewarding both personally and professionally. A series of letters, written by him during the summer of 1942, when he had to teach in Amherst and could come to their Woods Hole summer residence only for weekends, attests to his romantic love; he scattered phrases such as "I love you so much more" throughout his letters and usually concluded with the chemical mathematical symbols of their love: two side-linked hexagons in perspective, evoking her model of a fabric of hexagons, or with "love at the power of infinity." The letters also showed his devoted concern for her scientific preoccupations. Thus he would report to her on recent events in protein studies gleaned from both friends and colleagues, express anger at those who opposed or ignored her work, and continuously deploy his standing and connections as resources (ranging from information to technical services he was entitled to as chairman) for her work.[113] As clerk of the Marine Biological Corporation in Woods Hole, Glaser was a central member of the New England biological community, which regarded Woods Hole as its scientific and social summer resort. Through her marriage, Wrinch was thus propelled from her former misery and loneliness as a foreigner in Baltimore on the fringes of the American scientific community and a vulnerable target for the "ambushers" at the Gibson Island conference, to a solid position where she could enjoy her cherished dual goals of marital happiness and professional independence.

Yet, in view of her previous record of accumulated misery, she seems to have contemplated this step very carefully, weighing the advantages and disadvantages of the projected marriage almost in table form. Drawing up her life balance sheet, she noted the "positive net losses" that might result from remarrying: "dislike to companionship day and night; restraint of my actions; western hemisphere [later transferred to the column of positive net gains]; [relocation] + no job? I can't be a LU [lazy university?] wife; Somerville and Oxford in general certainly lost; D.C. [Dorothy Crowfoot?] LMH my God."

In the positive net gains column she noted: "sheer loneliness avoided; help in home [?] in bed financial gain in BLG int.; western hemisphere; moderation and self-control; after the lecture one can go home; holidays and recuperation in non-work periods; home and shelter for Pam." Adding that she felt she had suffered so many hurts that a protective love around her would be worth a great deal, she managed to produce an additional list of advantages or, rather, avoidances of hurts that had prevailed during her life as a single person: "1.2.3 [unclear or illegible]; 4. Night ending at midnight and then I go home; 5. Hanging alone after lectures; 6. Going to parties alone; 7. Giving parties alone; 8. Trying to have a home atmosphere for Pam & yet being ever so nervous about making the money or the home; 9. [unclear]."[114]

Her initial appointment of 1941–42 was renewed throughout the 1940s, when she enjoyed the bliss of happily married life, teaching, and research. She produced a monograph in 1946 for the American Society for X-ray and Electron Diffraction, *Fourier Transforms and Structure Factors,* which, unlike her theory of protein structure, has always been considered a solid contribution to knowledge. Her main research effort, however, remained an obsession with resurrecting her theory from the blows of 1939; it almost happened in the 1950s when the cyclol bond, pronounced unstable by Pauling and others in 1939, was discovered both in nature and in the laboratory. By then, however, the frontier of molecular biology, which she had cultivated from a transdisciplinary and structural-geometrical viewpoint in her seminars in the 1940s, shifted from proteins to nucleic acids and the "secret of life" came to be associated with the double helix.

Both Pauling and Bernal missed the discovery of the structure of DNA despite their veteran status as molecular structurists of biological compounds, in part because of their long-standing attachment to the problem of protein structure. Unlike Pauling and Bernal, however, who admitted their failure to anticipate the importance of DNA structure and even provided interesting rationalizations for missing it,[115] Wrinch remained adamant in her opposition to both alpha helix and double helix, entrenched as she was in her conviction that her hexagon-based model was the mathematical master key to the secrets of the biological universe:

> Well anyway, you see how one realizes that it is a work of superogation to go beyond a certain point in co-opting partners for the initially lonely little point on a cube face . . . this closure principle lies at the HEART [of] all the lovely things in

> proteins (according to me). . . . I can't be wrong about this otherwise no flower would be recognizable and one would never know an ash from an oak in 3 seconds at 100 yards etc. . . . do you see what I am talking about here, cos [because] this is the HEART not only just of proteins but of me and my ideas and my passions for the last few decades . . . oh dear, Do you see what it is about??[116]

Following Glaser's death of nephritis in 1950 , Wrinch moved to a residence on the Smith campus and continued her battle against the prevailing scientific consensus from her office in the physics department, where she held the position of visiting research professor until her retirement in 1971. Her daughter, Pamela, meanwhile had earned a Ph.D. from Yale in 1954 in international relations, one of the first women to do so; had married a Cambridge, Massachusetts, publisher; and was a lecturer in political science at Milton College. Wrinch died in 1976 shortly after Pamela had perished in an accidental fire.[117]

IN VIEW OF the widely perceived contradiction between career and marital life at the turn of and well into this century by both college-educated women and social authorities at large (as evidenced by both custom and legislation), the synergy between Wrinch's disciplinary and marital trajectories as documented here may come as a surprise. Not only was she able to produce an unusually large, versatile, and wholly independent body of research throughout the duration of both marriages, but she did so in two different disciplines and countries. Moreover, each time she was able to signal the compatibility and partnerlike relationship of the marriage with her research topic by occasional coauthorship of papers with each husband.

In her marriage to John Nicholson in the 1920s and to Otto Glaser in the 1940s, Wrinch's spouses shared her interest in applied mathematics and molecular biology, respectively. Yet this sharing allowed Wrinch to pursue her own unique mix of research topics without being assimilated into either husband's research work. For example, in her first marriage, Wrinch benefited from Nicholson's greater experience in applied mathematics, but this topic accounted for only one-third of her research output, the rest being work done independently in pure mathematics and the philosophy of science. Similarly, Glaser was a crucial biological (and social) resource for her work in molecular biology, yet most of that work revolved around a mathematical model of her own devising, one outside the research expertise of the classical biologist he was. Her capacity to maintain an independent credit line in research was not compromised, but rather enhanced, by each marriage.

Perhaps of greater importance in accounting for the synergy between Wrinch's career and marriages were her husbands' active and ingenious initiatives in securing her teaching positions, thus acting as the mentors she had never had. Both used their established positions as full professors of men's colleges, with additional administrative power as director of studies of Balliol and vice-president at Amherst, respectively, to exert influence on nearby women's colleges (Lady Margaret Hall and Smith College, respectively) so they would help Wrinch gain a professional foothold. In both cases, the husbands' insider position vis-à-vis the collegiate system at Oxford and the Connecticut Valley enabled them to "parachute" their outsider wife into teaching positions she may have never obtained or sought on her own, first temporarily and later on a more permanent basis.

The husbands' willingness to promote the career of their wife, though stemming primarily from their interest in her as a charming woman, not only entailed a rejection of the patriarchal ideal, but required actual intervention to enable an independent career for her at locations determined by their own prior careers. Their attitude is understandable to some extent in view of their age—both were thirteen years her senior. This not only meant that they were already established and need not feel threatened by her ambitiousness (especially since job segregation by sex, and in any case their positions at men's colleges, restricted their efforts on her behalf at the women's colleges), but also that they were susceptible to the social and sexual advantage of marrying a younger and attractive woman.

At the same time, their favorable response to her career was influenced by her insistence that she would not relocate to her husband's institutional setting unless a suitable teaching position was available for her, or as she put it, that she could not be a "lazy university wife." Nevertheless, the type of part-time positions for which she twice settled because of their compatibility with her marriages, though respectable enough under those circumstances, were not the real, full-time, or long-term secure positions an unattached woman of comparable standing would have sought.

Indeed, the limitations of her positions, obtained primarily for reasons of compatibility with marriage rather than with career per se, became obvious throughout the duration of each marriage, but especially once the marriage ended. As long as a marriage lasted, those limitations chiefly involved occasional embarrassments to which she was subjected as a result of not having a full faculty position (which were then held mostly by single women). Once a marriage was over, however, Wrinch found herself stranded in institutional contexts that were not optimal for her career. This situation was exacerbated by the fact that the relative compatibility between career and marriage had provided little incentive or opportunity for her to actively seek better positions during marriage.

As it happened, once each marriage ended, in 1930 and 1950, the job opportunities that her sudden new geographical flexibility made possible were bleak, both objectively and subjectively. In 1930 the economic depression made jobs difficult to find. This objective situation was reinforced by the sexual discrimination she encountered, by her lack of mentors in the several fields in which she pioneered, and by her socially anomalous status in the rigidly stratified and inbred Oxford milieu where a woman, for most of the interwar period, could be either a resident don or a faculty wife. Furthermore, her own sense of outstanding intellectual capacity and modest social background compelled her to stick to indicators of upward mobility such as her tenuous Oxford affiliation, rather than seek an actual job outside Oxbridge. The situation was similar in 1950 when Glaser died, but further compounded by her age; she was fifty-seven, not the best age at which to seek a new position.

One wonders how Wrinch's two career-compatible marriages affected the major scientific event of her life: her proposal of the first theory of protein structure, her subsequent conduct during the controversies it precipitated, and her obsessive defense of it for the rest of her life. Both marriages served as important social and scientific resources for her career in terms of generating teaching positions, technical information for her research, and upward social mobility in the academic community. In both cases, however, the marriage induced patrilocality, or shift of residence to the husband's location, a fact that had more than simple geographical significance; for she came to depend on institutional contexts chosen for compatibility with marriage rather than with long-range career goals. Initially, her decision to make this compatibility a top priority was a good one, given the lack of choice of independent positions that derived from both objective and subjective factors (most notably, discrimination against women in academic employment and her own desire to remain an Oxbridge scholar rather than seek jobs outside this sphere). In addition, her own social values, such as a desire for the experience of motherhood and a perception of marriage as an opportunity for upward social mobility, further compelled her to consider marriage, especially when it proved compatible with an independent career.

At the same time, the residential and institutional inertia induced by each marriage blurred the fact that she remained debarred from obtaining real positions, access to advanced students, decision-making power, and an insider's knowledge of institutions and departments. In this sense, the marriages reinforced her outsider status, a status already affected by her arrivist stance in the social, geographical, and institutional loci of her husbands, as well as by her multidisciplinary research record. Her research record meant that she was marginal to all the scientific communities she addressed. The Russellian legacy that stayed with her long after he himself left academia and thus could no longer serve as a mentor made her split her effort between mathematics and philosophy, with the result that she would not be a suitable protégée for authorities in either field.

This outsider status—reinforced though not wholly caused by her marriages and the false sense of security she got from her husbands' established positions—may explain not only why she stayed on at husband-induced locations after the termination of each marriage, but also how she behaved throughout the controversy over her theory of protein structure.

First, there was an initial displacement between the ideas underlying her theory and the social context into which the theory landed, a displacement that accounted for much of the later resistance she encountered and that can be traced to her superficial outsider's understanding of the various disciplinary and social traditions from which she sought approval. Thus, her theory revolved around certain geometrical notions of symmetry and logical notions of consistency both so abstract that, as she prophetically put it in her essay on the relationship between science and philosophy (1927), no empirical facts could possibly undermine such constructions. Yet she did not offer her theory to mathematicians and logicians accustomed to such standards, but to natural scientists who were primarily concerned with the technicalities of producing empirical facts while possessing an instrumental view of theory as guidance to practice. Though initially she proposed the theory as a working hypothesis capable of ordering a large set of facts, formerly isolated or meaningful only in terms of distinct disciplinary traditions, she soon elevated it to the status of an absolute theory that could only be increasingly corroborated but never refuted.

This position made sense in view of Wrinch's technical resources in mathematics and philosophy and her formative experience with the logical reductionism of mathematics, but was alien to the communities of protein workers who, she expected, would accept her theory as no less than the mathematical master key with which to unlock the secrets of the molecular-biological universe. In addition, not being an experimentalist, Wrinch had no direct access or control over the various empirical "facts" she regarded as supporting or even proving her theory. She was, thus, stuck with artifacts once the nonabstract facts of natural science, unlike the eternal facts of logic, occasionally dissolved.

In addition, she seemed to have constantly overestimated the role of theory in science, especially in chemistry and biology, as a result of her experiences with classical mathematical physics and with the theory of relativity, both of which she saw from the viewpoint of the mathematical logician rather than that of the physicist. This confusion may have been reinforced by her participation in the meetings of the Biotheoretical Gathering in the 1930s, a group that included atypical, philosophically and theoretically minded scientists who accepted her projected contribution to utopian projects (such as topological descriptions of embryological facts), as well as her own persona, on egalitarian and favorable terms. As she had done in Russell's circles in the 1910s, she came to confuse her social and intellectual acceptance based on her potential brilliance and shared avant-garde social ideals with a true scientific acceptance, which could only come from solid scientific record and an insider's mastery of scientific traditions, attitudes toward innovation, the subtle forms of legitimation of new ideas, and the authority of veteran practitioners in science.

This background of marginality to various scholarly traditions and their social worlds, derivative of both her social trajectory in the 1910s and 1930s when she was single, and in the 1920s and 1940s when she was married, explains her proposal of a transdisciplinary theory of protein structure and also her subsequent obsessive defense of it. Caught in institutional contexts where she was deprived of both the technical and human resources possessed by her rivals—all men with standard professional careers—she drew on her militant, utopia-inspired past and resorted to risky strategies of confrontation. These further alienated her enemies as well as her potential supporters as she became more and more concerned with the "siege" of her theory than with the shifting frontier of molecular-biological studies. This fixation with a theory that held meaning for her, not only in terms of her particular, unusual scientific trajectory, but also in terms of her social convictions, may explain the paradox of her lifelong refusal to abandon her theory and follow the changes in scientific consensus. Though her rivals had to drop their own theories once the "secret of life" turned out to lie beyond the boundaries of their common pursuit of protein structure in the 1930s, for Dorothy Wrinch the theory was no mere transient and discardable contribution, but the embodiment of all her ideas and passions, a mark of her unique trajectory as a woman theoretician in twentieth-century molecular life science.

Notes and References

Foreword

1. Anne Firor Scott, "Epilogue," in *Making the Invisible Woman Visible* (Urbana, Ill., 1984), 371.

Introduction

1. See, e.g., M. W. Rossiter, *Women Scientists in America: Struggles and Strategies to 1940* (Baltimore, 1982); D. Richter, ed., *Women Scientists: The Road to Liberation* (London, 1982); E. Fox Keller, *A Feeling for the Organism: The Life and Work of Barbara McClintock* (San Francisco, 1983); S. Gregory Kohlstedt, "In from the Periphery: American Women in Science, 1830–1880," *Signs* 4 (1978):81–96; M. L. Aldrich, "Women in Science," *Signs* 4 (1978):126–135; "Les femmes et la science," *Penelope pour l'histoire des femmes* 4 (1981); "Des femmes dans les sciences et des sciences sur les femmes," *Cahiers de recherche sociologique* 4 (1986). L. Schiebinger, "The History and Philosophy of Women in Science: A Review Essay," *Signs* 12 (1987): 305–332. On women's education and careers in other fields, see B. Solomon, *In the Company of Educated Women* (New Haven, Conn., 1985); H. Lefkowitz Horowitz, *Alma Mater: Design and Experience in the Women's Colleges from Their Nineteenth-Century Beginnings to the 1930s* (New York, 1984); A. Roland, ed., *Career and Motherhood: Struggles for a New Identity* (New York, 1979); H. Callan and S. Ardener,

eds., *The Incorporated Wife* (London, 1984); P. M. Glazer and M. Slater, *Unequal Colleagues: The Entrance of Women into the Professions, 1890–1940* (New Brunswick, N.J., 1986); K. Hausen and H. Nowotny, eds., *Wie Mannlich ist die Wissenschaft?* [How Masculine Is Science?] (Vienna, 1986). The last two books also include some material on women in science.

2. For a detailed argument, see D. Outram, "Politics and Vocation: French Science, 1793–1830," *British Journal for the History of Science* 13 (1980):27–43.

3. For many concrete examples, see D. Outram, *Georges Cuvier: Vocation, Science and Authority in Post-Revolutionary France* (Dover, N.H., 1984), chap. 9. See also S. Reynolds, *Women, State and Revolution: Essays on Power and Gender in Europe since 1789* (London, 1986).

4. On the British tradition, see J. Morrell and A. Thackray, *Gentlemen of Science: Early Years of the British Association for the Advancement of Science* (Oxford, 1981). On the American tradition, see Kohlstedt, "In from the Periphery"; idem, *The Formation of the American Scientific Community: The American Association for the Advancement of Science, 1848–1860* (Urbana, Ill., 1976).

5. See, e.g., M. Alic, *Hypatia's Heritage: Women and Science from Antiquity to the Present* (London, 1985); M. Bailey Ogilvie, *Women in Science, Antiquity through the Nineteenth Century: A Biographical Dictionary with Annotated Bibliography* (Cambridge, Mass., 1986).

6. See, e.g., R. Hubbard et al., eds., *Women Look at Biology Looking at Women* (Cambridge, Mass., 1979); H. and S. Rose, eds., *Ideology of / in the Natural Sciences* (Boston, 1980); J. Rotschild, ed., *Machina ex Dea: Feminist Perspectives in Technology* (New York, 1983); E. Fox Keller, *Reflections on Gender and Science* (New Haven, Conn., 1985); A. Fausto-Sterling, *Myths of Gender in Biological Theories about Women and Men* (New York, 1985); S. Harding, *The Science Question in Feminism* (Ithaca, N.Y., 1986); R. Bleier, ed., *Feminist Approaches to Science* (New York, 1986).

7. N. Elias, *The History of Manners* (New York, 1978), originally published in German as *Uber den Prozess der Zivilisation* (Basel, 1939); R. Sennett, *The Fall of Public Man* (London, 1978); J. B. Elshtain, *Public Men, Private Women: Women in Social and Political Thought* (Princeton, N.J., 1981).

8. P. G. Abir-Am, "The Rockefeller Foundation's Use of Advisory Systems in British Science in the 1930s: Policy or Patronage?" Presented at the Conference on Patronage, British Society for the History of Science, July 14–16, 1986, Oxford, forthcoming in *Minerva*; idem, "The Discourse on Physical Power and Biological Knowledge in the 1930s: A Reappraisal of the Rockefeller Foundation's Policy in Molecular Biology," *Social Studies of Science* 12 (1982):341–382.

9. E. Shorter, *The Making of the Modern Family* (London, 1976).

10. For preliminary information on the prevalence of collaborative marriages in various disciplines in the United States prior to 1940, see Rossiter, *Women Scientists*, Table 10.4 (pp. 293–294). About a dozen husband-and-wife scientist couples active in socialist politics in Britain in the interwar period, mostly in the fields of biochemistry and physics, are briefly discussed in G. Werskey, *The Visible College* (London, 1978), 211; not all these couples engaged in collaborative work, though almost

always they practiced the same discipline. For information on collaborative marriages in British astronomy, see P. A. Kidwell, "Women Astronomers in Britain, 1780–1930," *ISIS* 74 (1984):534–546. On the impact of gender and marriage on the scientific productivity of both men and women, see B. Reskin, "Sex Differentiation and Social Organization of Science" in J. Gaston, ed., *Sociology of Sciences* (San Francisco, 1978): 6–37; N. Heckman et al., "Problems of Professional Couples: A Content Analysis," *Journal of Marriage and the Family* 39 (1977):323–330; J. R. Cole and H. Zuckerman, "Marriage, Motherhood and Research Performance in Science," *Scientific American* 257 (Feb. 1987):119–125.

11. On the relationship between scientific research and universities, see J. Ben-David, "The Universities and the Growth of Science in Germany and the United States," *Minerva* 7 (1968–69):1–35; idem, *The Scientist's Role in Society: A Comparative Study* (Englewood Cliffs, N.J., 1971).

12. See, e.g., B. Latour and S. Woolgar, *Laboratory Life: The Social Construction of Scientific Facts* (London, 1979); K. Knorr et al., eds., *The Social Process of Scientific Investigation* (Boston, 1980); K. D. Knorr-Cetina and M. Mulkay, eds., *Science Observed* (London, 1983); B. Latour, *Science in Action* (Cambridge, Mass., 1987).

13. For the impact of gender on recognition, see J. R. Cole and H. Zuckerman, "The Productivity Puzzle: Persistence and Change in Patterns of Publication of Men and Women Scientists," in *Advances in Motivation and Achievement* (New York, 1984), 2:217–258.

14. On the relationships between action and structures in social theory, see A. Giddens, *Central Problems in Social Theory* (Berkeley, Calif., 1982); idem, *The Constitution of Society: Outline of the Theory of Structuration* (Cambridge, 1984); R. Boudon, *The Logic of Social Action* (London, 1981); R. J. Bernstein, *The Restructuring of Social and Political Theory* (London, 1985).

Chapter 1: Before Objectivity

1. A.E.M. Grétry, *La vérité, ou ce que nous fûmes, ce que nous sommes, ce que nous devrions être,* 3 vols. (Paris, 1807), 1:218; quoted in J. Simon, *Une académie sous la directoire* (Paris, 1885), 89.

2. See D. Outram, *Georges Cuvier: Vocation, Science and Authority in Post-Revolutionary France* (Dover, N.H., 1984), 169–202; idem, "Politics and Vocation: French Science, 1793–1830," *British Journal for the History of Science* 13 (1980):27–43.

3. For new work along these lines for a later period, see Françoise Mayeur, "Woman and Elites from the Nineteenth to the Twentieth Century," in *Elites in France: Origins, Reproduction and Power,* eds. J. Howorth and P. G. Cerny (London, 1981), 57–65. The importance of family groupings to cultural innovation and reproduction is also discussed in A.F.C. Wallace, *The Social Context of Innovation: Bureaucrats, Families and Heroes in the Early Industrial Revolution* (Princeton, N.J., 1982).

4. See Robert Mauzi, *L'idée de bonheur, dans la littérature et la pensée françaises au dix-huitième siècle* (Paris, 1960), 44–91, 147, 268–275, 355–485.

5. For technical philosophy supporting this view, see the theories of perception described in C. Van Duzer, *Contributions of the Idéologues to French Revolutionary Thought* (Baltimore, Md., 1935). The philosophy of Immanuel Kant only became influential in France after around 1800; see M. Vallois, *La formation de l'influence kantienne en France* (Paris, 1927); François Picavet, *La philosophie de Kant en France de 1773 à 1814* (Paris, 1888); K. Figlio, "Theories of Perception and the Physiology of Mind in the Late Eighteenth Century," *History of Science* 13 (1975):177–212. D. Outram, "The Language of Natural Power: The Funeral *Éloges* of Georges Cuvier," *History of Science* 16 (1978):153–178.

6. Ramond wrote in 1777: "Je regarde le marriage comme le lien le plus naturel, celui où la société a moins ajouté, celui dans lequel elle a le moins perverti le voeu de la nature, conséquemment celui dont le bonheur est le plus proche"; quoted in Cuthbert M. Girdlestone, *Poésie, politique, Pyrénées: L. Ramond 1755–1827: Sa vie, son oeuvre littéraire et politique* (Paris, 1968), 75.

7. D'Alembert wrote to Lagrange in 1767: "J'apprends que vous avez fait ce qu'entre nous, philosophes, on appelle 'le saut périlleux'"; quoted in Antoine Guillois, *La Marquise de Condorcet: sa famille, son salon, ses amis, 1764–1822* (Paris, 1897), 65.

8. Guillois, *Marquise,* 40–67.

9. Modern sociologists, unlike historians of science, take it for granted that such are the functions of family relations; e.g. P. Bourdieu, "Champ de pouvoir, champ intéllectuel et habitus de classe," *Scolie* (i) (1971):7–26; on p. 10: "Il faut se rappeler que parmi l'ensemble des privilèges qui sont l'instrument et le produit du pouvoir, il n'en est sans doute pas de plus important que le capital des relations; par l'intermédiaire du réseau de relations familiales et amicales s'opère un nombre important de transactions objectivement politiques et objectivement économiques."

10. Georges Cuvier, *Éloges historiques,* 3 vols. (Paris, 1819), 1:168: Éloge of Darcet. Such ruptures with the biological family as the price of personal authenticity had long been part of the formation of religious vocation; see Laurent Theis, "Saints sans famille? Quelques remarques sur la famille dans le monde franc à travers les sources hagiographiques," *Revue historique* 255 (1976):3–20; on p. 6: "Le saint n'est lui-meme qu'après que sa famille a été annihilié éventuellement au profit d'un groupe familial métaphorique." See also S. M. Silverman, "Parental Loss and the Scientist," *Science Studies* (1974):259–264; W. R. Woodward, "Scientific Genius and the Loss of a Parent," *Science Studies* (1974):265–279.

11. Cf. Laurent Theis, "Saints sans famille," 6.

12. For further historical parallels, which point to the deep historical roots of scientific vocational ideals, see A. D. Nock, "Conversion and Adolescence," in *Essays on Religion and the Ancient World,* 2 vols. (London, 1972), 1:469–480.

13. Isidore Geoffroy St. Hilaire, *Vie, travaux, et doctrine scientifique d'Etienne Geoffroy St. Hilaire* (Paris, 1847), 39; other examples include Lacepède's remarks on Buffon: "Il me traîta comme son fils"; Roger Hahn, "L'autobiographie de Lacepède retrouvée," *Dix-huitième siècle* 7 (1975):49–85, esp. 58; Dupont de Nemours of

Quesnay: "Je n'étais qu'un enfant lorsqu'il me tendait les bras. C'est lui qui m'a fait un homme," in H. A. Dupont de Nemours, ed., *L'enfance et la jeunesse de Dupont de Nemours* (Paris, 1906), 42.

14. Antoine Guillois, *Le salon de Mme. Helvétius: Cabanis et les idéologues* (Paris, 1894).

15. Charles Dupin, "Éloge de Mme. de Prony," *Mercure du dix-neuvième siècle* 2 (1822):203–215.

16. James F. Traer, *Marriage and the Family in Eighteenth-Century France* (Ithaca, N.Y., 1980); François Olivier Martin, *La crise du marriage dans la legislation intermédiaire, 1789–1804* (Paris, 1901); Roderick Philips, "Le divorce en France à la fin du dix-huitième siècle," *Annales* 24 (1979):252–287; J. Donzelot, *The Policing of Families* (New York, 1979).

17. E.g., Grétry, *Vérité*; J. A. Ségur, *Les femmes: leur condition, et leur influence dans l'ordre social,* 3 vols. (Paris, 1802), 3, 5–20; E. Legouvé, "La mérite des femmes," in *Oeuvres complètes,* 2 vols. (Paris, 1826), 2, 26–30.

18. Bonnie G. Smith, *Ladies of the Leisure Class: The Bourgeoises of Northern France in the Nineteenth Century* (Princeton, N.J., 1981), 4–51; Margaret H. Darrow, "French Noblewomen and the New Domesticity, 1750–1850," *Feminist Studies* 3 (1979):16–35.

19. J. B. Suard, "Éloge de Mme. d'Houdetot," *Journal des Débats* (February 6, 1813), 249. All translations from the French are my own, unless otherwise noted.

20. C. A. Lopez, *Mon Cher Papa: Benjamin Franklin and the Ladies of Paris* (New Haven, 1966), 243.

21. A. C. Kors, *D'Holbach's Coterie: An Enlightenment in Paris* (Princeton, N.J., 1976); Carolyn Lougée, *Le Paradis des Femmes: Women, Salons and Social Stratification in Seventeenth-Century France* (Princeton, N.J., 1976); Deborah Hertz, "Salonnières and Literary Women in Late Eighteenth-Century Berlin," *New German Critique* 1 (1978):97–108.

22. See the suggestions contained in D. Outram, "Politics and Vocation."

23. Dupin, "Éloge de Mme. de Prony," 213–214.

24. Henri Beyle, *Vie d'Henry Brulard,* ed. B. Didier (Paris, 1973), 380.

25. "Scientific elite" is a phrase that can only be used here anachronistically, since the period was distinguished by the social and intellectual interpenetratation of different areas of intellectual inquiry, few autonomous disciplines were yet in existence, and many individuals practiced both "humanistic" and "scientific" disciplines with success. In statistical studies, I have established that the demographic behavior of the "scientific elite," which I define here with necessary crudeness as the membership and the members of the "First Class" of the Institut de France, does not vary significantly from that of the 272 individuals composing the Institut as a whole between 1795 and 1814.

26. Louis Bergeron and Guy Chaussinand-Nogaret, *Les masses de granit: cent mille notables du premier Empire* (Paris, 1979). Demographic information for the Institut is based on: P. Franqueville, *Le premier siècle de l'Institut de France,* 2 vols. (Paris, 1895);

P. Gaujon, ed., *L'Académie des Sciences de l'Institut National de France* (Paris, 1934); *Institut de France: index biographique de l'Académie des Sciences du 22 décembre 1666 au 1 octobre 1978* (Paris, 1979); F. Michaud, ed., *Biographie universelle ancienne et moderne, nouvelle édition,* 45 vols. (Paris, 1843–48); C. C. Gillispie, ed., *Dictionary of Scientific Biography,* 14 vols. (New York, 1970–78); J. Balteau, M. Barroux, and M. Prevost, eds., *Dictionnaire de biographie française* (Paris, 1932); M. de Courcelles, *Dictionnaire universelle de la noblesse de France,* 5 vols. (Paris, 1822); Léonce de Brotonne, *Tableau historique des pairs de France, 1789–1848* (Paris, 1889); François Michel, ed., *Fichier stendhalien,* 2 vols. (Boston, Mass., 1964); Alfred Potiquet, *L'Institut national de France, . . . 20 novembre 1795–19 novembre 1869* (Paris, 1871); Romauald Szramkiewicz, *Les régents et les censeurs de la Banque de France, nommés sous le consulat et l'Empire* (Geneva, 1974); Auguste Jal, *Dictionnaire critique de biographie et d'histoire* (Paris, 1867).

27. Bergeron and Chaussinand-Nogaret, *Masses de granit,* 14–17.

28. For example, Lamarck (*Dictionary of Scientific Biography*); Cuvier (D. Outram, *Georges Cuvier,* 65, 217); Bosc (Claude Perroud, "Le roman d'un Girondin: le naturaliste Bosc," *Revue du six-huitième siècle* 2 [1916]:232–257, 348–367). See also R. Barroux ("Sebastien Mercier, le promeneur qui ne sait où il va," *Mercure de France* 64 [1960]:655–665); Jacques Payen (*Capital et machine à vapeur au dix-huitième siècle: les frères Perier et l'introduction en France de la machine à vapeur de Watt* [Paris, 1969], 56); S. Gillmor (*Coulomb* [Princeton, N.J., 1971], 75).

29. That is, those of Malus, Charles, Tessier, Daubenton, Delambre, Guyton de Morveau, Lacépède, Lagrange, Latreille, Lefèbvre-Ginau, Legendre, L. G. Lemonnier, Prony, Berthoud, Percy, Olivier (first marriage), Palisot, and Pinel (first marriage) produced no children. Marriages producing less than three children include those of Berthollet, Biot, Broussonnet, Cels, Desmarets, Deyeux, Girard, De Jussieu, Duhamel, Richard, Ramond, Silvestre, Deschamps, Mirbel, Messier, Buache, Coulomb, Defontaines, Fourcroy, Laplace, Geoffroy St. Hilaire, Carnot, Hallé, Monge, Bouvard, Claret de Fleurieu, Thenard, Ampère, Levêque, Adanson, Ventenat, and Pinel (second marriage).

30. For example, Bonnie G. Smith, *Ladies of the Leisure Class,* 5–6.

31. Jean Tudesq, *Les grands notables en France (1840–1849): étude historique d'une psychologie sociale,* 2 vols. (Paris, 1964), 1:458–459, 462–463.

32. For a specific example (the Brongniart family of naturalists), see D. Outram, *Georges Cuvier,* 174, 197–198; Louis Delaunay, *Une grande famille de savants: les Brongniart* (Paris, 1940).

33. For discussion of the role of the protégés, see Outram, *Georges Cuvier,* 189–202.

34. Such an approach would be typified by Pierre Bourdier, "Les stratégies matrimoniales dans le système de réproduction," *Annales* 27 (1972):1105–1127. For a commercial élite, marriage for capital was inescapable; see P. Leuillot, "Bourgeois et bourgeoises," *Annales* 11 (1956):87–101.

35. This failure is common to all recent studies in this area; e.g., J. Houdaille, "Les déscendants des grands dignitaires du premier Empire au dix-neuvième siècle,"

Population 29 (1974):263–274; L. Bergeron and Guy Chaussinand-Nogaret, eds., *Grands notables du premier Empire,* 7 vols. (Paris, 1978–). Technical manuals pay little attention to the problem; e.g., Louis Henry, *Manuel de demographie historique* (Geneva, 1967).

36. A first listing from the Bibliothèque Nationale, *Catalogue des imprimés,* would include the following works produced by wives of men of science, members of the Institut: Mme. Biot (Françoise Gabrielle Brisson), *La physique mécanique* (1813; reprint, Paris, 1830), translated from the German of Ernst G. Fischer; Mme. Daubenton (Marguerite Daubenton, 1720–1818), *Zélie dans le desert,* 2 vols. (Paris, 1787), and twenty-one subsequent editions, 1787–1861; Mme. Claret de Fleurieu (Aglae Deslacs d'Arcambal, 1776–1828), *Au Théâtre de la Nation: le siècle des ballons, satyre, nouvelle comédie en 1 acte* (Paris, 1784); *Stella, histoire anglaise,* 4 vols. (Paris, 1800). Mme. Guyton de Morveau (Claudine Poullet, 1735–1821), translations of Scheele and Stahl; Mme. de Lacépède (Anne Caroline Gauthier, née Jube [?–1801]), *Sophie ou mémoires d'une jeune religieuse* (Paris, 1790; 2d ed., 1792); Mme. de Lefrançois de Lalande (Marie Jeanne Emilie Harlay), *Tables horaires,* in J. J. Lefrançois de Lalande, *Abrégé de navigation* (Paris, 1793); Mme. Laplace (Marie-Anne Charlotte de Courty de Romanges [?–1862]), *Lettres de Mme de Laplace à Elisa Napoléon, Princesse de Lucques et de Piombino,* ed. P. Marmottan (Paris, 1897).

37. For Delambre and Deyeux, Szramkiewicz, *Régents et les censeurs,* 38, 268, 272; for Gay-Lussac, M.P.C. Crosland, *Gay-Lussac* (Cambridge, 1982); for Prony, Dupin, "Éloge de Mme. de Prony"; for Lemmonier, Michel Robida, *Les bourgeois de Paris: trois siècles de chronique familiale de 1675 à nos jours* (Paris, 1955), 45–46; for Leroy, G. Brusa and C. Allix, "Julien and Pierre LeRoy: Their Business, Their Relatives, and Their Namesakes," *Antiquarian Horology* 7 (1972):598–606; Lopez, *Mon Cher Papa,* 213–215.

38. Jean Tulard, "Problèmes sociaux de France impériale," *Revue d'histoire moderne et contemporaine* 18 (1970):639–663.

39. For example, Yves Durand, *Finance et mécenat: les fermiers-généraux au dix-huitième siècle* (Paris, 1976), 165.

40. For example, in 1792 the mathematician P. L. Lagrange married the daughter (Renée Françoise) of the botanist Pierre Charles Lemonnier (1715–1799), whose third daughter, Renée Michelle, married her uncle the horticulturalist Louis Guillaume Lemonnier in 1794. In 1800 the chemist A. L. Fourcroy married Adelaide née Belleville, widow of Institut member architect Charles de Wailly. See Robida, *Les bourgeois de Paris;* for Fourcroy, G. Kersaint, "Antoine François Fourcroy, 1755–1809: sa vie et son oeuvre," *Mémoires du Muséum National d'Histoire Naturelle,* sér. D, 2 (1966):1–296.

41. J. Schiller, "Physiology's Struggle for Independence in the First Half of the Nineteenth Century," *History of Science* 7 (1968):64–89; D. Outram, *Georges Cuvier,* 118–140; idem, "Uncertain Legislator: Georges Cuvier's Laws of Thought in Their Intellectual Context," *Journal of the History of Biology* 19 (1986), 323–368.

42. See the suggestions in D. Outram, "Politics and Vocation."

43. Wealth from, e.g., chemical industry was possessed by Chaptal: Roland

Peigeire, *La vie et l'oeuvre de J. C. Chaptal* (Paris, 1934). Multiple administrative positions are detailed in D. Outram, "Politics and Vocation."

44. An example here, passed to us with an unusual wealth of financial detail, is Paul Cottin's edition of the journal of Mme. Moitte, wife of sculptor Jean Guillaume (1746–1810); P. Cottin, ed., *Journal de Mme. Moitte* (Paris, 1932). The family only married into other "working" artistic families, in spite of Moitte's membership in the Institut from 1795.

45. For a detailed working out of family politics within one scientific group, see D. Outram, *Georges Cuvier,* 161–188.

Chapter 2: Botany in the Breakfast Room

1. *A Feeling for the Organism: The Life and Work of Barbara McClintock* (New York, 1983).

2. The best general account is David E. Allen, *The Naturalist in Britain: A Social History* (London, 1976); see also his *The Botanists: A History of the Botanical Society of the British Isles through 150 Years* (Winchester, 1986). For a rich historical and bibliographical account of the books that shaped eighteenth-century botanical culture, and for biographical information on individuals, see Blanche Henrey, *British Botanical and Horticultural Literature before 1800* (London, 1975), 2. On nineteenth-century culture, see Nicolette Scourse, *The Victorians and Their Flowers* (London, 1983). On botany and the development of one nineteenth-century cultural institution, see Jack Morrell and Arnold Thackray, *Gentlemen of Science: Early Years of the British Association for the Advancement of Science* (Oxford, 1981).

3. For example, Julius von Sachs, *History of Botany (1530–1830),* trans. Henry E. F. Garnsey (1890; reprint, New York, 1967), which seeks "to discover the first dawning of scientific ideas and to follow them as they developed into comprehensive theories" (p. vi); Edward Lee Greene, *Landmarks of Botanical History* (Stanford, Calif., 1983), which charts the life and work of twenty-six "fathers" and "forefathers" of botany; and A. G. Morton, *History of Botanical Science* (London, 1981), which traces "the evolution of botanical theory, of what men at different times have thought about plants as a class of beings" (p. v).

4. County floras since the mid-nineteenth century have included biographical information on local people whose findings were recorded. Women figure in the lists of eighteenth- and nineteenth-century contributors to the project of collecting and reporting on indigenous plants. See, e.g., G. C. Druce, *Flora of Bedfordshire* (1897), *Flora of Buckinghamshire* (1926), and *Flora of Northamptonshire* (1930); and A. R. Horwood and C.W.F. Noel, *Flora of Leicestershire and Rutland* (1933); and H. J. Riddelsdell, *Flora of Gloucestershire* (1948).

5. On women as flower painters, see Wilfrid Blunt, *The Art of Botanical Illustration* (London, 1950). Women did many plates over the years for Curtis's *Botanical Magazine.* One well-known flower painter was Marianne North (1830–90); see *A*

Vision of Eden: The Life and Work of Marianne North (New York, 1980). A recent historical find on a Victorian family of sisters and aunts who collected and drew plants in Gloucestershire is reported in Richard Mabey, *The Frampton Flora* (London, 1985).

6. François Delaporte, *Nature's Second Kingdom: Explorations of Vegetality in the Eighteenth Century*, trans. Arthur Goldhammer (Cambridge, Mass., 1982).

7. *The Naturalist in Britain; The Botanists;* "Life Sciences: Natural History," in *Information Sources in the History of Science and Medicine,* ed. Pietro Corsi and Paul Weindling (London, 1983), 349–360; and the earlier programmatic essay, "Naturalists in Britain: Some Tasks for the Historians," *Journal of the Society for the Bibliography of Natural History* 8 (1977):91–107.

8. "The Women Members of the Botanical Society of London, 1836–56," *British Journal for the History of Science* 13, no. 45 (1980):240–254; The Botanical Society of London had an inclusive and nondiscriminatory membership policy, yet women never made up more than 10 percent of the group. Allen's discussion of this would benefit from wider gender analysis and from inclusion of the voices of the women botanists themselves about their relationship to the scientific culture of their day.

9. Family links were formative for women in the history of art; see Ann S. Harris and Linda Nochlin, *Women Artists 1550–1950* (New York, 1977); and Germaine Greer, *The Obstacle Race* (New York, 1979), chap. 1. On family networks in the history of British natural history, see Allen, *The Naturalist in Britain,* passim. On family links specifically in botany during the mid-Victorian era, see Scourse, *The Victorians and Their Flowers,* 74–78. On family encouragement as part of the history of women's science study at Cambridge during the late nineteenth century, see Roy MacLeod and Russell Moseley, "Fathers and Daughters: Reflections on Women, Science and Victorian Cambridge," *History of Education* 8, no. 4 (1979):321–333. Family networks also were important in nineteenth-century French natural history, as Dorinda Outram has shown regarding work at the Paris Museum of Natural History; see her *Georges Cuvier: Vocation, Science and Authority in Post-Revolutionary France* (Dover, N.H., 1984), esp. chap. 7.

10. Ann B. Shteir, "Linnaeus's Daughters: Women and British Botany," in *Women and the Structure of Society,* ed. Barbara J. Harris and JoAnn K. McNamara (Durham, N.C., 1984).

11. Dr. T. S. Traill, *Memoir of Wm. Roscoe* (Liverpool, 1853), 41.

12. See, e.g., Rosalind C. Barnett and Grace K. Baruch, *The Competent Woman: Perspectives on Development* (New York, 1978).

13. Sir William J. Hooker was director of Kew Gardens from 1841 to 1865; his son Sir Joseph D. Hooker succeeded him, from 1865 to 1885. For a general account of the family, see Mea Allan, *The Hookers of Kew, 1785–1911* (London, 1967). Another notable example from the period under discussion is the family of James Sowerby (1757–1822), botanical artist and collaborator with Sir James E. Smith on *English Botany (1790–1814)*; his three sons all did botanical and natural history work. See Arthur De Carle Sowerby, *The Sowerby Saga* (Washington, D.C., 1952).

14. Eliza and Marianne Boswell, "Manuscript List of Plants, Vicinity of Balmuto, Fifeshire," 1820 (Linnean Society, London). They helped J. R. Scott and W. Jameson on *Herbarium Edinense,* 1820.

15. Mrs. Gatty wrote: "We are doing all we can over the boys' education, looking upon it as money laid out to interest. The girls are pretty well; they teach and visit the sick and are as good as four curates" (cited in Christabel Maxwell, *Mrs. Gatty & Mrs. Ewing* [London, 1949], 115).

16. *Journal of Botany* 32 (1894):205–207.

17. On Lydia Becker, see *Dictionary of National Biography* (London, 1921), hereafter DNB; and Helen Blackburn, *Women's Suffrage: A Record of the Woman's Suffrage Movement in the British Isles, with Biographical Sketches of Miss Becker* (1902; reprint, New York, 1971). On her BAAS botanical paper, see *Journal of Botany* 7 (1869): 291–292.

18. See letters to Lydia Becker from John Leigh, "The Becker Letters," Fawcett Library Autograph Collection, 33, pt. A, letters 21–23.

19. Mary Kirby, *"Leaflets from My Life": A Narrative Autobiography* (London, 1887); and A. R. Horwood and C.W.F. Noel, *The Flora of Leicestershire & Rutland,* 205–208. I am grateful to David E. Allen for bringing Mary Kirby's autobiography to my notice.

20. *Quarterly Review* 101 (1857):6. See also G. C. Druce, *Flora of Northamptonshire,* 88–89.

21. See Bea Howe, *Lady with Green Fingers: The Life of Jane Loudon* (London, 1961).

22. Linnean Society MSS 393.

23. Letter dated October 7, 1815; Smith MSS 19:219; Linnean Society Library.

24. For example, Frances Rowden's *A Poetical Introduction to the Study of Botany* (London, 1801), a verse exposition of Linnaean botany, based on Erasmus Darwin's "The Loves of the Plants"; Rowden's topics for moral exhortation include friendship, modesty, virtue, and maternal love.

25. Deborah Gorham, *The Victorian Girl and the Feminine Ideal* (Bloomington, Ind., 1982).

26. Women members of the upper-middle-class Paget family, e.g., attended "botanical lectures" and were urged to join a neighbor in studying botany. Botany was among the subjects that three generations of women in the Paget family studied; see M. Jeanne Peterson, "No Angels in the House: The Victorian Myth and the Paget Women," *American Historical Review* 89 (1984):677–708.

27. *The Journal of Emily Shore* (London, 1898). Thanks to Barbara Brandon Schnorrenberg for drawing Emily Shore to my attention.

28. For example, as recorded in the entry for May 27, 1833, regarding two sorts of bryony: "only one of which is the real bryony; the Latin name is the *Bryonia dioica*. . . . The other is the *Tamus communis*. . . . The first plant belongs to the class Triandria, monogynia; the other to Hexandria, monogynia; that is, according to the new arrangement of Thunberg and Withering" (pp. 54–55).

29. Helen Blackburn, *Women's Suffrage,* 30.

30. Letter to Mrs. Henry Fawcett, February 6, 1887, in "The Becker Letters," Fawcett Library Autograph Collection, 28, pt. A.

31. On the history of women's education in England during the 1860s and 1870s, see Joan N. Burstyn, *Victorian Education and the Ideal of Womanhood* (New York, 1980); and Margaret Bryant, *The Unexpected Revolution: A Study in the History of the Education of Women and Girls in the 19th Century* (London, 1979).

32. Lydia Ernestine Becker, "On the Study of Science by Women," *Contemporary Review* (1869):368–404.

33. Bea Howe, *Lady with Green Fingers,* 54.

34. For example, *Gardening for Ladies* (1840); *The Ladies' Flower-Garden* (1840–48); *The First Book of Botany* (1841).

35. Jane Louden wrote: "It is so difficult for men whose knowledge has grown with their growth, and strengthened with their strength, to imagine the state of profound ignorance in which a beginner is" (*Botany for the Ladies, vi).*

36. Ann B. Shteir, "Priscilla Wakefield's Natural History Books," in *From Linnaeus to Darwin: Commentaries on the History of Biology and Geology* (London, 1985), 29–36.

37. Kirby, *"Leaflets from My Life,"* 40–43.

38. Kirby records being delighted when Sir William Hooker praised her flora and expressed support for local efforts to spread interest in botany, but she was "constrained" by the eminence and manner of John Lindley, professor of botany, University College, London, who encouraged work only "of the most scientific kind" (p. 147).

39. Of particular note here are *Plants of the Land and Water* (1857) and *Chapters on Trees* (1873).

40. On Henslow's school and its curriculum, see Jean Russell-Gebet, *Henslow of Hitcham* (Lavenham, Suffolk, 1977), chap. 4; on Elizabeth Twining's educational project, see the preface to her *Short Lectures on Plants, for Schools and Adult Classes* (London, 1858).

41. One early and much reprinted example was the Thomas Martyn translation and enlargement of Rousseau's *Letters on the Elements of Botany* (1785; reprint, London, 1815).

42. For an example of the conversational format, see Sarah and Elizabeth Fitton's *Conversations on Botany* (1817); for the epistolatory format, see Priscilla Wakefield's *Introduction to Botany* (1796).

43. See Mitzi Myers, "Reform or Ruin: A Revolution in Female Manners," in *Studies in Eighteenth-Century Culture,* vol. 11, ed. Harry C. Payne (Madison, Wis., 1982), 199–216; idem, "Impeccable Governesses, Rational Dames, and Moral Mothers: Mary Wollstonecraft and the Female Tradition in Georgian Children's Books," *Children's Literature* 14 (New Haven, Conn., 1986).

44. Ray Desmond's *Dictionary of British and Irish Botanists and Horticulturalists* (London, 1977) is the starting point for any British biographical foray, for he and his precursors combed nineteenth-century journals and many a county flora for references to individuals and their botanical contributions.

45. Eliza Brightwen, *Glimpses into Plant Life: An Easy Guide to the Study of Botany* (1898); and W. H. Chesson, ed., *Eliza Brightwen: The Life and Thoughts of a Naturalist* (London, 1909).

46. There is a poignant tale to be reconstructed about the career of Agnes Ibbetson (1757–1823), avid student of plant physiology, who wrote periodical essays ca. 1809–1822. On her, see DNB (1921) 10, and *Gentleman's Magazine* (1823), 93, pt. 1. She sought a recommendation to a publisher from Sir James E. Smith, president of the Linnean Society, ca. 1816, but was rebuffed. (See her letter to him, dated May 26, 1816; Linnean Society, Smith MSS 23:70.) Other well-known botanical researchers such as George Bentham and Thomas Andrew Knight also were cool toward her work. Her sense of grievance is unmistakable in, e.g., *The Philosophical Magazine and Journal* 56 (1820):3–9, and 59 (1822):3–8, 243–244.

47. The case studies in Ian Inkster and Jack Morrell, eds., *Metropolis and Province: Science in British Culture, 1780–1850* (Philadelphia, 1983), are models of research into local contexts, but there is barely a mention of women. Jack Morrell and Arnold Thackeray offer slightly more in their account of the early history of the BAAS; see *Gentlemen of Science,* 148–157. The "Manchester model" thesis of social mobility as a reason for involvement in scientific culture is based only on men. Two examples from Manchester of women's involvement in cultural institutions of science would reward study. Circa 1811, wives, "some of whom were . . . Botanists," attended meetings of a working man's botanical society (as reported in Linnaean Society MSS 401; thanks to Gwen Averley for the reference). In 1867 Lydia Becker formed the Manchester Ladies' Literary Society because both the Manchester Literary and Philosophical Society and the Manchester Scientific Students' Association were closed to women. She emphasized scientific study, and Charles Darwin sent a paper on plant sexuality to be read at the inaugural meeting. For Lydia Becker's presidential address to the Manchester Ladies' Literary Society, see Helen Blackburn, *Women's Suffrage,* 31–39.

48. As a model of the new historiographical possibilities in using children's literature, see James A. Secord, "Newton in the Nursery: Tom Telescope and the Philosophy of Tops and Balls, 1761–1838," *History of Science* 23 (1985):127–151.

Chapter 3: The Many Faces of Intimacy

1. Charles Rosenberg, "The Therapeutic Revolution: Medicine, Meaning, and Social Change in Nineteenth-Century America," in *The Therapeutic Revolution: Essays on the Social History of American Medicine,* ed. Charles Rosenberg and Morris Vogel (Philadelphia, 1979), 10–11.

2. *Valedictory Address* (Philadelphia, 1872), 67.

3. For a more detailed analysis of these events, see Regina Markell Morantz-Sanchez, *Sympathy and Science: Women Physicians in American Medicine* (New York, 1985).

4. *Daughters of Aesculapius: Stories Written by Alumnae and Students of the Woman's Medical College of Pennsylvania* (Philadelphia, 1897).

5. Ibid., 66–79.

6. Ibid., 133–149.

7. Ibid., 53–65.

8. See, e.g., Caroline H. Dall, ed., *A Practical Illustration of Woman's Rights to Labor* (Boston, 1860), 2.

9. Gertrude Baillie, M.D., "Should Professional Women Marry?" *Woman's Medical Journal* 2 (February 1894):33–35.

10. See Morantz-Sanchez, *Sympathy and Science,* 134–135.

11. Rachel L. Bodley, *The College Story: Valedictory Address to the Twenty-ninth Graduating Class of the Woman's Medical College of Pennsylvania* (Philadelphia, 1881), 4–10.

12. "Inaugural Address at the Opening of the Woman's Medical College of the New York Infirmary, October, 1880," reprinted in the Women's Medical Association of New York City, eds., *Mary Putnam Jacobi: Pathfinder in Medicine* (New York, 1925), 390.

13. For more detailed information on marriage, see Morantz-Sanchez, *Sympathy and Science,* 134–143. Rosalie Slaughter Morton, *A Woman Surgeon* (New York, 1937), 143–148, 177.

14. Emily Dunning Barringer, *Bowery to Bellevue* (New York, 1950), 67.

15. See the Thomas Longshore manuscript biography of his wife and notes from an interview with Longshore's daughter, Mrs. Lucretia Blankenburg, Longshore MSS, Medical College of Pennsylvania. Also oral interview in 1979 conducted with Stacy May, Dr. Cohen's son. I am grateful to Ruth Abram for sharing this material with me.

16. Regina Morantz, oral interview with Dr. Pauline Stitt, December 9, 1977, 29–33, Oral History Project on Women in Medicine, Medical College of Pennsylvania; Alan Chesney, M.D., *The Johns Hopkins Hospital and the Johns Hopkins School of Medicine: A Chronicle* (Baltimore, Md., 1943), 1:10; Elinor Bleumel, *Florence Sabin: Colorado Woman of the Century* (Boulder, Colo., 1959), 62; Howard to her parents, February 18 and March 11, 1916. Ernestine Howard MSS, Radcliffe Women's Archives, Schlesinger Library.

17. Mendenhall autobiography, handwritten section, F, 1–10, typescript, 18. Dorothy Reed Mendenhall MSS, Sophia Smith Collection, Smith College.

18. Elizabeth Robinton, "Anna Wessels Williams," in *Notable American Women: The Modern Period,* ed. Barbara Sicherman and Carol Hurd Green (Cambridge, Mass., 1980), 737–739; Notes, Folder #61, Williams MSS, Radcliffe Women's Archives, Schlesinger Library.

19. Notes, Clelia Mosher MSS, Stanford University Library.

20. Harriet Belcher to Eliza Johnson, February 18, 1977, August 10, 1978; Belcher MSS; in private hands.

21. Belcher to Johnson, February 5, 1879, Belcher MSS; Elizabeth Cady Stanton, *Eighty Years and More* (New York, 1972), 172; *Marie Zakrezewska: A Memoir* (Boston, 1903), 22.

22. See Fullerton's diary, "Our Life in Dekra Dun" (1915 and 1916), in the hands

of her great-niece, Beatrice Beech MacLeod. For a fuller description of the nuances of this single female professional community, see Jane Hunter, *The Gospel of Gentility* (New Haven, Conn., 1984), 52–89.

23. Nancy Sahli, "Elizabeth Blackwell, M.D.: A Biography" (Ph.D. diss., University of Pennsylvania, 1974), 128–129; Elizabeth Putnam Gordon, *The Story of the Life and Work of Cordelia A. Greene, M.D.* (Castile, N.Y., 1925), 20–21; August 15, 1885, Mosher MSS, Michigan Historical Collections.

24. Letters not dated, but between 1871 and 1876. Helen Morton Papers, Radcliffe Women's Archives, Schlesinger Library.

25. See Fullerton diary, Belcher letters, passim. Barbara Sicherman, "Alice Hamilton," in *Notable American Women*, ed. Sicherman and Green, 303–306; Genevieve Miller, "Lilian Welsh," in *Notable American Women*, ed. Janet and Edward T. James (Cambridge, Mass., 1971), 3:567–568; Jacobi to Elizabeth Blackwell, December 25, 1888, Blackwell MSS, Library of Congress.

26. For historians' comments on these homosocial relationships, see Carroll Smith-Rosenberg, "The Female World of Love and Ritual: Relations between Women in Nineteenth-Century America," *Signs* 1 (1975):1–20; Nancy Sahli, "Smashing: Women's Relationships before the Fall," *Chrysalis* 2 (1979):17–27; Blanche Weisen Cook, "Female Support Networks and Political Activism: Lillian Wald, Crystal Eastman, Emma Goldman," *Chrysalis* 1 (1977):43–61; Leila Rupp, "'Imagine My Surprise': Women's Relationships in Historical Perspective," *Frontiers* 5 (Fall 1980):61–70.

27. *Boston Medical and Surgical Journal* 43 (1850):69–75; 53 (1855–56):292–294; 54 (1856):169–174.

Chapter 4: Field Work and Family

Acknowledgments: Research for this project has been supported in part by a Frank M. Chapman Memorial Grant of the American Museum of Natural History, and in part by the American Ornithologists' Union. I gratefully acknowledge the manifold support of my husband, David, and the encouragement and help of Mary Baldwin, Marcia Bonta, Maxine Benson, Michael J. Brodhead, Erica Dunn, Marjorie Nice Boyer, Mary Gilliland, Katherine Goodpasture, Frances C. James, Stewart Holohan, Mary and C. Stuart Houston, Louise de Kiriline Lawrence, Ann Hibner Koblitz, Harriet Kofalk, Robert Orr, Matthew Perry, Iola Price, Margaret W. Rossiter, Doris Huestis Speirs, and Barbara Blanchard De Wolfe. I am also grateful to the many contemporary women ornithologists, too numerous to mention, who shared their experiences with me, and to Smithsonian archivist Susan Westgate for her assistance.

1. In the history and sociology of science the professionalization of a discipline has often been considered a measure of the maturity of the field. Among the many useful sources on disciplines and professions, see J. Ben-David and A. Zloczower,

"Universities and Academic Systems in Modern Societies," *Archives européennes de sociologie* 3 (1962):45–82; J. D. Beer and W. D. Lewis, "Aspects of the Professionalization of Science," in *The Professions in America,* ed. K. S. Lynn et al. (Boston, 1965), 110–130; G. Lemaine, ed., *Perspective on the Emergence of Scientific Disciplines* (Paris, 1976); E. Mendelsohn, "The Emergence of Science as a Profession in Nineteenth Century Europe," in *Management of Scientists,* ed. K. Hill (Boston, 1964), 3–48; D. Outram, "Politics and Vocation: French Science, 1793–1830," *British Journal for the History of Science* 13 (1980):27–43; N. Reingold, "Definitions and Speculations: The Professionalization of Science in America in the Nineteenth Century," in *The Pursuit of Knowledge in the Early American Republic,* ed, A. Oleson and S. C. Brown (Baltimore, Md., 1976), 33–68; R. A. Stebbins, "The Amateur: Two Sociological Definitions," *Pacific Sociological Review* 20 (1977):583–605; R. Kargon, *Enterprise and Expertise: Science in Victorian Manchester* (Baltimore, Md., 1978).

2. Warren Hagstrom, in *The Scientific Community* (Carbondale, Ill., 1965), emphasized that in science the acceptance by scientific journals of contributed manuscripts establishes the author's status as scientist and this assures him or her a place in the scientific community. In ornithology, where positions were scarce, publication was the only way to measure the extent of scientific activity.

3. M. G. Ainley, "D'assistantes anonymes à chercheures scientifiques: une retrospective sur la place des femmes en sciences," *Cahiers de recherche sociologique* 4 (April 1986):55–71.

4. To my knowledge *only* the following women were employed as ornithologists during the 1860–1950 period: May T. Cooke (1885–1963), who in 1916 took over her late father, W. W. Cooke's (1858–1916) work on bird migration and bird distribution for the Bureau of Biological Survey; in 1927 Phoebe Knappen (1905–1979), a Cornell M.Sc. in ornithology, was employed by the survey. Apparently the director, Dr. Clarence Cottam, prevented her professional advancement because she was a woman. I am grateful to M. Katz and M. Perry for providing me with information on Knappen's career. Dr. Hildegarde Howard (b. 1901) is an avian paleontologist (Ph.D., 1928) who apparently had no career problems. She cannot properly be considered an ornithologist. Dr. Theodora Nelson (1894–1981) did not publish anything else besides her Ph.D. thesis on the spotted sandpiper. After her graduation from the University of Michigan in 1939, she was employed at Hunter College, but had no time for research. In the 1930s Ruth Trimble (b. 1902) was assistant curator of birds at the Carnegie Museum, Pittsburgh. Of the early women graduates in ornithology only Mary M. Erickson (1905–1983) and Barbara Blanchard De Wolfe (b. 1912) had successful careers as teachers and researchers. After having taught in private colleges in the late 1930s, they both found employment at the University of California, Santa Barbara, where they encountered no discrimination. I thank Dr. De Wolfe for this information.

5. A. Marguerite Heydweiller, a Cornell Ph.D. in ornithology (1935), could not find employment as an ornithologist after her marriage to fellow scientist F. Baumgartner. He, however, found paid positions, and she both helped his research and carried out some projects on her own as an independent researcher. After Ruth

Trimble's marriage in 1940 to ornithologist James P. Chapin of the American Museum of Natural History, she had to give up work in ornithology; antinepotism regulations prevented her employment at the museum, and no other positions were available.

6. Margaret W. Rossiter, "Women's Work in Science, 1880–1910," *Isis* 71 (1980):381–399.

7. For instance, F. M. Chapman and T. S. Palmer, eds., *Fifty Years' Progress in American Ornithology, 1883–1933* (Lancaster, Pa., 1934); E. Mayr, "The Role of Ornithological Research in Biology," *Proceedings of the Thirteenth International Ornithological Congress, 1962* (Washington, D.C., 1963), 27–38; E. Stresemann, *Ornithology, from Aristotle to the Present* (Cambridge, Mass., 1975). Paul L. Farber, in his recent book *The Emergence of Ornithology as a Scientific Discipline, 1760–1850* (Boston, 1982), mentions only one woman, Mme. De Bandeville, a French collector. There were others. Recent historical works on American women ornithologists are Marianne Ainley, "The Involvement of Women in the American Ornithologists' Union," in *A Centennial History of the American Ornithologists' Union, 1883–1983,* ed. K. B. Sterling and M. G. Ainley (Washington, D.C., 1987 [in press]), and idem, "Women in North American Ornithology during the Last Century," in *Proceedings of the First International Conference on the Role of Women in the History of Science, Technology and Medicine in the 19th and 20th Centuries* (Veszprem, Hungary, 1983), 3–7.

8. Robert Ridgway of the Smithsonian Institution and Frank M. Chapman of the American Museum of Natural History were among these. Regular college courses in ornithology were first given during the 1909–1911 period at the University of California, Berkeley, the State College of Washington, and at Cornell University. The University of California and Cornell later developed into graduate centers of ornithological education.

9. Of the twenty-five founders of the American Ornithologists' Union, only three were employed as ornithologists: Ridgway and S. F. Baird at the Smithsonian, and J. A. Allen (1838–1921) at the Museum of Comparative Zoology, Harvard. The rest were civil servants, army officers, and businessmen. See M. G. Ainley, "The Contribution of the Amateur to American Ornithology: A Historical Perspective," *Living Bird* 18 (1979–80):161–177.

10. D. J. Warner, *Graceanna Lewis: Scientist and Humanitarian* (Washington, D.C., 1979); Maxine Benson's biography of Martha A. Maxwell, *Rocky Mountain Naturalist* (Lincoln, Nebr., 1986); Harriet Kofalk's work on F. M. Bailey, *No Woman Tenderfoot* (forthcoming).

11. Grace Anna Lewis (1821–1912), a Pennsylvania Quaker, was a typical nineteenth-century museum ornithologist and exemplified early women scientists who aspired to professional employment. See Warner, *Graceanna Lewis.* Martha Ann Maxwell (1831–81) was a naturalist, taxidermist, museum builder, and an early student of animal behavior; see Benson, *Rocky Mountain Naturalist.* F. M. Bailey (1863–1946) was a conservationist, naturalist, and field ornithologist.

12. For an excellent discussion concerning the importance of female relation-

ships, see Carroll Smith-Rosenberg, "The Female World of Love and Ritual: Relations between Women in Nineteenth-Century America," *Signs* 1 (1975):1–29.

13. She shared her home with her unmarried older sister, Amelia Sherman, M.D. Another sister, this one married, lived in the southwestern United States. See H. J. Taylor, "Iowa's Woman Ornithologist: Althea Rosina Sherman, 1853–1943," *Iowa Bird Life* 13 (1943):19–35; also Marcia Bonta, "The Chimney Swift Lady," *Bird Watcher's Digest* (March–April 1985):36–40.

14. Taylor, "Iowa's Woman Ornithologist," 27.

15. Although the AOU had admitted women as associates since 1886 (the first was Florence M. Bailey), in 1900 they constituted only 11.4 percent of the total membership. In 1901 three women became elective members (Bailey and conservationist–nature writers Olive Thorne Miller and Mabel Osgood Wright). In 1910 women constituted 18 percent of the total membership, but only 3 percent of the elective members; this number rose to 5 percent only in 1950.

16. A. R. Sherman to M. M. Nice, December 8, 1925, Margaret Morse Nice Papers, #2993, Olin Library, Cornell University Archives, Ithaca (hereafter Nice Papers).

17. Ibid.

18. A. R. Sherman to William Rowan, June 21, 1927, William Rowan Papers, University of Alberta Archives, Edmonton.

19. A. R. Sherman to M. M. Nice, October 11, 1929, Nice Papers.

20. Ibid., March 10, 1926. The domestic "problem" in America is explored by Ruth S. Cowan in *More Work for Mother: Ironies of Household Technology from the Open Hearth to the Microwave* (New York, 1983), 119–127.

21. A. R. Sherman, *Birds of an Iowa Dooryard,* ed. F. J. Pierce (Boston, 1952), 60–61. Between 1918 and 1932 at least one hundred visitors per year saw the swifts in the tower.

22. A. R. Sherman to M. M. Nice, November 13, 1924, Nice Papers.

23. M. M. Nice, review of Sherman, *Birds of an Iowa Dooryard, Bird Banding* 23 (1952):135.

24. M. M. Nice, "Some Letters of Althea Sherman," *Iowa Bird Life* 22 (1952):55.

25. M. M. Nice, *Research Is a Passion with Me,* ed. Doris H. Speirs (Toronto, 1979), 33.

26. Ibid., 14.

27. Ibid., 15.

28. Ibid., 21. Although Morse did not like the approach to zoology, she did like her teachers, including Dr. Cornelia Clapp.

29. Ibid.

30. Ibid., 28.

31. Ibid., 32.

32. Ibid., 33.

33. Ibid., 34.

34. Ibid., 41.

35. Ibid.

36. Bailey's book (1902) combined original field observations with much technical data provided by museum ornithologists. It remained a classic for over half a century.

37. Dr. Marjorie Nice Boyer, daughter of Margaret M. Nice, personal communication.

38. Nice Papers. Amelia Laskey Papers at the Cumberland Museum and Science Center, Nashville, Tennessee. Private papers of Louise de Kiriline Lawrence, Pimisi Bay, Ontario, and Doris H. Speirs, Pickering, Ontario.

39. Boyer, personal communication.

40. Ainley, "D'assistantes"; Hanna Papanek, "Men, Women and Work: Reflections on the Two-Person Career," in *Changing Women in a Changing Society,* ed. Joan Huber (Chicago, 1973), 90–111.

41. J. Chapin to P. A. Taverner, March 3, 1942, National Museum of Natural Sciences, Vertebrate Zoology Division, Ottawa (hereafter Taverner Papers).

42. In 1929 Bailey became the first woman fellow of the AOU; two years later she was the first recipient of the prestigious Brewster Medal, given biannually since 1921.

43. Dr. Doris H. Speirs founded this club in Toronto, Ontario, because the Toronto Ornithological Club would not admit women as members.

44. K. A. Goodpasture, "In Memoriam: Amelia Rudolph Laskey," *Auk* 92 (1975):254.

45. Ibid.

46. A. R. Laskey to M. M. Nice, December 29, 1943, Laskey Papers.

47. A. R. Laskey to Roy Ivor, August 4, 1944, Laskey Papers. Ivor was a Canadian independent researcher working on comparative life history studies. He also benefited from Nice's network of correspondents.

48. Ibid.

49. Goodpasture, "Amelia Rudolph Laskey," 257.

50. D. E. Allen, "The Women Members of the Botanical Society of London, 1836–1856," *British Journal for the History of Science* 13 (1980):238–254.

51. See Warner, *Graceanna Lewis.*

52. Maxwell, at odds with her husband and only child, could not realize her scientific potential.

53. M. S. White, "Psychological and Social Barriers to Women in Science," *Science* 170 (1970):414.

Chapter 5: Nineteenth-Century American Women Botanists

1. Emanuel D. Rudolph, "Women in Nineteenth Century American Botany: A Generally Unrecognized Constituency," *American Journal of Botany* 69 (1982): 1346–1355.

2. Almira H. Lincoln [Phelps], *Familiar Lectures on Botany* (Hartford, Conn., 1829).

3. Almira H. L. Phelps, *Botany for Beginners* (Hartford, Conn., 1833), preface.

4. J.F.A. Adams, "Is Botany a Suitable Study for Young Men?" *Science* 9 (1887):117–118. See comment in Margaret W. Rossiter, *Women Scientists in America: Struggles and Strategies to 1940* (Baltimore, Md., 1982), 86.

5. Harry Baker Humphrey, *Makers of North American Botany* (New York, 1951); seven were added by R. H. Humphrey after his father's death in 1955. C. Earle Smith, Jr., "A Century of Botany in America," *Bartonia* (1954–56):1–30, appendix.

6. Rossiter, *Women Scientists in America,* index.

7. In order to try to answer these questions I have used a variety of sources. Rossiter's book was both a source of information and of further valuable sources. *American Men and Women of Science* and particularly Edward T. James, Janet Wilson James, and Paul W. Boyer, eds., *Notable American Women,* 3 vols. (Cambridge, Mass., 1971), and its supplement, Barbara Sichermond and Carol Hurd Green, eds., *Notable American Women: The Modern Period* (Cambridge, Mass., 1980), were useful for biographical information. I have used the botanists in the NAW volumes as a basic sample for this paper.

Full-length biographies of women botanists are few. Bolzau's biography of Almira Lincoln Phelps (Emma L. Bolzau, *Almira Hart Lincoln Phelps: Her Life and Work* [Philadelphia, 1936]) was valuable, as were the Phelps Archives at the Emma Willard School, Troy, New York. Ethel M. McAllister, *Amos Eaton: Scientist and Educator, 1776–1842* (Philadelphia, 1941), contains additional material about Phelps. Her own published writings are perhaps the best source, particularly in terms of her views on woman's sphere in society. The best source for Eliza Sullivant, characteristically, is the biography of her husband, William Sullivant (Andrew D. Rodgers III, *"Noble fellow": William Starling Sullivant* [New York, 1968]). Edith S. Clements's autobiographical *Adventures in Ecology* (New York, 1960) sheds light on her own work and its relationship to her husband's work.

I have found the accomplishments of women botanists who published in botanical journals in articles and obituaries in those journals; the *Bulletin of the Torrey Botanical Club,* the *Journal of the New York Botanical Garden,* and the *Bryologist* have been particularly useful. Other printed sources were often more local or more obscure—local historical society bulletins or botanical publications, such as that of the Boston Mycological Club. A biographical account of the immigrant botanist Katherine Esau, for example, was found in an American Botanical Society publication for members.

Letters have often been the best source of information. There is as yet no biography of Elizabeth Britton, but the New York Botanical Garden Archives contain hundreds of letters to her, as well as many copies of her own letters. In addition, there are letters of several women botanists to the first New York State botanist, Charles Horton Peck, in the New York State Museum Archives, Albany, New York. Discussion of marriages even in letters is unusual. Spouses are rarely mentioned in the many letters between botanists that I have read except in terms of sending regards, trips necessitating the writer's absence (e.g., "Dr. Britton and I will be in Porto Rico until March 17"), or condolences at the death of a spouse. Letters between husbands and wives, John and Abigail Adams and other well-known exceptions notwithstanding, are often not kept or are kept by only one of the two. I have found letters of Nathaniel to Elizabeth Britton, but not hers to him. Primary materials elucidating relationships between single women botanists are even harder to find. Oral history is still possible even for nineteenth-century botanists; I have interviewed a few people who knew Elizabeth Britton, although she died over fifty years ago. Specific references to all these sources are given where appropriate.

8. There is considerable literature on the professionalization of science in nineteenth-century America, some of which relates specifically to botany. See, e.g., Nathan Reingold (New York, 1976); and Douglas Sloan, "Science in New York in America in the Nineteenth Century," in *The Pursuit of Knowledge in the Early American Republic,* ed. Alexandra Oleson and Sanborn C. Brown (Baltimore, Md., 1976), 33–69; George H. Daniels, "The Process of Professionalization in American Science: The Emergent Period, 1820–1860," in *Science in America Since 1820,* ed. Nathan Reingold (New York, 1976), and Douglas Sloan, "Science in New York City, 1867–1907," *Isis* 71 (1980):35–76, the latter particularly including botany. See also Nancy G. Slack, "Charles Horton Peck and the Professionalization of Botany in New York State" (forthcoming). The amateur in general, including the amateur scientist, is discussed in Robert A. Stebbins, "The Amateur: Two Sociological Definitions," *Pacific Sociological Review* 20 (1977):582–606; and more specifically in the current context by Sally Kohlstedt, "The Nineteenth-Century Amateur Tradition: The Case of the Boston Society of Natural History," in *Science and Its Public: The Changing Relationship,* ed. G. Holton and W. Blanpied (Dordrecht, Holland, 1976), 173–190. Also relevant is Elizabeth Barnaby Keeney, *The Botanizers: Amateur Scientists in Nineteenth-Century America* (Ph.D. diss., University of Wisconsin at Madison, 1985).

9. Frank Lamson-Scribner, "Southern Botanists," *Bulletin of the Torrey Botanical Club* 20 (1893):315–334. This geographical bias probably has to do in large part with the development of botany in Philadelphia, New York, and New England, and in the latter part of the nineteenth century at the midwestern land grant universities. Another factor was the earlier establishment in the northeastern United States of female seminaries providing at least some science education. The discovery of further primary material may perhaps remove some of this geographical bias. We know nothing about the young women who were introduced to botany in the Arkansas Territory by Almira Lincoln Phelps's book. Perhaps no one has yet really looked. It is a recent search; the nineteenth-century letters, including a number from women, to Charles Horton Peck that started me on this project were found only about ten years ago in a scrapbook in Peck's granddaughter's attic.

10. Ruth Schwartz Cowan, *More Work for Mother: The Ironies of Household Technology from the Open Hearth to the Microwave* (New York, 1983).

11. Rossiter, *Women Scientists in America,* 238, 241.

12. Edna Yost, *American Women of Science* (Philadelphia, 1943).

13. *Notable American Women* and *Notable American Women: The Modern Period,* botany and horticulture entries in both.

14. Ann B. Shteir, "Linnaeus's Daughters: Women and British Botany," in *Women and the Structure of Society: Selected Research from the Fifth Berkshire Conference on the History of Women,* ed. Barbara J. Harris and Jo Ann K. McNamara (Durham, N.C., 1984).

15. Jane Colden, *Botanical Manuscript,* ed. Harold W. Ricket and Elizabeth C. Hall (Garden Clubs of Orange and Dutchess Counties, N.Y., 1963). The quote is from Elizabeth Hall's introduction, "The Gentlewoman, Jane Colden, and Her Manuscript on New York Native Plants," 21. It is interesting to note that Cadwallader Colden, whose botanical works were certainly more important than his daughter's, was not included by Humphrey in *Makers of American Botany,* whereas Jane, exceptional among eighteenth-century women, was. Elizabeth Hall is herself an interesting botanist; born in 1898, she was a graduate of Radcliffe College and the Ambler School of Horticulture for Women. From 1937 she was librarian for the New York Botanical Garden; and from 1963, associate curator of education. At eighty-eight, she is still actively working there, although she retired in 1967. She knew Elizabeth Britton and many other earlier botanists.

16. Geraldine Kaye, "Violetta S. White: A Mycologist Who Got Away," *Boston Mycological Club Bulletin* 39, no. 2 (1984):10–11. White's monographs were published in the *Bulletin of the Torrey Botanical Club* in 1901 and 1902.

17. Violetta White was herself the daughter of a prominent lawyer, John Jay White, and had been born in Florence. The clubs to which she belonged as Mrs. Delafield indicate an upper-class New York City life-style. Her obituary, quoted by Kaye, noted her Fenwick Medal at the National Flower Show. Garden clubs and flower arrangement were no doubt more acceptable activities for an upper-class matron with botanical interests than research on fungi.

18. Charlotte Haywood, "Shattock, Lydia White," in *NAW,* 3:273–274. Mildred E. Mathias, "Sessions, Kate Olivia, in *NAW,* 3:262–263. It is surprising that these descriptions should be included at all in the biographies of eminent women botanists.

19. Goldring was a paleontologist and is sometimes included among botanists (Rossiter, *Women Scientists in America,* index). She was the first woman to be appointed state paleontologist for New York, a position she held for fifteen years.

20. Sally Gregory Kohlstedt, "Goldring, Winifred," in *NAW* supplement, 282–283.

21. Rossiter, *Women Scientists in America,* 15–16.

22. Kohlstedt, "Goldring, Winifred," in *NAW* supplement, 283. Botanical collecting was a male profession throughout much of the nineteenth century. These men were financially supported by such botanists as William J. Hooker in England and Asa Gray in America. The great exploring expeditions to botanically unknown areas of the American West, as well as later exploration of Mexico, kept this tradition alive in the United States even in the 1880s, when other positions for male bota-

nists were available. C. C. Parry, a famous collector, "proved unsuited to the life of a civilized botanist" in a government position and went back to being a roving collector in Mexico. See A. Hunter Dupree, *Asa Gray* (Cambridge, Mass., 1959), 204–210, 388–390. Taxonomic botany continued as a major component of American botanical research even after the rise of experimental biology and of physiological, ecological, and genetic studies of plants. This was less true in Europe during the same period. Rodgers, after a lengthy discussion of American botanical exploration, quoted a letter from William Hooker to Asa Gray in 1878, noting that "all the world is mad after Physiology and Histology, and Morphology *pure* and classification are despised on the Continent, and Britain is fast following suit," Andrew D. Rodgers III, *American Botany 1873–1892: Decades of Transition* (Princeton, N.J., 1944), 144.

I surveyed the *Bulletin of the Torrey Botanical Club* for 1936 and 1937, which published results of research. (A second club journal published botanical notes and news; early issues of the *Bulletin* before "professionalization" had included both). For those two years there were thirty-one taxonomic papers—twenty-one in physiology and development, three in paleobotany, three in ecology, four in cytology, two in anatomy, and two in genetics. Women contributed papers in genetics, cytology, development, and paleobotany. See also Joel Hagen, "Experimentalists and Naturalists in Twentieth Century Botany: Experimental Taxonomy, 1920–1950," *Journal of the History of Biology* 17(1984):249–270, on the interaction of experimental and more classical methods.

23. A. Hunter Dupree and Marian L. Gade, "Brandegee, Katherine Curren," in *NAW,* 1:228–229; and William A. Setchell, "Townsend and Mary Brandegee," *University of California Publications in Botany* 13 (1926):156–158. Michael T. Stieber, "Chase, Mary Agnes," *NAW* supplement, 146–148; and F. R. Fosberg and J. R. Swallen, "Agnes Chase," *Taxon* (1959):145–151.

24. Joseph Ewan, "Mexia, Ynes Enriquetta Julietta," in *NAW,* 2:533–534.

25. There were defenders of a woman's right to limit childbearing as early as the 1840s. They argued that a woman "must have the right to control her own body; without this she was slave not only to the sexual impulses of her husband but also to endless childbearing." In this controversy one physician could write, "It is her absolute right to determine when she will not be exposed to pregnancy"; but another wrote of the ills caused by condoms and diaphragms, including not only gynecological lesions, but also nymphomania and insanity. See Carroll Smith-Rosenberg and Charles E. Rosenberg, "The Female Animal: Medical and Biological Views of Women," in *No Other Gods, On Science and American Social Thought,* ed. C. E. Rosenberg (Baltimore, Md., 1976). The physicians' quotes are from that article, which provides many other references on this controversy.

26. Almira Lincoln Phelps, *The Female Student; or Lectures to Young Ladies on Female Education* (New York, 1836).

27. Almira Phelps, "Additional Sketch of John Willard," from an unpublished scrapbook owned by Frank Phelps, as quoted in Bolzau, *Almira Hart Lincoln Phelps,* 27–28.

28. Bolzau, *Almira Hart Lincoln Phelps,* 44. Bolzau described both of Phelps's marriages in the chapter "The Busy Bride."

29. These courses went on until at least 1834. A "botanical retreat for women" at the institute (Rensselaer Polytechnic Institute by that date) was set for that year (Bolzau, *Almira Hart Lincoln Phelps,* 73). A letter (in the Emma Willard School Archives) from Emma Willard to Amos Eaton that year reads, "I have been expecting to see you on the subject of your winter's course of lectures," and refers to "settling that business," presumably payment for the lectures.

Eaton thought that girls "should be taught like boys" and gave the lecture part of the courses to them himself. In some years the lecture audience included both the girls and Rensselaer students, as recorded by a Troy Seminary student in a letter home in 1825: "I have been to the Rennselaer School to attend the philosophical lectures. They are delivered by the celebrated Mr. Eaton who has several students, young gentlemen. I hope they will not lose their hearts among twenty or thirty pretty girls." Eaton himself wrote, when asked about separate but equal instruction for the two sexes: "They should always be educated together. The usual separation arose in the monkish policy of the dark ages and has been continued with some other absurdities of our colleges and boarding schools. I can conceive of no greater exposure of female delicacy at a school than in church. A change in this particular would greatly improve the state of society . . . change the attachment [between the sexes] which savors rather of appetite than esteem into rational and durable friendship" (Bolzau, 74).

30. Eaton referred to Phelps in an 1838 letter to William Darlington as "scientific associate"; see also the Eaton letter to Silliman, n. 35, below. (Both quoted in McAllister, *Amos Eaton.*)

31. Phelps, private note, quoted in Bolzau, *Almira Hart Lincoln Phelps,* 77.

32. "There is not a female educational institution in America in which her textbooks are not known and there are few of the best similar European institutions to which their fame is not attended" (a quote by Phelps from an editor, Dr. Sears, who wrote a preface to her article "Duties and Rights of Women," published in the *National Quarterly Review* [June 1874]).

33. In the first edition of *Familiar Lectures on Botany* she wrote: "The descriptions of the genera and species have been furnished me by Professor Eaton, to whom my thanks for this and other kind offices are justly due." Later, in an 1832 letter to Eaton quoted by Arnold, she acknowledged: "Whatever success I have met with as a writer is wholly owing to your encouragement at the outset." Lois Barber Arnold, *Four Lives in Science: Women's Education in the Nineteenth Century* (New York, 1984), 41, n. 43.

34. In an 1830 letter to Silliman, quoted in McAllister, *Amos Eaton,* Eaton wrote: "As the Troy Female Seminary needs a similar work I prevailed on Mrs. Lincoln to undertake the translation . . . and do not hesitate to say that it is well translated."

35. Eaton did not think as well of all women who undertook the writing of textbooks. Emma Willard wanted to write a "kitchen chemistry" and asked Eaton's aid.

Eaton wrote to Silliman: "Mrs. Willard has been harrassing [*sic*] me these two years about a Kitchen Chemistry. She thinks she could write a good one, and is asking me for suitable books to aid her. My repeated reply is in my usual abrupt manner with her. I tell her she is totally incompetent, and that I can direct her to nothing" (1829 letter, quoted in McAllister, *Amos Eaton*).

36. Two Phelps daughters and Jane Lincoln were at the Troy Seminary. Bolzau discussed the Phelpses' courtship and Almira Phelps's relationships with her stepchildren in *Almira Hart Lincoln Phelps*, 50–56.

37. "Mother is much pleased with the ladies of that place [Brattleboro]. . . . We have attended many parties this winter. . . . We had a very great party . . . last evening and a very gay one too, there was about 30 here and they staid till about 11 o'clock" (quoted in Bolzau, *Almira Hart Lincoln Phelps*, 56–57, from a letter of Lucy Phelps to Helen M. Phelps [sister], February 6, 1832).

38. Letter of Almira Phelps to Mrs. Samuel Hart, her sister-in-law, quoted in Ruth Galpin, "Mrs. Almira Hart Lincoln Phelps" (unpublished manuscript, 1914); a copy of the manuscript is in the Emma Willard School Archives.

39. John Phelps wrote in the Phelps family Bible: "This lady was eminent as the authoress of various useful and popular works on general literature, science and education." Almira Phelps wrote in the same Bible that he encouraged her to take the West Chester position because "he had noticed the literary success of his wife." She noted further: "That in the decline of his life he was desirous of aiding his wife in literary and educational pursuits is proof of a noble and liberal spirit towards woman. . . . He was gratified in seeing his wife successful and honored, never imagining that this could detract from any distinctions to which he felt himself entitled" (Bolzau, *Almira Hart Lincoln Phelps*, 66–67). From her vantage point at least, he was a very unusual husband for that time—or any other time.

40. Bolzau, *Almira Hart Lincoln Phelps*, 118, from a letter of Almira and John Phelps to Hon. T. B. Dorsey, January 10, 1842.

41. Quoted from a letter of Emma Willard to Lydia Sigourny, September 23, 1841. A newspaper announcement by Phelps also noted that "she has accepted the charge of that institution, and will open her school there the first Wednesday in November" (Bolzau, *Almira Hart Lincoln Phelps*, 117–118).

42. Reading her own comments in succeeding editions of *Familiar Lectures on Botany* (which abound in Troy in the archives of both the Emma Willard School and Russell Sage College), one can see that Phelps was conversant with the new "natural" system of classification and did not deny its importance. She did, however, deny its value as a system for teaching botany as compared with the simpler, but artificial, sexual system of Linnaeus. Botanical manuals used today by college students include many "artificial keys," because in some plant families keys based on natural relationships are difficult to use for identification. Thus Phelps's views are to some extent vindicated. A. Hunter Dupree discussed the change from the Linnaean to the natural system of classification in his excellent account of the internal and external history of botany in the mid-nineteenth century in *Asa Gray*, 27–29, 51–53. Phelps's *Familiar Lectures* was also discussed in this context by Dupree, 51. Peter Ste-

vens has recently discussed the early ideas of de Candolle and Correa de Serra on natural classification of plants in "Haüy and A.-P. Candolle: Crystallography, Botanical Systematics and Comparative Morphology, 1780–1840," *Journal of the History of Biology* 17 (1984):49–82.

43. Sally Gregory Kohlstedt, "In from the Periphery: American Women in Science, 1830–1880," *Signs* 4 (1978):81–96.

44. Men as well as women used her textbook, as gray-haired male scientists told her at the AAAS meeting in 1874. There is a letter from daughter Myra Phelps to Ruth Galpin indicating that these may have included Asa Gray himself as a young man (Bolzau, *Almira Hart Lincoln Phelps,* 450). Her methods of teaching were the hands-on experiential methods popularized by her mentor Amos Eaton. Not only did students take apart flowers instead of just reading about them, they experimented with potent chemicals, as recommended in her *Familiar Lectures on Chemistry*. Many of these students, whether or not they themselves moved into Kohlstedt's third generation of women scientists, the scientific investigators, did move to other schools and later women's colleges. There they taught these methods of science to future women investigators. An annotated copy of Phelps's chemistry book was recently found at Russell Sage College. The original owner was a well-known chemistry professor in the early days of the college; it was probably her secondary school text.

45. See Smith-Rosenberg and Rosenberg, "The Female Animal." Physicians from the 1820s and throughout the rest of the century documented the supposed relationship between a woman's restricted place in society and her anatomy and physiology. The problems were primarily with her uterus and her delicate nerves. In addition, a young girl was thought to have a limited amount of "vital energy," which if overused on educational activity at puberty or thereafter would thwart development of her reproductive organs, endangering the production of the next generation. That this rhetoric was widely accepted is shown by a quote by the authors from the Regents of the University of Wisconsin in 1877: "Education is greatly to be desired but it is better that the future matrons of the state should be without a University training than that it should be produced at the fearful expense of ruined health; better that the future mothers of the state should be robust, hearty, healthy women, than that, by over study, they entail upon their descendants the germs of disease."

46. Phelps, "Duties and Rights of Women," *National Quarterly Review* (June 1874). Natural science as a means of appreciating and glorifying the creation was the establishment view in the early nineteenth century and the one she still clung to as Darwinism gained adherents in the latter part of the century. Her address to the AAAS at Hartford in 1870, at the age of eighty-seven, read: "Let us honor Darwin for all the good he has done in the search of truth; but may no member . . . ever be found willing to sell his birth-right, as a child of God, created by Him in His own image, for a miserable mess of *potage* 'evolved from a primordial cell'" (quoted in Bolzau, *Almira Hart Lincoln Phelps,* 449–450). Just before the meeting she had written to Henry Barnard (July 31, 1874): "You will have 'Evolution' ad nauseum, and I

should like to bear my testimony against this." She was not alone in these views. One of the two greatest biologists of her time, Louis Agassiz, who had died the previous year, denounced evolutionary ideas. By the turn of the century, however, evolution had become the establishment view for the religious as well as for the agnostic.

47. Almira Phelps's letter to Arianna Trail, quoted by Bolzau, *Almira Hart Lincoln Phelps.*

48. See n. 23. There is also a recent article by Frank S. Crossgate and Carol D. Crosswhite, "The Plant Collecting Brandegees, with Emphasis on Katherine Brandegee as a Liberated Woman Scientist of Early California," *Desert Plants* 7 (1985): 128–139, 158–163; this article and its sources may not be entirely reliable. Katherine Brandegee was a fascinating and controversial character.

49. Shteir, "Linnaeus's Daughters."

50. David E. Allen, "The Women Members of the Botanical Society of London," *British Journal for the History of Science* 13 (1980):240–254. An American counterpart, the New England Botanical Club, founded in 1895, was not open to women, though there was apparently no objection to amateurs. Richard A. Howard, "A Partial History of the New England Botanical Club," *Rhodora* 75 (1973):493–513. The charter members, according to Howard, were "seventeen gentlemen . . . seven professional botanists and ten amateurs." Mary Anna Day (1852–1924), a botanical librarian, aided the club in various ways, but the NEBC "was a gentlemen's organization and Miss Day was not eligible for membership." Women were first admitted in 1968, "after much agonizing debate." Ironically, the national elite society, the Botanical Society of America, did admit women from its founding in 1893, or at least one woman, Elizabeth Britton.

51. Rodgers, *"Noble Fellow,"* 110.

52. Gray wrote in the journal of his European trip in 1839: "Sullivant wants, I suppose, a microscope of single lenses—a good working instrument. . . . This last I shall procure for him in London, where they produce a more perfect instrument than the French" (Rodgers, *"Noble Fellow,"* 111). Gray's search for Sullivant's microscope in London is also discussed by Dupree in *Asa Gray.*

53. Rodgers, *"Noble Fellow,"* 153. Eliza Wheeler Sullivant was probably born in 1817, since Rodgers stated (p. 99) that she was only seventeen at the time of her marriage, November 29, 1834. He also stated that she was thirty at the time of her death, August 23, 1850, which would make her fourteen at the time of her marriage! I have not found another source. In 1850 Sullivant married Eliza's niece, Caroline Eudora Sutton. In addition to producing six more children, she also did botanical work with him. She claimed, through her husband, the title of "assistant muscologist" to the United States Exploring Expedition and had a Hawaiian moss named for her (Rodgers, 212). She survived Sullivant, but no birth or death dates were given for her.

54. Rodgers, *"Noble Fellow,"* 129.

55. "Within a little more than a decade, Sullivant and his wife conceived an arrangement, which because of its clarity and truthfulness, became a model to Bryolo-

gists the world over. Not only was their arrangement of specimens in bound form emulated but also their plan and mode of drawing botanical illustrations on plates were followed. Their drawings were analytically accurate, illustrating botanical specimens in detail. . . . Undoubtedly the artistic judgment of Eliza Sullivant lent a finished excellence" (Rodgers, *"Noble Fellow,"* 140).

56. William Sullivant to Asa Gray, quoted in Rodgers, *"Noble Fellow,"* 208 (written after Eliza Sullivant's death in 1850, but not sent until a year later, with the following letter).

57. William Sullivant to Asa Gray, October 12, 1851, quoted by Rodgers, *"Noble Fellow,"* 208–209.

58. Asa Gray to Charles Wright, professional plant collector, January 23, 1852, quoted by Rodgers, *"Noble Fellow,"* 209.

59. Rossiter, *Women Scientists in America,* 208–209. Edith and Frederick Clements were chosen as a botanical couple to determine how "conjugal collaboration" in the 1920s compared with the Sullivants' collaboration in the 1840s and to examine the recognition question.

60. Eugene Cittadino referred to the "many theoretical works that established Clements's reputation as the preeminent American ecologist" and also commented on the concept of ecological succession, the regular replacement of one plant community by another, introduced in one of his early books, as "a concept that was to remain central to Clements's theoretical work and also in plant ecology for the next half century." Cittadino, "Ecology and the Professionalization of Botany in America, 1890–1905," *Studies in the History of Biology* 4(1980):171–198. Henry Chandler Cowles was the other important theorist in plant ecology of this period. His classical studies of sand dune succession strikingly illustrated the concept of succession.

61. Rossiter, *Women Scientists in America,* 80, table 4.1.

62. Anna Botsford Comstock did important work in her husband, J. H. Comstock's entomology laboratory at Cornell. Unlike Edith Clements, she did not have a doctorate. Comstock became the first woman assistant professor at Cornell University in 1899. She was demoted to lecturer in 1900 because of trustee objections, but was finally reappointed assistant professor (of nature study) in 1913 and professor in 1920 (James G. Needham, "The Lengthened Shadow of a Man and his Wife, II," *Scientific Monthly* [March 1946]).

63. Edith S. Clements, *Adventures in Ecology: Half a Million Miles . . . from Mud to Macadam* (New York, 1960). This is a largely chronological account of their life and travels. A list of her publications is at the front of the book, as well as a list of nine books by Frederick Clements.

64. Clements, *Adventures in Ecology,* 222.

65. Rossiter, *Women Scientists in America,* 85.

66. John Hendley Barnhart, "Elizabeth Britton as a Scientist," *Journal of the New York Botanical Garden* 41 (1940):142–143.

67. General information on Britton and her publications can be found in William Campbell Steere, "Britton, Elizabeth Gertude Knight," in NAW, vol. 3. Steere did not know her personally, but on the basis of interviews with those who did he de-

scribed her as "a woman of great physical and mental energy and the possessor of a keen intellect," and in her later years as a "strong-minded, outspoken, formidable woman."

68. Henry A. Gleason, *The New Britton and Brown Illustrated Flora of the Northeastern United States and Adjacent Canada* (New York, 1968). Notice the title; Addison Brown, Britton's original coauthor, was more a financier for the project than a botanist.

69. Marshall A. Howe, "Elizabeth Gertrude Britton," *Journal of the New York Botanical Garden* 35 (1934):97–105.

70. Douglas Sloan, "Science in New York City, 1867–1907," *Isis* 71 (1980): 35–76. Sloan noted that "While Britton [N. L.] had been able to amass only a modest fortune, he had the friendship of Vanderbilt, Carnegie, and Morgan, and his garden was the pet philanthropic project of more than 350 of New York's highest social elite" (p. 59).

71. Immigrant women are important in American science, but they were in other fields, such as chemistry, or they appear later. See P. Thomas Carroll, "Immigrants in American Chemistry," in *The Muses Flee Hitler: Cultural Transfer and Adaptation, 1930–1945*, ed. Jarrel C. Jackman and Carla M. Borden (Washington, D.C., 1983). An eminent exception among women botanists is Katherine Esau, who was born in the Ukraine in 1898. After studies in Germany, she immigrated to California in 1922. She eventually earned a Ph.D. in botany, and after serving the maximum number of years in each rank she became a full professor at the University of California at Davis. She has done important research in plant anatomy. Ray F. Evert ("Katherine Esau," *Plant Science Bulletin* [of the Botanical Society of America] 31, no. 5[1985]:33–37) called her today's "grande dame of American Botany" and reported that she was currently preparing a new text on a personal computer purchased at age eighty-six.

72. My informants, all at the New York Botanical Garden, were Elizabeth Hall (see n. 15); Thomas H. Everett, retired director of horticulture; and Lilian Weber, former secretary to Elizabeth Britton and T. H. Everett. More oral history is urgently needed, both from women scientists still living and from the friends and coworkers of those who have died.

73. Evelyn Fox Keller (*A Feeling for the Organism: The Life and Work of Barbara McClintock* [New York, 1983], 53) mentioned that Margaret Ferguson in her many years of teaching at Wellesley had trained more women botanists and wives of botanists than anyone else. At Cornell in McClintock's time there were four such botanical wives all working in their (professor) husbands' laboratories.

74. A letter of May 29, 1912, addressed to "Mrs. Elizabeth Britton, New York Botanical Garden," reads: "Dear Madam:—I have the honor to inform you that at a meeting of the Board of Managers of the New York Botanical Garden held May 23, 1912, you were appointed Honorary Curator of Mosses. Yours truly, [signed] N. L. Britton, Secretary" (in Elizabeth Britton Archives, New York Botanical Garden; hereafter NYBG).

75. Letter from Nathaniel L. Britton to Elizabeth G. Britton, August 10, 1891. Elizabeth Britton Archives, NYBG.

76. Elizabeth Britton, "The Jaeger Moss Herbarium," *Bulletin of the Torrey Botanical Club* 20 (1893):235–236.

77. Letter from Nathaniel L. Britton to Elizabeth G. Britton, September 1, 1908. Elizabeth Britton Archives, NYBG.

78. Letter from Nathaniel L. Britton to Elizabeth G. Britton, August 9, 1891. Elizabeth Britton Archives, NYBG.

79. Steere, *Notable American Women,* vol. 3. See also John Hendley Barnhart, "The Published Work of Elizabeth G. Britton," *Bulletin of the Torrey Botanical Club* 62 (1935):1–17.

80. Some of these positions were sex-segregated, as was true also in astronomy in the 1880s, particularly at the Harvard University Observatory. In federal government employment from the late 1880s until 1920, women were hired as scientific assistants, particularly in plant pathology, a recent segregate from botany. E. F. Smith at the USDA hired at least twenty such assistants, several of whom did outstanding research. See Margaret Rossiter, "'Women's Work' in Science," *Isis* 71 (1980): 381–391; and idem, *Women Scientists in America,* 60–63.

81. Rossiter, *American Women in Science,* 140, table 6.3.

82. Saul D. Feldman, "Impediment or Stimulant? Marital Status and Graduate Education," in *Changing Women in a Changing Society,* ed. Joan Huber (Chicago, 1973), 220–232.

83. E. D. Rudolph, "Women in Nineteenth Century American Botany: A Generally Unrecognized Constituency," *American Journal of Botany* 69 (1982): 1346–1355.

84. Dorinda Outram, *Georges Cuvier: Vocation, Science and Authority in Post-Revolutionary France* (Dover, N.H., 1984), esp. chaps. 3, 4, and 5. Cuvier's colleagues and rivals, all male, used similar strategies.

85. George H. Daniels, "The Process of Professionalization in American Science: The Emergent Period, 1820–1860," in *Science in America Since 1820,* ed. Nathan Reingold (New York, 1976).

86. Dupree, *Asa Gray,* 107–112.

87. The fields of genetics and cytogenetics were both developed in the decade before 1920 in T. H. Morgan's *Drosophila* laboratory at Columbia. McClintock received her Ph.D. in botany from Cornell in 1927, but her work then and thereafter was in the genetics and cytogenetics of corn (maize). Distinctions between botany and other fields in not always clear-cut. Cytogenetics remained in the botany department at Cornell in the 1950s, whereas genetics was in the plant breeding department.

88. Keller, *A Feeling for the Organism,* 63–77; and idem, *Reflections on Gender and Science* (New Haven, Conn., 1985), 158–159. Cornell and Iowa State universities both refused her a faculty position; the University of Missouri finally did hire her, but failed to promote her. Her only strategy for many years was one of temporary

fellowships. Harriet Creighton, who did important work on corn cytogenetics with McClintock at Cornell, left her research for a teaching position at a woman's college shortly after finishing her Ph.D.

89. To resort to autobiography as evidence, I can report such obstacles. As a young married graduate student in botany at Cornell many years after McClintock, I was told that a graduate teaching assistantship in botany would only be offered to a woman if a man could not be found. In the late 1960s at another university, I was told that married women with children could not successfully carry out a Ph.D. program in biology. In the early 1970s, when I applied for a faculty position, future colleagues seriously doubted that a married woman with three children would have the time and energy to be a professor.

90. Until recently many women, well trained in science, settled for the "two-person career"; as described by Papanek, this varied from a wife's role as fosterer of her husband's career by status maintenance to actual collaboration, usually publicly unrecognized, in the husband's laboratory (Hanna Papanek, "Men, Women, and Work: Reflections on the Two-Person Career," in *Changing Women in a Changing Society,* ed. Huber). One could extend that concept to the still fairly common tenure-track faculty member (usually male)–research associate (usually female) couple at the same university and to the couple in the same field who share a faculty position. In the fifteen years since that article was written, "two-person careers" and even the sharing of faculty positions, in which both often work full-time for half-pay, have gone out of favor, especially with young women scientists.

Two-career families have replaced two-person careers to the extent that Winfield's recent book on commuter marriages describes the problems of employers who counted on a wife (unpaid) as well as the employee—but she is commuting to her own position in another city. "'Commuter Marriages' a Growing Necessity for Many Couples in Academe" reads a recent headline in the *Chronicle of Higher Education* 31, no. 19 (January 22, 1986). It reports that wives who formerly did part-time teaching or research on "soft money," now pursue tenure-track positions, even in distant cities. The result is a weekly or less frequent commute for one or both partners. Two books discuss the advantages and disadvantages of commuter marriages; both point out that of the 700,000 American couples that are currently doing it, commuting often works out best for academic couples with flexible work schedules, work that can be done at home, and long vacations. See Fairlee E. Winfield, *Commuter Marriage: Living Together, Apart* (Irvington, N.Y., 1985); and Naomi Gerstel and Harriet E. Gross, *Commuter Marriage: A Study of Work and Family* (New York, 1984). What neither book points out is that for commuting botanists and other scientists, although teaching hours may be flexible if chairpersons and administrators are sympathetic, research laboratories are not flexible. They cannot easily move with the commuter. Other problems involve families, especially young children, whether or not it is the woman who commutes. Nevertheless, the willingness of many couples to try to solve these career-and-marriage problems in a way that considers the careers of both partners equally is a new and encouraging development.

Chapter 6: Marital Collaboration

1. Pierre Curie, "Radioactive Substances, Especially Radium" (Nobel lecture, June 6, 1905), in *Nobel Lectures. Including Presentation Speeches and Laureates' Biographies. Physics. 1901–1921* (Amsterdam, 1967), 73–78.

2. Traditionally, collaboration with a male has provided a back door through which women could enter science unobtrusively. For example, Caroline Herschel's love for her astronomer brother, William, inspired her to work primarily on his projects, but secondarily on her own enthusiasms. Ostensibly, she appeared incensed by the idea that she had anything to offer science apart from helping William. "I did nothing for my brother," she wrote, "but what a well-trained puppy-dog would have done: that is to say, I did what he commanded me. I was a mere tool which he had the trouble of sharpening" (Mary Cornwallis Herschel, *Memoir and Correspondence of Caroline Herschel* [New York, 1876], 142). In spite of her protestations, Caroline began to enjoy her own successes and became annoyed when her sweeping for comets "was interrupted by being employed with writing down my Brother's observations with the large 20-feet" (Constance A. Lubbock, ed., *The Herschel Chronicle: The Life-Story of William Herschel and His Sister Caroline Herschel* [Cambridge, 1933], 149–150).

Collaboration, however, was only one of several ways in which a married woman might enter a scientific career. Marriages in which the husband, though not a scientist himself, was indulgent toward his wife's scientific interests also produced women scientists. Jane Marcet (1769–1858) and Mary Somerville (1780–1872) represent this group. Marcet, whose husband preferred his hobby, chemistry, to his profession, medicine, encouraged his wife to indulge the nineteenth-century public taste for popular science. She produced a number of introductory science books, including the popular *Conversations on Chemistry*. Mary Somerville's first marriage, to Samuel Grieg, a captain in the Russian navy with a low opinion of the intellectual capabilities of women, was a disaster for her mathematical studies. After Grieg died and she married her first cousin, William Somerville, Mary's scientific career flourished. Unlike his predecessor, William's encouragement made his wife's scientific work possible in spite of her active social life and family responsibilities.

Other married women became interested in science because their husbands were scientists. They probably would have participated in any activity in which their husbands were engaged, and their involvement in science seldom long survived their husbands. Elizabeth Agassiz (1822–1907) and Mary Anne Palmer Draper (1839–1914) exemplify this group. Elizabeth Agassiz's interest in science only surfaced after her marriage to Louis Agassiz. Her younger sister, Emma, reported that Elizabeth "was a very attractive girl, but one for whom I should not have predicted a career that would give her a wide reputation" (Lucy Allen Paton, *Elizabeth Cary Agassiz: A Biography* [Boston, 1919], 25). Totally without scientific training, her information came from her association with her husband. She became important in conserving, elucidating, and popularizing his ideas. Mary Anne Palmer Draper, wife of Henry Draper, professor of physiology and chemistry at the University of the City of New

York (later New York University), became involved in her husband's interest in astronomy after their marriage. During his lifetime, she participated in astronomical projects, ably assisting her husband. After Henry Draper died, her active participation as an investigator ended. Her major contribution to astronomy was in the institution of a memorial to her husband in the form of a bequest to the Harvard College Observatory.

3. "Dietrich," *Neue deutsche Biographie,* Herausgegeben von der historischen Kommission bei der bayerischen Akademie der Wissenschaften, 3 (Berlin, 1957), 694–695.

4. In a natural system of classification it is assumed that the categories reflect relationships between members of the taxa. In an artificial system, the categories are considered convenient ways of considering animals and plants, not necessarily reflective of actual relationships.

5. Amalie Dietrich, *Australische Briefe,* ed. Augustin Lodewyck (Melbourne, 1943), 1–120.

6. Charitas Bischoff, *The Hard Road: The Life Story of Amalie Dietrich, Naturalist, 1821–1891* (London, 1931), 47.

7. Ibid., 42.

8. Ibid., 146–147.

9. Ibid., 261.

10. Ibid., 141.

11. Ibid., 146.

12. Dietrich, *Australische Briefe,* 64.

13. Bischoff, *The Hard Road,* 244.

14. Ibid., 315–316.

15. William Huggins and Margaret Huggins, *An Atlas of Representative Stellar Spectra. From 4870 to 3300. Together with a Discussion of the Evolutional Order of the Stars, and the Interpretations of Their Spectra. Preceded by a Short History of the Observatory and Its Work* (London, 1899), 21.

16. Ibid., 3.

17. Ibid., 6–7.

18. Ibid., 12.

19. William Huggins, "Further Observations on the Spectra of Some of the Stars and Nebulae with an Attempt to Determine Therefrom whether These Bodies Are Moving towards or from the Earth, Also Observations on the Spectra of the Sun and of Comet II," *Philosophical Transactions of the Royal Society of London* 158, pt. 2 (1868):529–564.

20. William and Margaret Huggins, *An Atlas of Representative Stellar Spectra,* 17.

21. "Lady Huggins: Her Work in Astronomy," *Times,* March 25, 1915.

22. Ibid.

23. William Huggins, "On a New Group of Lines in the Photographic Spectrum of Sirius," *Proceedings of the Royal Society* 48 (1890):216–217.

24. Huggins, "On the Spectrum, Visible and Photographic of the Great Nebula in Orion," *Proceedings of the Royal Society* 46 (1889):40–61.

25. Huggins, "On Wolf and Rayet's Bright-Line Stars in Cygnus," *Proceedings of the Royal Society* 49 (1890):33–46.

26. Huggins, "On the Photographic Spectra of Uranus and Saturn," *Proceedings of the Royal Society* 46 (1889):231–233; "On the Atmospheric Bands in the Spectrum of Mars, 1895," *Astrophysical Journal* 1 (1895):380.

27. Huggins, "Preliminary Note on Nova Aurigae," *Proceedings of the Royal Society* 50 (1892):465–466; "On Nova Aurigae," *Proceedings of the Royal Society* 51 (1892):486–495; "On the Bright Bands in the Present Spectrum of Nova Aurigae," *Proceedings of the Royal Society* 54 (1893):30–36.

28. William Huggins and Margaret Huggins, eds., *The Scientific Papers of Sir William Huggins, K.C.B., O.M.* (London, 1909), 393–458.

29. "Sir William and Lady Huggins on Stellar Spectra," *Times,* April 20, 1900.

30. Margaret Lindsay Huggins, obituary notice, *Monthly Notices of the Royal Astronomical Society* 76 (February 1916):278–284.

31. Sarah Marks (Hertha Ayrton), "The Uses of a Line-Divider," *London, Edinburgh, and Dublin Philosophical Magazine and Journal of Science* 19, 5th ser. (April 1885):280–285.

32. Evelyn Sharp, *Hertha Ayrton, 1854–1923: A Memoir* (London, 1926), 107–109.

33. Unconventional Barbara Bodichon's father was radical, Unitarian, and wealthy. He provided Barbara, who was illegitimate, with a liberal education. She became a good friend of George Eliot, married a French Algerian doctor at thirty, and authored *Women and Work,* in which she asserted women's rights to all kinds of work, before and after marriage. In 1863 Emily Davies, who had organized schoolmistresses into associations, convinced officials at Cambridge University to open its local examinations (equivalent to entrance examinations) to girls. Davies opened a women's college at Hitchin, near Cambridge, in 1869, and in 1873 the college moved to Girton, on the outskirts of Cambridge. She insisted that girls should study the same subjects and take the same examinations as men. Bodichon was involved in the establishment of the college; she took an active part in fund raising, but was left off the committee, since it was feared that her unconventional reputation, particularly her association with the suffrage movement, would antagonize potential supporters. A second women's college established at Cambridge, Newnham, operated under a different philosophy. Anne Clough and others involved in this college insisted that women's education should not be cast in the anachronistic pattern of male institutions—classics and mathematics—but, instead, should be made relevant to women's needs.

34. Sharp, *Hertha Ayrton,* 88. Philanthropy, not simply almsgiving, was considered well suited to woman's nature. Education became one focus of this philanthropy. Barbara Bodichon, although her reputation made her participation on school boards unlikely, delighted in helping individual students obtain an education.

35. Ibid., 92.

36. Ibid., 113.

37. Ibid., 133.

38. Ibid., 154.
39. Ibid., 129.
40. Ibid., 132; Hertha Ayrton, "On the Relations between Arc Curves and Crater Ratios with Cored Positive Carbons," *Report of the British Association for the Advancement of Science* [held at Toronto in August 1897] (London, 1898), 575–577.
41. Hertha Ayrton, "The Drop of Potential at the Carbons of the Electric Arc," *Report of the British Association for the Advancement of Science"* [held at Bristol in September 1898] (London, 1899), 805–807.
42. Sharp, *Hertha Ayrton,* 138–139.
43. Ibid.
44. Ibid.
45. Ayrton, *The Electric Arc,* 1902.
46. R. Appleyard, *The History of the Institution of Electrical Engineers (1871–1931)* (London, 1939), 167.
47. Ayrton, "The Mechanism of the Electric Arc," *Philosophical Transactions of the Royal Society of London,* ser. A, 199 (1901–1902):299–336.
48. Carpenter, H.C.H., "Mrs. Hertha Ayrton," *Nature* 112 (December 1, 1923):800–801.
49. Ibid., "The Origin and Growth of Ripple-Marks," *Proceedings of the Royal Society of London,* ser. A, Containing Papers of a Mathematical and Physical Character, 84 (October 21, 1910):285–310.
50. Ibid.
51. Sharp, *Hertha Ayrton,* 177.
52. Ibid., 254.
53. Ibid., 225–235.
54. Ibid., 233.
55. Sharp, *Hertha Ayrton,* 183.
56. Carpenter, "Mrs. Hertha Ayrton," 800.
57. Ibid., 801.
58. Ibid.

Chapter 7: Maria Mitchell and the Advancement of Women in Science

Acknowledgments: This essay is reprinted, with permission and with minor changes and updated notes, from "Maria Mitchell: The Advancement of Women in Science," *New England Quarterly* 51 (1978):39–63. The author thanks Janet Bogdan, Margaret Rossiter, Robert Post, and Carol Gregory Wright for their comments and assistance.

1. This activity is discussed in Sally Gregory Kohlstedt, "In from the Periphery: American Women in Science, 1830–1880," *Signs* 4 (1978):39–63. A comprehensive study of women in science is in Margaret Rossiter, *Women Scientists in America:*

Struggles and Strategies to 1940 (Baltimore, Md., 1982). The few books previous to Rossiter that dealt with women and science recount often heroic efforts without much attention to the mutual support networks; among the best are Edna Yost, *American Women of Science* (Philadelphia, 1943); Madeline B. Stern, *We the Women* (New York, 1962); and Phoebe Kendall, *Maria Mitchell: Her Life, Letters and Journals* (Boston, 1896).

2. Quoted in Joan N. Burstyn, "Sex and Education: The Medical Case against Higher Education for Women in England, 1870–1900," *Proceedings of the American Philosophical Society* 17 (April 10, 1973):79–89. Contemporary John Zahn (under pseudonym H. J. Mozans) in *Women in Science* (New York, 1913) wrote: "Perhaps the greatest change is that woman now does thoroughly what before she did only as amateur" (p. 105). Standard accounts of the process of professionalization tend to ignore the subgroups excluded for reasons other than ability: George Daniels, "The Process of Professionalization in American Science: The Emergent Period, 1820–1860," *Isis* 58 (1967):151–166; and Daniel H. Calhoun, *The American Civil Engineer: Origins and Conflict* (Cambridge, 1960).

3. Mitchell's biographers, especially her sister Phoebe Kendall, tend to deemphasize her women's rights activism. An important exception is Dorothy J. Keller, "Maria Mitchell: An Early Woman Academician" (Ph.D. diss., University of Rochester, 1974). Only a few scholars—Ellen Du Bois, for example—have begun to explore the reform activities of women between the Civil War and the emergence of a second suffrage movement and a rise in socialist feminism.

4. The academy's *Report* for 1848 is quoted in Keller, "Maria Mitchell," 3.

5. Eleanor Flexner, *Century of Struggle: The Woman's Rights Movement in the United States* (Cambridge, 1959), 71–77.

6. Mitchell's diary has no comments about the role and status of women until about 1855. Most of the Mitchell manuscripts are at the Maria Mitchell Science Library on Nantucket Island (hereafter MMM) and available on microfilm at the American Philosophical Society in Philadelphia; some correspondence is preserved in the Vassar College Archives (hereafter VCA).

7. The trip and the meetings with the Herschels and Somerville seemed to redirect her attention toward the role of women in science. Throughout her career, Mitchell would use Caroline Herschel, in particular, as an example of women's underutilized potential. See her "Reminiscences of the Herschels," *Century Magazine* 38 (1889):903–909.

8. Helen Wright, *Sweeper in the Sky: The Life of Maria Mitchell, First Woman Astronomer in America* (New York, 1949), 104, 120. A late notebook entry (undated, item 35, MMM) by Mitchell acknowledges three crucial aids to her success as a scientist: the first woman teacher who taught her to love learning, the man who loaned her $100, and the women who gave her the telescope.

9. Ibid., 69.

10. "The Female World of Love and Ritual: Relations between Women in Nineteenth-Century America," *Signs* 1 (1975):1–30.

11. Mitchell to Rufus Babcock, March 18, 1964, VCA.

12. "Maria Mitchell at Vassar," unidentified clipping, item 35, MMM. Students considered Mitchell "sternly just yet marvelously kind." See comments in Catherine Weed Barnes and Ella Dietz Clymer, "Report to the Delegates from Sorosis," in Association for the Advancement of Women, *Report of the Fifteenth Women's Congress held in New York City, October, 1887* (Fall River, Mass., 1890), 26; news clipping from the *Toronto Mail* (October 16, 1890); and an undated *Toronto Globe* in the Julia Ward Howe Scrapbooks, 2:5, 52, Schlesinger Library on the History of Women in America, Radcliffe College (hereafter SLR).

13. Kendall, *Maria Mitchell,* 31.

14. Julia M. Pease to "Momma," May 2, 1875, Student Letter File, VCA. The series of letters from January 1872 to May 1875 suggest the importance a concerned faculty member might have in the life of her students. Also see David F. Allmendinger, Jr., "Mount Holyoke Students Encounter the Need for Life-Planning, 1837–1857," *History of Education Quarterly* 19 (1979):27–46.

15. Martha Leb. Goddard to [Sarah J.] Spaulding, Boston, September 24, 1875, Mitchell MSS, VCA. Goddard was incensed at some "libel" by a "coarse, disloyal, treacherous" Vassar student who "told the secrets of her sex, and told them in so false a way that she lied—even if her facts were true." Goddard expressed none of the fears of homosexuality that emerged in Victorian America when she wrote: "Almost every girl has, for her first love, an older girl or some young woman, often her school-teacher, to whom she gives a pure and adoring affection, and to whom she offers gifts as she dares—Haven't you many a time saved your sweetest flowers, and done your finest handwork for some woman whom you loved?" This period may well have been a turning point away from the acceptance of the close woman-to-woman relationships discussed in Smith-Rosenberg's "Female World of Love and Ritual."

16. Julia Pease to Carrie [Pease], Vassar, January 17, 1875, Student Letter File, VCA; and John H. Raymond to Mitchell, Vassar, Monday [1875], Mitchell MSS, VCA.

17. Gertrude Mead to Mary Whitney, Vassar, September 29, 1870, Astronomical Observatory Notebooks, VCA. A letter to Miss Hopson dated only November 25 indicates that Mead might be willing to sell her Clark telescope, thus indicating that she did drop her astronomical study. Mitchell MSS, VCA. The number of unsuccessful aspirants to science careers can never be tabulated, but various evidence exists about thwarted hopes; see, e.g., the letters from Alice Bache Gould to Elizabeth Agassiz, Agassiz MSS, SLR.

18. Miscellaneous notes, October 18, 1868, item 35, MMM.

19. European notes, item 49, MMM. Mitchell interviewed Cobbe and discussed the role of women, concluding: "The world would never know how much it had lost by Mrs. Somerville's not having been educated."

20. June 3, 1878, item 29, MMM. She attended meetings of the New England

Women's Club and on one occasion sat impatiently through a lecture by Harvard mathematician Benjamin Pierce, who argued "that women were the educators of men and should be satisfied with that" (notebook, item 29, MMM).

21. Journal, item 48, MMM. The journal for the second trip is filled with detailed notes of Mitchell's interviews with educators in Edinburgh, Cambridge, and London.

22. Wright, *Maria Mitchell,* 197–204.

23. These included Mrs. Horace Mann, Elizabeth Peabody, and Mrs. Charles Pierce (J. J. Croly, *Sorosis: Its Origins and History* [New York, 1886]).

24. *Papers [of the AAW] . . . 1876* (Washington, D.C., 1877), 121–122.

25. Mitchell was simply listed as an honorary member in Marguerite Dawson Winant's, *A Century of Sorosis, 1868–1968* (Uniondale, N.Y., 1968), 113.

26. *Papers . . . 1876,* 123. Altogether the planning committee estimated that 1,620 letters and notices had been mailed to women in the United States and Europe.

27. Lita Barney Sayles, *History and Results of the Past Ten Congresses of the Association for the Advancement of Women,* published separately and as part of the *Papers Read before the Association for the Advancement of Women at Its Tenth Annual Congress* (1882), 6–7. There is only a chapter on the AAW, whose history has been eclipsed by the more long-lived, activist General Federation of Women's Clubs, in Karen J. Blair, *The Clubwoman as Feminist: True Womanhood Redefined, 1868–1914* (New York, 1980). Also see Julia Ward Howe, *Reminiscences*; Laura E. Richards and Maude Howe Elliott, *Julia Ward Howe, 1819–1910* (Boston, 1916); and Howe, *Historical Account of the Association for the Advancement of Women, 1873–1893, World's Columbian Exposition, Chicago, 1893* (Dedham, 1893), although this last introduces some inaccuracies repeated in subsequent accounts. The AAW published reports from 1873 to 1893, although those to 1881 are only lists of papers and some summary of business items. Detailed reports of committees and state vice-presidents began in 1882. Selected papers, although numbered in sequence from 1873 to 1891, were apparently published from 1873 to 1876 and from 1881 to 1891. The most complete series found by this author are housed in the Vassar College Archives. There is apparently no manuscript record of the AAW, although the Howe papers at the Schlesinger Library have considerable material relating to the late 1880s and 1890s. A series of letters from Howe to Emily Howland between 1888 and 1894 indicate Howe's unsuccessful efforts to find younger members, a more efficient governing system, and money for the organization; see Sophia Smith Collection, Smith College, Northampton, Massachusetts.

28. Ibid., 9. Local AAW officers themselves printed disclaimers in local newspapers to ally fears of the potential audience. "It has been announced that women's right to suffrage, marriage, divorce, free love, and Bloomerism will not [be] considered in the deliberations of the Congress" (*Fayetteville Recorder,* October 14, 1875); and "It will be seen from the list of topics printed above, that the Congress will have nothing to do with the discussion of women's suffrage, or other topics connected

with what is popularly termed the 'women's rights' movement" (*Syracuse Journal,* September 18, 1875). Clippings file of the Onondaga Historical Society, Syracuse, New York.

29. *Papers and Letters Presented at the First Woman's Congress of the Association for the Advancement of Women* (New York, 1974), 89–98.

30. Note to Miss Lapham, undated, attached to a reprint of "Notes on the Satellites of Jupiter," Mitchell MSS, VCA; also see Mitchell to Antoinette Brown Blackwell, January 19, 1878, Blackwell Family MSS, SLR.

31. "Address of the President," *Papers . . . 1875,* 3. Commonplace Book, November 2, 1873, item 50, MMM. Mitchell wrote: "I am amazed that they considered me a good presiding officer. I made my little speech boldly and fearlessly," and she emerged from the meeting feeling "fifteen years younger" (Howe, *Reminiscences,* 387).

32. Confronted with an almost immediate controversy about whether to have an opening prayer, the pragmatic Mitchell compromised on the spot with a minute of silent prayer; it was over, according to one reporter, "in half that time" (*Woman's Journal,* October 23, 1875).

33. "Address of the President," *Papers . . . 1875,* 3.

34. *Syracuse Journal,* October 18, 1875.

35. Journal for 1873, item 48, MMM. Higginson gave an address on the higher education of women and Agassiz took exception to critical comments about Harvard's treatment of women, insisting that he made no distinction among his assistants. Mary Livermore asked why her daughter could not get a regular college eduction in Boston or Cambridge. A bemused Mitchell recorded the debate in detail and concluded optimistically in her journal: "It is clear that learning is to be made possible for women; that they shall accept it and value it as the duty for them, and that they shall use it wisely. For myself it seems to me my duty is to prepare a Lecture on 'Science for Women' and in the claim to their consideration which I think it has."

36. *Papers . . . 1876* (Washington, D.C., 1877), 9–10.

37. The meeting in Philadelphia was not pleasant, the assigned hall had problems, and the Philadelphia women threatened to cancel their sponsorship if suffrage was mentioned. When Howe was nominated for president, the western representatives shouted out the names of western women, including controversial temperance leader Frances Willard, and ultimately Mitchell was accused of "carrying things with a high hand" (Copybook, November 15, 1876, item 50, MMM). Mitchell confessed privately: "I admire the courage of people who can work with all kinds of people, but I cannot do good work in inharmonious relations & must keep free of them." (Mitchell to Alice Stone Blackwell, January 19, 1878, Blackwell Family MSS, SLR).

38. Participants reported on summer school sessions on Penikese Island under Agassiz and at Harvard's Botanic Garden in which women were given unprecedented opportunity to study ("Our Museums and Our Investigators," *Papers . . . 1876,* 26–28).

39. "Lectures on Women," undated, item 17, MMM.

40. To Antoinette Brown Blackwell, March 22, [1879], Blackwell Family MSS, SLR.

41. When too busy to work on the committee reports, Mitchell solicited papers on education and science (Mitchell to Blackwell, January 19, 1878, Blackwell Family MSS, SLR). Upon leaving the presidency, she resolved "to push the Scientific Dept. going a little. It is all very uphill work—mainly because women have so little money or control" (Copybook, item 50, MMM).

42. Item 47, MMM.

43. Ibid. Although the circulars themselves are missing, a final total of seventy-nine responses are recorded in a copybook apparently complete and organized by state.

44. Ibid. Mitchell's own tabulations are at the end of the copybook.

45. The Dana Society, composed largely of social leaders concentrated on self-help and on local park projects; there seemed to be little interest in careers for women in science. The society's manuscripts are deposited at the Albany Institute and Historical and Arts Society, Albany, New York.

46. Dolley apparently wrote to every woman about whom she had heard in up-state New York, but none felt she was a contributor to science. Within three years, however, Dolley again wrote to Mitchell, describing the newly founded Rochester Society of Natural History, which was composed chiefly of women (item 46, MMM). Also see the society's manuscripts and annual reports for 1880 and 1882 at the Rush Rhees Library, University of Rochester, Rochester, New York.

47. Delaware similarly reported an active study group, but no women physicians.

48. Item 46, MMM. A note from J. J. Burleigh to Grace Anna Lewis pointed out that Reed had applied unsuccessfully to the summer school under the Agassiz family on Penikese Island, but had subsequently enrolled at Cornell.

49. Ibid. Richards wrote a long note to her former teacher, explaining the situation in Boston.

50. Ibid. F. E. Foote also noted the variety of educational opportunities in the greater Boston area, including the special courses at Harvard, the scientific course at the Lowell Institute, the summer course in Salem, and the students at the Technical Institute.

51. Ibid.

52. Sally Gregory Kohlstedt, "The Nineteenth-Century Amateur Tradition: The Case of the Boston Society of Natural History," in *Science and Its Public: The Changing Relationship,* ed. G. Holton and W. A. Blanpied (Dordrecht, Holland, 1976), 185–186.

53. Annotations in her copybook for 1876 indicate that the second mailing in 1880 was much smaller. Mitchell had only nineteen replies by the time of her report. A list dated August 20, 1880, contains the names of twenty-one women in various states (most of whom had sent replies in 1876) and five high schools; responses out-

side this initial group indicate that Mitchell must have sent additional letters and that not all inquiries were answered.

Walcott was a close friend of Howe and active in the AAS, sometimes making the report in Mitchell's absence. Blackwell, whose book of essays *The Sexes Through-out Nature: An Inquiry into the Dogma of Woman's Inferiority to Man* (New York, 1875) contained a "new scientific estimate of feminine nature," had relinquished her ministerial work in favor of scientific and philosophical study.

54. Mitchell suggested that "when any town has a [woman] physician in 1876, it has several in 1881." Draft copy of Mitchell's report for 1881 is in item 24; the published reports began in 1882.

55. Crocker had described the associate status of members in the Boston Society of Natural History and expressed her hope that women would soon be made regular members as well. Lewis submitted a list of women admitted to the Academy of Natural Sciences since 1876. Although there was a movement toward specialized societies in the 1870s, these women were of the generation that still relied on local acceptance.

56. Item 46, MMM. Endowment was a major theme of her address "The Collegiate Education of Girls," read in 1880 before the AAW and published separately in Boston in 1881.

57. Item 21, MMM. On April 26, 1882, Mitchell recorded a visit from Thompson in her diary during which Thompson offered to pay Mitchell's expenses as a summer travel companion in Europe. Although Mitchell declined the offer, she was taken with the cheerful, unconventional woman who was not "weak-minded." For a discussion of Thompson's scientific philanthropy, see Howard S. Miller, *Dollars for Research: Science and Its Patrons in Nineteenth-Century America* (Seattle, Wash., 1970), 127–129.

58. See Rossiter, *Women Scientists,* chap. 1.

59. Notes and diary, August 20, 1881, item 24, MMM.

60. The debate was taken up with vigor in the pages of Edward Youman's *Popular Science Monthly* in the 1870s. Also see Rosalind Rosenberg, "In Search of Woman's Nature, 1850–1920," *Feminist Studies* 3 (Fall 1975):141–154.

61. Diary for 1883, item 21, MMM. Mitchell quoted E. B. *[sic]* Stanton as stating there were "470 women physicians with university diplomas in the United States."

62. *Report . . . 1887,* 34–35. For discussion of this optimism and subsequent disillusionment in the twentieth century, see Mary Roth Walsh, *"Doctors Wanted: No Women Need Apply," Sexual Barriers in the Medical Profession, 1835–1975* (New Haven, Conn., 1977).

63. *Report . . . 1884,* 25.

64. *Report . . . 1883,* 23. Mitchell recorded privately, "How slowly women work into science," but her public comments were always encouraging: "If we cannot show rapid advancement of the culture of science among women, we can claim that there is slow and steady advancement" (*Report . . . 1882,* 25).

65. Mitchell apparently did not write the reports for 1878 and 1886, in the latter year because she was working hard to find financial support for the observatory at Vassar College (Mitchell to Antoinette Brown Blackwell, August 22, 1886, Blackwell Family MSS, SLR).

66. Her personal concern with women in science did not diminish and her notebooks contain numerous clippings and anecdotes on the subject (item 21 and others, MMM).

67. According to Alice Stone Blackwell, "Wherever it met, it left behind a trail of new women's clubs and other organizations then considered novel and dangerous" (quoted in Inez Irwin, *Angels and Amazons: A Hundred Years of American Women* [New York, 1934], 229). In her *Reminiscences,* Howe wrote: "Talk of the man with the hoe! We were the women of the plough. To us it was given to draw the furrow and drop the seed from which have sprung up clubs and federations scarcely to be numbered."

The General Federation of Women's Clubs was created as an alternative to the AAW and reflected an ongoing competition between the Boston area women involved with New England Women's Club (which did not join the federation when it was founded in 1889) and the New York Sorosis. The AAW persisted under Howe until she "laid it to rest" in 1900. Howe's correspondence with Ednah Dow Cheney is particularly useful on these later years. Jane Cunningham (Jennie June) Croly's *The History of the Woman's Club Movement in America* (New York, 1898) dismisses the AAW's two decades of work in one paragraph, calling it the "John the Baptist" of women's clubs.

68. Winant, *A History of Sorosis, 1868–1968,* 20.

69. Robert Clark, *Ellen Swallow: The Woman Who Founded Ecology* (Chicago, 1973), 86–90.

70. Mitchell conducted an officer's meeting of the AAW at the home of Phoebe Hanaford and recorded with disappointment that most of the executive committee were women "of business" rather than learning (Commonplace notebook, June 20, 1875, item 50, MMM). Mitchell gave a paper to the collegiate group in Boston; see her copybook, January 1880, item 50, MMM.

71. Copybook, August 27, 1879, item 50, MMM.

72. Her eye was always on the women who would succeed her, and she wrote in her diary that "the continuance of scientific study seems to be on the increase," with women more involved in serious study and "less apt to take up science as a pastime" (August 14, 1883, item 21, MMM).

73. She apparently had successfully opposed, for example, a presentation by Mary Livermore on writer Harriet Martineau (Livermore to Antoinette Brown Blackwell, June 27, 1877, Blackwell Family MSS, SLR.

74. Phoebe Ann (Coffin) Hanaford, *Daughters of America: Or Women of the Century* (Augusta, Maine, 1882), 252.

Chapter 8: "Strangers to Each Other"

1. Most recent studies of Clémence Royer have focused on her evolutionary or social evolutionary ideas. See, for a recent example, Linda Clark, *Social Darwinism in France* (Tuscaloosa, Ala., 1984). Sara J. Miles is preparing a Ph.D. thesis for the University of Chicago that will examine in detail her evolutionary theories by focusing on her Darwin prefaces. A recent French intellectual biography written by Geneviève Fraisse, *Clémence Royer, philosophe et femme de sciences* (Paris: 1985), focuses on her feminism, her economic theories, and her novel. See also Claude Blanckaert's illuminating piece "Clémence Royer (1830–1902): l'anthropologie au féminin," *Revue de synthèse* 109 (1982):23–39; and Joy Harvey, "Doubly Revolutionary: Clémence Royer before the Société d'Anthropologie de Paris," *Proceedings of the Sixteenth International Congress History of Science* (Bucharest, 1981), *Symposia B*, 250–256. Joy Harvey, "Races Specified, Evolution Transformed: The Social Context of Scientific Debates Originating in the Société d'Anthropologie de Paris, 1859–1902" (Ph.D. diss., Harvard University, 1983), looks at Clémence Royer in the context of the Société d'Anthropologie debates. Yvette Conry's pioneering study *Introduction du Darwinisme en France au XIX siècle* (Paris, 1974) examined Clémence Royer's Darwinism outside these debates.

2. Clémence Royer. "Sur la natalité," suppressed communication before the Société d'Anthropologie de Paris (exists in page proofs), July 16, 1874, Archives Société d'Anthropologie de Paris (uncatalogued materials deposited, Musée de l'Homme, Paris). This interesting and unknown manuscript was located by Claude Blanckaert and myself in 1979 in these archives.

3. Clémence Royer, "Préface," in Charles Darwin, *L'origine des espèces ou des lois du progrès chez les êtres organisés* (Paris, 1862).

4. Many of these statements come from her unpublished autobiography "Clémence Royer par elle-même," written for Pascal Duprat in the 1870s, but added to in the late 1880s and 1890s (cited as "Autobiography" hereafter). Dossier Clémence Royer, Marguerite Durand Library, Paris (hereafter MDL).

5. As she claimed in her *Introduction à la philosophie des femmes* (Lausanne, 1859). (Translations from the French are my own.)

6. *Congrès internationale sur la condition et des droits des femmes (1900)* (Paris, 1901), 304.

7. Clémence Royer, "Autobiography," 41.

8. Clémence Royer to Avril de Sainte-Croix, undated letter, [1899?], 5. Dossier Clémence Royer.

9. "Autobiography" and "Rectifications biographiques" (typed manuscript by Clémence Royer [1899] in Dossier Clémence Royer).

10. "Autobiography," 15, 39.

11. Ibid., 15.

12. Ibid., 17.

13. Though she suffered from her father's paternal despotism, she clearly ad-

mired him greatly. Her depiction of the origins of tyrannical government in family authority is in *L'origine de l'homme et des sociétés* (Paris, 1870), 491–492.

14. Her anger at the poor education she had received as an adolescent during eighteen months in a convent school and the sense that she had been lied to by the church is very strong in "Rectifications biographiques" (pp. 5–6).

15. In the course of a depopulation debate, she gave extensive comparisons of the child-raising techniques in France and England. *Bulletin Société d'Anthropologie* (hereafter *BSAP*) 1, ser. 4 (1890):680–700.

16. "Rectifications biographiques," 4. Here she discussed adopting a Channing style of Deist Unitarianism before becoming a complete freethinker.

17. Ibid., 6.

18. "Autobiography," 57.

19. Clémence Royer to Avril de Sainte-Croix, 1.

20. Ibid., 2–3.

21. Clémence Royer, "La vie politique de François Arago," *Revue International* (1886), cited in Albert Millice, *Clémence Royer et sa doctrine de la vie* (Paris, 1926), 180.

22. Clémence Royer to Avril de Sainte-Croix, 1–5. See also discussion of this in Geneviève Fraisse, *Clémence Royer,* and Albert Millice, *Clémence Royer et sa doctrine de la vie.*

23. *Les jumeaux d'Hellas* (Belgium, 1864). This book was placed on the Vatican's Index and prevented from distribution in France. Two biographers have examined the novel from different points of view: Albert Millice, *Clémence Royer,* 59–72, and Geneviève Fraisse, *Clémence Royer,* 53–60. Although set in an earlier time, Clémence Royer incorporated some of the contemporary excitement over Mazzini and Garibaldi's concept of a revolutionary, unified Italy in her novel.

24. A. Moufflet, "L'oeuvre de Clémence Royer," *Revue internationale de socologie* (1911):659–693.

25. Clémence Royer to Avril de Sainte-Croix, 3.

26. Ibid., 2.

27. Ibid, 3–4. Divorce was finally legalized in 1884, in France. The motion was brought by Alfred Nacquet, who sat on the radical left, a member of the same political and intellectual circle as Duprat.

28. Clémence Royer, "Sur la natalité," 613.

29. Clémence Royer, *Introduction à la philosophie des femmes,* 11–12.

30. Ibid., 11.

31. Ibid., 5. She adds that if men reject this feminine philosophy, "we will keep it for ourselves, for our own use" (p. 27).

32. Clémence Royer, *L'origine de l'homme et des sociétés,* 391.

33. Clémence Royer, *Théorie de l'impôt,* 2 vols. (Paris, 1862). She emphasizes in her biography the attention this gave her outside Switzerland. G. Fraisse has analyzed her economic theories in this book.

34. Clémence Royer, "Rectifications biographiques." See also her preface to *Le bien et la loi morale* (Paris, 1881).

35. Clémence Royer, "Préface," *L'origine des espèces*. See also her articles on Lamarck for Emile Littré's positivist journal: "Lamarck, sa vie, ses travaux, et son système," *La philosophie positive* 3 (1868):173–209, 343–372; 4(1869):5–30.

36. Claparède had expressed annoyance to Darwin about Clémence Royer's insistence on adding corrective notes to Darwin's expressions of scientific doubt. C. Darwin to J. D. Hooker, September 22, 1862, in Francis Darwin and A. C. Seward, *More Letters of Charles Darwin* (New York, 1903), 2:288.

37. Clémence Royer, "Préface," *L'origine des espèces* (reprinted in 3d ed.), 23.

38. Ibid., 65.

39. Charles Darwin to Asa Gray, June 1862, cited in footnote. Francis Darwin, *Life and Letters of Charles Darwin,* 2:79, footnote.

40. Charles Letourneau, ms. of speech, "Banquet, Clémence Royer," March 10, 1897 (Dossier Clémence Royer).

41. Mme. le Docteur (Madéleine) Brès, ms. of speech, "Banquet, Clémence Royer," March 10, 1897 (Dossier Clémence Royer).

42. Clémence Royer, "Preface," *L'origine des espèces* (reprinted in 3d ed.).

43. Ibid., 65.

44. Aristide Pratelle, "Clémence Royer, notice biographique," *Revue anthropologique* 28 (1918):263–276.

45. Clémence Royer to Avril de Sainte-Croix. It is interesting to note that Duprat's book was on the eighteenth-century Encyclopedistes from whom Royer derived her philosophy.

46. Clémence Royer, "Avant propos" to her second edition of Charles Darwin, *De l'origine des espèces par sélection naturelle ou des Lois de transformations des êtres organisés* (Paris, 1866). She discusses some of the changes in language requested by Darwin, xii–xiii. These letters are no longer extant. (See also nn. 64, 65.)

47. The subtitle changes in the French editions from "or the laws of *progress* of organized beings" (1862) to "by *natural selection,* or the laws of *transformation* of organized beings" [italics mine] (1866, 1870).

48. Clémence Royer, "Lettre à M. le President de l'Académie des Sciences Morales et Politiques," *Journal officiel,* August 2, 1873, separately printed with a published concession to Clémence Royer by the vice-president of the academy and a note of thanks from her dated August 3, 1873.

49. Clémence Royer to Paul Broca, unpublished letter, December 22, 1865. Archives at the Société d'Anthropologie de Paris, F_1, no. 796.

50. Clémence Royer, "Avant propos," 2.

51. Paul Broca, "Sur la mâchoire de Naulette," *BSAP* 1, ser. 2 (1866):288–296, and again to Carl Vogt, expresses his inability to accept Darwin as a polygenist since Darwinism was based on a common ancestor, not on parallel lines of descent. See Paul Broca to Carl Vogt, May 24, 1867, Carl Vogt Correspondence MS 2188, no. 154, Bibliothèque Universitaire et Publique, Geneva, Switzerland. His comment about Clémence Royer's preface is in this same letter to Vogt.

52. Paul Broca, "Sur le transformisme," *BSAP* 5, ser. 2 (1870). For an extended

description of this debate and Clémence Royer's participation in the Société d'Anthropologie debates, see Joy Harvey, "Races Specified, Evolution Transformed" (47–50, 170–182, 233), and in a shorter form, "Evolution Transformed, Positivists and Materialists before the Société d'Anthropologie de Paris," in *The Wider Domain of Evolutionary Thought,* ed. D. Oldroyd and I. Langham (Dordrecht, Holland, 1983), 289–310.

53. René's existence and Duprat's presence were both freely acknowledged by Royer in her social life, as shown in her letters to Lenoir (see no. 93, below).

54. Millice, *Clémence Royer.*

55. Clémence Royer, "Lamarck, sa vie, ses travaux et son système."

56. Clémence Royer, *L'origine de l'homme et des sociétés.*

57. Ibid., 547.

58. Clémence Royer, "Autobiography," 77.

59. Clémence Royer, "Rémarques sur le transformisme," BSAP 5, ser. 2 (1870):265–312. (See also n. 50, above.)

60. Charles Letourneau, "Banquet, Clémence Royer."

61. J. T. Bury, *Gambetta and the Making of the Third Republic* (London, 1973), 229. See also "Pascal Duprat," *La grande encyclopédie* (Paris, [1890]), 92.

62. For the links between politics and science in the Société d'Anthropologie de Paris during the Third Republic, see Harvey, "Races Specified, Evolution Transformed," chap. 1: "Positivists and Materialists" (pp. 1–112).

63. These articles were collected into a book: Paul Bert, ed., *Revues scientifiques pour la republique française,* 7 vols. (Paris, 1879–1885).

64. Darwin had written to C. Reinwald, who had just published his *Descent of Man* for a new translation. Francis Darwin, ed., *Life and Letters of Charles Darwin* (London, 1887), 3:110. J. J. Moulinié's translation in 1873 was the result. The new translation bore the notation "requested and authorized by (Darwin) with (his) revisions and notes." This was so unsatisfactory in style that Reinwald issued another translation by Barbier in 1880.

65. Francis Darwin cites a letter of Charles Darwin to J. D. Hooker in *Life and Letters of Charles Darwin,* 110.

66. See, e.g., J. Azezat's comments on J. J. Moulinié's style in his review of *Descent of Man, Revue d'anthropologie* 3 (1874):331–332. He quotes Clémence Royer's disparagement of the style as a "roman-patois." See also Paul Broca's discussion of Royer's translation in "Les sélections," *Revue d'anthropologie* 1 (1872):691.

67. Clémence Royer, "Lettre à M. Le Président de l'Académie des Sciences Morales et Politiques."

68. "Deux hypothèses sur l'hérédité," *Revue d'anthropologie* 6 (1877):443–484.

69. Clémence Royer contributed more than 130 times in articles and discussions to the Société d'Anthropologie, according to the index of the bulletins.

70. Quoted in J.P.T. Bury, *Gambetta and the Making of the Third Republic,* 229. The effect of this attitude on feminism has been discussed recently by Claire Goldberg Moses, *French Feminism in the 19th Century* (Albany, 1984).

71. Clémence Royer's discussion is not printed in the Congress reports, although her later contributions that year are. Bertillon's talk opened the congress. Louis-Adolphe Bertillon, "La population de France," *Comptes rendus Association Français pour l'Avancement des Sciences,* 2d Congress (Lyon, 1873).

72. Delausiave's discussion, which began the session, was the only contribution retained in the published bulletins for July 16, 1874 (*BSAP* 9, ser. 2 [1874]). The debate on population had begun at the previous session (July 2) with a discussion of Malthus by Clémence Royer (*BSAP* 9, ser. 2 [1874]:585–590). Galleys and page proofs were retained in the archives, along with her letter to Broca and the publication committee's decision not to publish. The comments replying to her paper were also deleted from publication.

73. Clémence Royer, "Sur la natalité," 603.

74. Ibid., 609.

75. Ibid. James Reed has shown how easily available abortion was in the United States at about this time; see his *From Private Vice to Public Virtue: The Birth Control Movement and American Society since 1830* (New York, 1978).

76. "Sur la natalité," 613. Her belief in colonialism echoes that expressed in *L'origine de l'homme* four years before. She continued to decry European wars, however, whereas she felt colonial aggression served an evolutionary purpose.

77. Ibid., 611.

78. Ibid., 614.

79. Ibid.

80. Clémence Royer, *L'origine de l'homme,* 381.

81. Clémence Royer (letter), "Diminution de la population de France," *BSAP* 6, ser. 4 (1895):653–656.

82. Geneviève Fraisse, *Clémence Royer,* 49.

83. Paul Broca, "Les sélections," 707.

84. Henri Thulié, "L'Ecole d'Anthropologie depuis sa fondation," *L'Ecole d'Anthropologie de Paris,* 1876–1906 (Paris, 1907). For a look at the politics of this, see Harvey, "Races Specified, Evolution Transformed."

85. Clémence Royer to Paul Broca, May 15, 1875, Archives Société d'Anthropologie, no. B_1 1890. This discussion occurs almost a year after the debates. The note of suppression is dated first May and then July 1875.

86. Ibid.

87. Ibid. Her additional notes are not extant. They must have been returned to her, as she requested, in order to write a projected (but unknown) long work on this topic.

88. Patrick Kay Bidelman has recently discussed the civil code and its later challenge from the French feminists. Patrick Kay Bidelman, *Pariahs Stand Up: The Founding of the Liberal Feminist Movement in France* (Westport, Conn., 1982).

89. Louis Lartet to Ernest Hamy, February 1874, Correspondence à Ernest Hamy, Muséum d'Histoire Naturelle (Paris) MS 2254, no. 230–231.

90. Ibid., no. 246–248, February 1875.

91. A notebook from the Laboratory of Anthropology records her presence,

along with that of other enrolled students and colleagues (Archives, uncatalogued, in Laboratoire d'Anthropologie de Paris, Juvisy-sur-Orge).

92. "Ils admirent avant tout la battérie de cuisine des savants, leur instruments, leurs outils" (Clémence Royer, "Preface," *Le bien et la loi morale,* 24).

93. Clémence Royer, "Banquet, Clémence Royer."

94. Some of his regular contributions on evolution published as books were *L'evolution de la morale* (Paris, 1887), *L'evolution de marriage et de la famille* (Paris, 1888), *L'evolution de la propriété* (Paris, 1889), and *L'evolution religieuse* (1892).

95. Even when Broca, Mortillet, Quatrefages, and others criticized Clémence Royer for her "easy hypotheses," they did so in a friendly manner. Her contributions were usually answered in discussion and printed in the journal.

96. Clémence Royer to François Lenoir, Correspondence à François Lenoir, Bibliothèque Nationale, Paris, *n.a.f.* 21481, nos. 158–160. (Other material runs from 1872 to 1883 and includes both sides of the correspondence, along with notices of meetings, talks, and announcements of Clémence Royer's own society.)

97. Clémence Royer to François Lenoir, December 1, 1872, "Correspondence à M. Lenoir." Mme. Lenoir seems to have also been a member of some of these organizations (Bibliothèque Nationale, n.a.f. 21481 [through 1883]).

98. Quoted by A. Pratelle in his biographical article. The newspaper *La fronde* mentions it in a 1900 article by Marie Louise Neron, asking, "Why *almost,* Why a *man*?" (Dossier Clémence Royer). The source of Renan's quote is not given.

99. Clémence Royer, "Rectifications biographiques."

100. Darwin to J. D. Hooker, *More Letters,* 2:288.

101. A. Moufflet, "Clémence Royer," 660.

102. Ibid.

103. Alexandre Dumas, fils, *L'homme-femme* (Paris, 1872). The French feminist Maria Deraismes had written a scathing reply to this in *Eve contre M. Dumas, fils* (Paris, 1872).

104. Gustave Le Bon, "Récherches anatomiques et mathématiques sur les lois des variations du volume des cerveaux," *Revue d'anthropologie* 2, ser. 2 (1879): 27–104. Stephen Jay Gould has discussed this in *Mismeasure of Man* (New York, 1982).

105. For a lengthy discussion of this, see Harvey, "Races Specified, Evolution Transformed," 288–292.

106. Central Committee, Procès Verbaux, Archives Société d'Anthropologie de Paris, vol. 2, meeting for March 27, 1879.

107. Paul Broca insisted that education would enlarge and increase brain size as well as knowledge (P. Broca, "Les sélections"). Charles Letourneau related this directly to women's need for education to "masculinize" their brain, whereas Manouvrier rejected this entire argument, along with Le Bon's "objective" measurements (see n. 102, above).

108. L. Manouvrier, "L'anthropologie des sexes, III," *Revue de l'Ecole d'Anthropologie* 9 (1909):41–61.

109. Clémence Royer, *Le bien et la loi morale.*

110. Clémence Royer, "Autobiography."

111. See Clémence Royer letter to Avril de Sainte-Croix, and the presidential address to the Société d'Anthropologie, BSAP 5, ser. 3 (1885):595.

112. Linda Clark has shown that this was a good salary in 1885 and 1886, but in spite of regular appeals from friends and colleagues, this drastically dwindled over the years (Clark, *Social Darwinism in France,* 57–58).

113. Geneviève Fraisse has seen her request but is ignorant of the result.

114. Clémence Royer to Avril de Sainte-Croix.

115. Clémence Royer, "Autobiography," but see also Millice, *Clémence Royer,* where this is described in great detail.

116. Correspondence à François Lenoir.

117. Linda Clark, *Social Darwinism in France,* 7. See also Royer's own summary of articles or books published, or in manuscript, some of which, as she carefully noted, had received payments or monetary awards; Clémence Royer, "Autobiography," 83–93.

118. Clémence Royer to Armand de Quatrefages, April 1891, Correspondence à Quatrefages, Muséum d'Histoire Naturelle, Paris, MS 2258. Yvette Conry cites only part of this letter in the appendix to her book *Introduction du Darwinisme,* 437.

119. See, e.g., the letters by Ernest Hamy over many years in his attempts to gain a place at the Institut de France. Ernest Hamy Correspondence, Muséum d'Histoire Naturelle, MS 2252–2254.

120. This was published in two parts: "Facultés mentales et instincts sociaux des singes," *Revue scientifique* 38, ser. 3 (1886):257–270, and "Les notions de nombre chez les animaux," *Revue scientifique* 40, ser. 3 (1887):649.

121. These appeared in English as "Faculties of Monkies: Mental Evolution," *Popular Science Monthly* 30 (1887):17–24, and "Animal Arithmetic," *Popular Science Monthly* 34 (1889):252–262. Since Clémence Royer had received some American attention, a short biography on her was published in 1899 by Jacques Boyer (*Popular Science Monthly* 54 (1899):690–698).

122. René Verneau, "Discours aux obsèques de Mme. Clémence Royer," BSAP 3, 5e ser. (1902), 75.

123. Central Committee (Bureau Centrale, Société d'Anthropologie), Procès verbaux, 2 (1900), MS, Archives Société d'Anthropologie.

124. For example, the letter on depopulation (1895) cited in n. 78, above.

125. The connection with Maria Deraismes apparently dated back to the Second Empire. See Geneviève Fraisse on Clémence Royer's ties to feminism (*Clémence Royer,* 95–101). Her strong support of freemasonry as the "army of science" echoed that of many of the political and scientific men and women around her. See Harvey, "Races Specified, Evolution Transformed," 35.

126. In December 1872, Léopold Lacour wrote an article on her attempts to enter the Institut de France: "Les femmes à l'Institut," *L'evénement,* December 18, 1892 (copy in Dossier Clémence Royer).

127. Many of these have been excerpted and are collected in the Dossier Clémence Royer, MDL.

128. On the organization of this banquet, see Millice, *Clémence Royer,* 201–202. For the letters requesting assistance, including speeches given at the banquet on March 10, 1897, see Dossier Clémence Royer, MDL.

129. Clémence Royer, "Banquet, Clémence Royer."

130. Charles Letourneau, "Banquet, Clémence Royer."

131. For his life, see discussion in Moufflet, "L'oeuvre de Clémence Royer," Millice, *Clémence Royer,* and Fraisse, *Clémence Royer.*

132. Clémence Royer to Felix Nadar, March and November 19, 1899, Bibliotèque Nationale, n.a.f. 24285, no. 691,694. Nadar photographed Clémence Royer in 1865; the portrait has been used recently on the cover of Fraisse, *Clémence Royer.*

133. René was awarded a posthumous Legion of Honor medal. Some biographies list his death as from liver cancer (Millice, *Clémence Royer*).

134. A file of the letters soliciting this honor are also in MDL.

135. *Congrès sur la condition et les droits des femmes,* 304.

136. Georges Clémenceau, "Madame Clémence Royer," *L'illustration* (March 13, 1897):194–195. This is accompanied by a lithograph of her, looking heavy and sad.

137. She had written this for the same positivist journal for which she had written her Lamarck articles (edited at this time by the biologist Charles Robin): "Attraction et gravitation d'après Newton," *La philosophie positive* (1883):206–226. This article carried a disclaimer by the editors.

138. *La constitution du monde* (Paris, 1900).

139. *Science* 2 (1900):785. The reviewer in the English journal *Nature* was far kinder, insisting that "some of the ideas give one the idea that there is much to be said in their favor." He concludes: "But is it new?" G.H.B., *Nature* 63 (1901): 534.

140. Clémenceau, "Mme. Clémence Royer." She had begun this work in 1873, giving lectures on the fluid atom. Claude Blanckaert contributes insights into her attempts at encyclopedic knowledge ("L'anthropologie au feminin"). J. D. Saulze saw her, in 1912, as one of the major French monists (Saulze, *Le monisme matérialiste en France* [Paris, 1912], 129–151). Moufflet also discusses this ("L'oeuvre de Clémence Royer," 661–670).

141. A special six-page issue of *La fronde* was devoted to Clémence Royer on June 19, 1930, including the program of the centennial celebration of her birth. This commemoration included male scientists and male and female scholars, as well as artists and writers.

142. Nikolai Sergevich Rusanov, *O Raznopravnosti zhenschin* [On the Equal Rights of Women] (Saint Petersburg, 1905). I am indebted to Dr. Anne C. Hughes for the reference and translation.

143. Millice embraced this interpretation in his discussion of "Royerian philosophy" (*Clémence Royer).*

144. Although Linda Clark sees this work as espousing a more sympathetic look at the underclass than her earlier writings (Clark, *Social Darwinism,* 57), it should be

remembered that Spencer also shared the view of population growth as a motor for progress.

145. See, e.g., the bitter tone in which she spoke of her fame primarily as a translator in "Autobiography," 77. One of the translations for which she was suggested was Karl Marx's *Capital*! See Claude Blanckaert, "L'anthropologie au feminin."

Chapter 9: Career and Home Life in the 1880s

Acknowledgments: I am grateful to the School of Social Science of the Institute for Advanced Study in Princeton for its support and hospitality during the writing of this paper; and also to Bert Hansen, Fritz Schütze, Susan Mosher Stuard, and other members of the 1984–85 Biography Seminar at the Institute for many helpful suggestions.

1. Robert Hermann, "The Geometric Foundations of the Integrability Property of Differential Equations and Physical Systems. II. Mechanics on Affinely Connected Manifolds and the Work of Kowalewski and Painlevé," *Journal of Mathematical Physics* 25, no. 4 (April 1984):782.

2. For details of Kovalevskaia's life, see Ann Hibner Koblitz, *A Convergence of Lives. Sofia Kovalevskaia: Scientist, Writer, Revolutionary* (Boston, 1983).

3. T. A. Bogdanovich, *Liubov' liudei shestidesiatykh godov* (Leningrad, 1929); Chernyshevskii's remark is quoted on p. 95. Both Chernyshevskii and Shelgunov remained true to this belief throughout their lives; Shelgunov raised his wife's two sons by her lovers as if they had been his own, and always referred to his wife as his "one true friend." For information on "free love" movements in America around this time, see Hal D. Sears, *The Sex Radicals: Free Love in High Victorian America* (Lawrence, Kans., 1977).

4. See J. M. Meijer, *Knowledge and Revolution: The Russian Colony in Zürich (1870–1873)* (Assen, Holland, 1955); Barbara Alpern Engel, *Mothers and Daughters: Women of the Intelligentsia in Nineteenth Century Russia* (London, 1983). See also Blanche Wiesen Cook, "Female Support Networks and Political Activism," *Chrysalis* 3 (Autumn 1977), for other examples of this phenomenon.

5. The translation is Beatrice Stillman's, in Sofya Kovalevskaya, *A Russian Childhood* (New York, 1978), 23.

6. E. F. Litvinova, *S. V. Kovalevskaia* (Saint Petersburg, 1893), 50.

7. Ibid., 51.

8. Kovalevskaia, October 1880 letter to A. O. Kovalevskii, in *Vospominaniia i pis'ma* (Moscow, 1951), 254.

9. Kovalevskaia's diaries for this period are in the archives of the Institut Mittag-Leffler in Djursholm, Sweden; some of her letters are also there, and others were published in *Vospominaniia i pis'ma*.

10. For information on the family and women in tsarist Russia, see Dorothy Atkinson, Alexander Dallin, and Gail Warshofsky Lapidus, eds., *Women in Russia* (Stanford, Calif., 1977); and Engel, *Mothers and Daughters*.

11. The exceptions were the German number theorist E. E. Kummer, who opposed Kovalevskaia's admission to Berlin University in 1870, and the Russian A. A. Markov, who included Kovalevskaia (along with Chebyshev and other prominent specialists) among the people whose works he attacked. Kummer's opposition to Kovalevskaia appears to have been based purely on misogyny and antifeminism, and as such struck a responsive chord among most professors at Berlin University. Markov's antagonism toward Kovalevskaia was more complicated. Partly, he had an irascible personality and was known for attacks on his colleagues' work. Partly, he was jealous that Kovalevskaia's election to full membership in the Saint Petersburg Academy of Science was being discussed while he was still an adjunct member. Also, he resented Kovalevskaia's German degree and her position as the favorite student of the great Weierstrass. In any event, two years after Kovalevskaia's death the Moscow Mathematical Society investigated Markov's allegations, found them "unsubstantiated and without foundation," and refused to accept any more of his attacks on Kovalevskaia and other mathematicians into their records. (See p. 227 of Koblitz, *Convergence* for more details.)

12. Kovalevskaia, May 1884 letter to Theresa Gyldén, in *Vospominaniia i pis'ma,* 283–284.

13. Sofia Vladimirovna Kovalevskaia (Fufa), "Vospominaniia o materi," in *Vospominaniia i pis'ma,* 360–361; see also her expanded memoirs in P. Ia. Kochina, *Sofia Kovalevskaia* (Moscow, 1982).

14. Kovalevskaia, September 1885 letter to Gustav Hansemann, in Maria von Bunsen, "Sonja Kowalevsky: Eine biographische Skizze," *Illustrierte Deutsche Monatschefte,* no. 82 (1897):226.

15. Ellen Key, "Sofia Kovalevskaia," in *Vospominaniia i pis'ma,* 411.

16. Jessica Tovrov, "Mother-Child Relationships among the Russian Nobility," in *The Family in Imperial Russia,* ed. D. L. Ransel (Urbana, Ill., 1978), 15–43.

17. Kovalevskaia, "Vospominaniia detstva," in *Vospominaniia i pis'ma,* 12–13, and passim.

18. "Vospominaniia o materi," 362–363; see also the appendix to Kochina, *Sofia Kovalevskaia.*

19. Ibid., 367.

20. "M. M. Kovalevskii o S. V. Kovalevskoi," in *Vospominaniia i pis'ma,* 388–407; 393, n. 2; 530.

21. See, e.g., Don H. Kennedy, *Little Sparrow* (Athens, Ohio, 1983), 283ff.; *Sonya Kovalevsky: Her Recollections of Childhood with a Biography by Anna Carlotta Leffler, Duchess of Cajanello* (New York, 1895); Gunilla Tengvall, "Om förnuftsdyrkan och kärleksnöd," *Dagens Nyheter,* July 31, 1982.

22. Formulations taken from L. F. Austin, "At Random," (London) *Sketch,* March 24, 1897, 390. Isabel F. Hapgood, "Notable Women: Sonya Kovalevsky," *Century Magazine* 50 (1895):536–539. See also n. 21 and Koblitz, "Changing Views of Sofia Kovalevskaia," to appear in L. Keen, ed., *The Legacy of Sonia Kovalevskaya* (American Mathematical Society Series in Contemporary Mathematics, Providence, R.I., 1987), 53–76.

Chapter 10: Marie Curie's "Anti-natural Path"

Acknowledgments: The author wishes to thank the archives of the Laboratoire Curie, Paris, and the Institut International de Physique Solvay (Courtesy AIP Niels Bohr Library), which provided the illustrations in this chapter.

1. For consistency, all future references will be to Marie Curie, although her original name was Maria Sklodowska. There are two recommended biographies of Curie: Eve Curie, *Marie Curie,* trans. Vincent Sheean (Garden City, N.Y., 1940); and Robert Reid, *Marie Curie,* copyright © 1974 by Robert Reid; reprinted by permission of E. P. Dutton, a division of New American Library, and the author. An autobiography appears in Marie Curie, *Pierre Curie,* trans. Charlotte and Vernon Kellogg (1923; reprint, New York, 1963). The present essay relies on these works for the details and general themes of Curie's life. But a particular biography is cited only where details or themes peculiar to it are presented, or where it is directly quoted.
2. Marie Curie, *Pierre Curie,* 78.
3. Ibid., 79.
4. Quoted in Eve Curie, *Marie Curie,* 36.
5. Marie Curie, *Pierre Curie,* 78.
6. Ibid., 83. There are some striking similarities between the Sklodowska sisters' close relationship and that of Sofia Kovalevskaia and her sister, Aniuta, not the least of which is the strong influence of progressive ideologies on the two pairs of sisters. On Kovalevskaia's early relationship with Aniuta and their interest in Russian nihilism, see Ann H. Koblitz, *A Convergence of Lives. Sofia Kovalevskaia: Scientist, Writer, Revolutionary* (Boston, 1983), esp. 31–79.
7. Marie Sklodowska to Kazia Przyborovska, October 25, 1888, in Eve Curie, *Marie Curie,* 78.
8. Marie Sklodowska to Henrietta Michalovska, December 1886, in Eve Curie, *Marie Curie,* 72.
9. Ibid.
10. Bronia Sklodowska to Marie Curie, March 1890, in Eve Curie, *Marie Curie,* 83–84.
11. Marie Sklodowska to Bronia Sklodowska, March 12, 1890, in Eve Curie, *Marie Curie,* 84.
12. Marie Curie, *Pierre Curie,* 83.
13. Marie Sklodowska to Joseph Sklodowski, September 15, 1893, in Eve Curie, *Marie Curie,* 115.
14. For details on Pierre Curie's life and work through 1894, see Marie Curie, *Pierre Curie,* 11–33; and Jean Wyart, "Pierre Curie," *Dictionary of Scientific Biography* (New York, 1970–80) 3:503–508.
15. This is Pierre Curie's description of the collaboration, quoted from one of his letters to Marie Sklodowska shortly before their marriage (Marie Curie, *Pierre Curie,* 23).

16. Eve Curie, *Marie Curie,* 120.

17. The terms "anti-natural path" and "anti-natural thought" were Pierre Curie's. As he explained, they referred to a life so dominated by science that, although he was "obliged to eat, drink, sleep, laze, love," he would not "succumb" to these distractions. (See Eve Curie, *Marie Curie,* 120, 126.)

18. The only indication that Pierre shared some domestic responsibilities is Marion Cunningham's claim that visitors to the young couple's apartment "found Pierre sweeping the floor and his wife cooking the meals" (Cunningham, *Madame Curie (Sklodowska) and the Story of Radium* [London, n.d.], 32).

19. Marie Curie, *Pierre Curie,* 86.

20. Eve Curie, *Marie Curie,* 207.

21. Marie Curie to Joseph Sklodowski, November 23, 1895, in Eve Curie, *Marie Curie,* 145–146.

22. Ibid., 144–145.

23. Marie Curie taught her laboratory students "that a research bench should always be neat and tidy, that 'a good researcher' did not wait until the end of an experiment to clear up, but did so as he went along" (Maurice Goldsmith, *Frédéric Joliot-Curie: A Biography* [London, 1976], 35).

24. Marie Curie, *Pierre Curie,* 88.

25. Ibid. Marie Curie, however, seems to have spent more than her usual amount of time at home immediately before and after the birth of her second daughter, Eve (see Eve Curie, *Marie Curie,* 225–227).

26. Lawrence Badash, "Radioactivity before the Curies," *American Journal of Physics* 33 (1965):128.

27. G. C. Schmidt independently made the same discovery in 1898. See Lawrence Badash, "The Discovery of Thorium's Radioactivity," *Journal of Chemical Education* 43 (1966):219–220.

28. Marie Curie's 1903 thesis contains a lucid account of her early work on radioactivity. See Marie Curie, *Radioactive Substances* (Westport, Conn., 1971). This edition is a reprint of the translated version of the thesis that appeared in *Chemical News* of 1903.

29. For Marie Curie's own account of the extraction of radium salt, see Marie Curie, *Radioactive Substances,* 19–31.

30. James Christie, "The Discovery of Radium," *Journal of the Franklin Institute* 167, no. 5 (May 1909):361.

31. Marie Curie, "Note," *Comptes rendus . . . de l'Académie des Sciences* 126 (April 12, 1898):1101.

32. See, e.g., Marie Curie, *Radioactive Substances,* 93–94. Also, when Gösta Mittag-Leffler (who had been so supportive of Sofia Kovalevskaia) informed Pierre of his consideration for the Nobel Prize of 1903, he responded with a private letter emphasizing Marie's part in the couple's research on radioactivity and thus the appropriateness of her sharing the prize. (See Elisabeth Crawford, *The Beginnings of the Nobel Institution: The Science Prizes, 1901–1915* [Cambridge, 1984], 141.)

33. Marie Curie, *Pierre Curie,* 94.

34. Ibid., 95.

35. Ibid.

36. André Debierne (1874–1949), a student of Pierre Curie, was an early collaborator and close friend of both Curies and, after 1906, of Marie Curie. He was one of the select group admitted to Curie's home, where he apparently spent many Sundays trying to entertain the Curie girls; he also saw Curie off on her trips abroad. Upon Curie's death, Debierne succeeded her as director of the Laboratoire Curie. (For details of the Curie-Debierne friendship, see Eve Curie, *Marie Curie,* 269, 280, and Reid, *Marie Curie,* 278–279.)

37. For a basic discussion of this work, see Reid, *Marie Curie,* 137–140.

38. David Wilson, *Rutherford: Simple Genius* (Cambridge, Mass., 1983), 253. Copyright © 1983, by David Wilson. Reprinted by permission of Literistic, Ltd.

39. Marie Curie to Ernest Rutherford, September 1910, Cambridge University Library, Add. 7653. Material from the Rutherford Collection courtesy of the Syndics of Cambridge University Library.

40. Reid, *Marie Curie,* 144.

41. Wilson, *Rutherford,* 253.

42. Besides Gleditsch (and Curie's famous daughter Irène), Curie's women students included May Leslie and Marguerite Perey. The Norwegian Gleditsch, who began a fruitful career devoted to radioactivity in Curie's laboratory in 1907, eventually returned to her native country as a professor of chemistry at the University of Oslo. Leslie worked with Curie during 1910–11, after which time Curie arranged for her admission to Rutherford's laboratory from which she published a long paper on the coefficients of diffusion of thorium and actinium emanations. Marguerite Perey (1909–1975), who served as an assistant at the Laboratoire Curie beginning in 1929, eventually became a professor of nuclear chemistry at the University of Strasbourg and, like her mentor, discovered a new element, francium.

43. Bertram Boltwood to Ernest Rutherford, September 12 [1913], the Bertram Boltwood Papers, courtesy Yale University Library; Mary Rutherford to Boltwood, October 6 [1913], Cambridge University Library. The first passage and all others from Boltwood to Rutherford are reproduced by permission of Yale University Library; the second and all other passages from the Rutherfords to Boltwood are reproduced by permission of the Syndics of Cambridge University Library. The cited letters are also found in *Rutherford and Boltwood: Letters on Radioactivity,* ed. Lawrence Badash, Yale Studies in the History of Science and Medicine, 4 (New Haven, Conn., 1969), 285–287.

44. Of non-French scientists, Rutherford—a major pioneer of radioactivity along with the Curies—seems to have been Marie Curie's closest and most supportive peer. For discussions of the Rutherford-Curie relationship, see Reid, *Marie Curie,* 141–145, and Wilson, *Rutherford,* 254–267. Both take very positive views of Rutherford's treatment of Curie.

45. A. S. Eve, *Rutherford: Being the Life and Letters of the Rt. Hon. Lord Rutherford,*

O.M. (New York, 1939), 213, courtesy Cambridge University Press.

46. Lise Meitner, "Looking Back," *Bulletin of the Atomic Scientists* 20, no. 9 (November 1964):5.

47. *Otto Hahn: My Life. The Autobiography of a Scientist,* trans. Ernst Kaiser and Eithne Wilkins (New York, 1970), 89.

48. Rutherford to Boltwood, January 11, 1909, in Badash, *Rutherford and Boltwood,* 206.

49. Ernest Rutherford, "Radium Standards and Nomenclature," *Nature* 84 (1910):430.

50. Rutherford to Boltwood, November 20, 1911, in Badash, *Rutherford and Boltwood,* 258.

51. Rutherford to Boltwood, April 22, 1912, in Badash, *Rutherford and Boltwood,* 270.

52. Eve Curie, *Marie Curie,* 266.

53. Marie Curie to Ernest Rutherford, April 18, 1919, Cambridge University Library, Add. 7653.

54. Adrienne R. Weill-Brunschvicg, "Paul Langevin," *Dictionary of Scientific Biography,* 8:9–14.

55. For a reconstruction of the Curie-Langevin relationship and details of the scandal, see Reid, *Marie Curie,* 162–182.

56. Henry Smith Williams, "The Case of Madame Curie," *World To-Day* 21 (1912):1632–1635. Williams was a distinguished American physician who specialized in mental diseases and pioneered nonspecific protein therapy ("Dr. H. S. Williams Dies on the Coast," *New York Times,* July 5, 1943, p. 15, col. 3).

57. "Editors in Duel over Mme. Curie," *New York Times,* November 24, 1911, p. 3, cols. 4–5.

58. For some details of the address, see Reid, *Marie Curie,* 184.

59. Drawing these two women together were their shared experiences of rejection by major scientific societies. Whereas the Royal Society refused Ayrton admission on the grounds that she was married, in early 1911 the French Academy of Sciences had turned down Curie's bid for membership primarily because she was a woman.

60. Rutherford to Boltwood, November 20, 1911, in Badash, *Rutherford and Boltwood,* 258–259.

61. Reid, *Marie Curie,* 167; Eve, *Rutherford,* 224. This material is copyrighted by and reprinted with the permission of Cambridge University Press.

62. Marie Curie to Ernest Rutherford, February 25, 1919, Cambridge University Library, Add. 7653.

63. See, e.g., Marie Meloney, "The 'New Woman,'" *Delineator* (November 1922):1; and Marie Meloney to Charles Eliot, December 9, 1920, Marie M. Meloney Papers, Columbia University.

64. For an insightful discussion of the effects of Curie's visit and the "Madame Curie strategy" on American women scientists, see Margaret W. Rossiter, *Women*

Scientists in America: Struggles and Strategies to 1940 (Baltimore, Md., 1982), esp. 122–128, 156–159.

65. Charles Eliot to Marie Meloney, December 18, 1920, Meloney Papers, Rare Book and Manuscript Library, Columbia University. Reprinted with permission of Columbia University.

66. Meloney to Charles Eliot, December 24, 1920, Meloney Papers, Columbia University.

67. Charles Eliot, "The Woman That Will Survive," *Delineator* (August 1914):5.

68. "Mme. Curie in Boston," *New York Times,* June 19, 1921, sec. 2, p. 1, col. 2.

69. Boltwood to Rutherford, July 14, 1921, in Badash, *Rutherford and Boltwood,* 346.

70. Ibid.

71. Eve Curie, *Marie Curie,* 337.

72. Marie Curie to Bronia Dluska [née Sklodowska], April 12, 1932, in Eve Curie, *Marie Curie,* 358.

Chapter 11: Cecilia Payne-Gaposchkin

Acknowledgments: This paper was begun under a Smithsonian Fellowship and took its present form under the gentle prodding of Pnina Abir-Am. It owes a semblance of structure to Deborah J. Warner and editorial improvements to Mark Kidwell. It would have been impossible without the insights of those I interviewed, including Owen Gingerich, Nancy Roman, Barbara Welther, Kenneth Janes, Stephen Kahler, Katherine Haramundanis, and Frances W. Wright, as well as diverse correspondents. Other helpful people include Clark Elliott, Margaret Rossiter, Karl Hufbauer, Spencer Weart, and Pamela Mack. I thank the archives of the American Association of University Women, the record keepers of the FBI, Jane Knowles and others of the Schlesinger Library, the staff of the Harvard Archives, Emilie Belserene and the staff of the Maria Mitchell Association, hardworking archivists of the Institute Archives at MIT, the Center for History of Physics at the American Institute of Physics and the Manuscripts Division of the Princeton University Library. Rhoda Ratner advised on *Little Dorrit,* Uta Merzbach provided cheerful oversight, and Juanita Morris came through in the clutch.

1. On women in astronomy, see Deborah J. Warner, "Women Astronomers," *National History* 88, no. 5 (1979):12–26; and Pamela Mack, "Women in Astronomy in the United States" (bachelor's thesis, Harvard University, 1977). On the work of British women, see P. Kidwell, "Women Astronomers in Britain, 1780–1930," *Isis* 74 (1984):534–546, as well as sources cited there.

2. Williamina Fleming, "Journal of Williamina Paton Fleming, Curator of Astronomical Photographs, Harvard College Observatory," Harvard University Archives. I thank Margaret Rossiter for bringing this journal to my attention.

3. On the lodgings of female college faculty, see Helen Lefkowitz Horowitz, *Alma Mater: Design and Experience in Women's Colleges from Their Nineteenth-Century Beginnings to the 1930s* (New York, 1984), 179–197.

4. For a more general discussion of marriage and scientific careers, see M. Rossiter, *Women Scientists in America: Struggles and Strategies to 1940* (Baltimore, Md., 1982), 15–16. On A.S.D. Maunder, see M. A. Evershed, "Mrs. Walter Maunder," *Journal of the British Astronomical Association* 57 (1947):238. On Priscilla Fairfield Bok, see the obituary in *Sky and Telescope* 51 (1976):25; and the fourth edition of *American Men of Science* (1927). On the possibility of her teaching during World War II, see Harlow Shapley to Ada Comstock, December 24, 1941, Harlow Shapley Directorial Papers, Harvard University Archives (hereafter Shapley Directorial Papers, HUA). Priscilla Bok was not listed in the Radcliffe catalogue as actually teaching any astronomy courses. This and other letters from the Shapley Directorial Papers are cited by permission of the Harvard University Archives.

5. Cecilia Payne-Gaposchkin's account of her life appears in *Cecilia Payne-Gaposchkin: An Autobiography and Other Recollections,* ed. Katherine Haramundanis (Cambridge, 1984); hereafter *Autobiography.*

6. *David Hannay,* "Edward John Payne," *Dictionary of National Biography* (London, 1912), 2d suppl., 3:85–86.

7. Payne-Gaposchkin, *Autobiography,* 89.

8. Norbert Wiener to Constance Wiener, July 21, 1925, Norbert Wiener Papers, Institute Archives, Massachusetts Institute of Technology. This and other letters from the Wiener Papers are cited by permission of the Institute Archives.

9. Dilys Powell, *The Traveller's Journey Is Done* (London, 1943), 24.

10. C. Payne-Gaposchkin, *Autobiography,* 82, 123–124.

11. L. J. Comrie to H. Shapley, March 7, 1923, Shapley Directorial Papers, HUA. Payne-Gaposchkin describes her female-dominated childhood, her college years, and her admiration for Eddington in *Autobiography.*

12. Payne was not the only one in her family who settled outside of England. Humfry also used funds provided by his college to do postgraduate studies; he went to Greece. According to his wife, wishing to marry, he became director of the British School of Archeology in Athens rather than accept the bachelor status of an Oxford don. (See Powell, *Traveller's Journey,* 90.)

13. Interview of Katherine Haramundanis, August 11, 1981. Interview of Helen Sawyer Hogg by David DeVorkin, August 17, 1979, Niels Bohr Library, American Institute of Physics, New York City. Interview of Dorrit Hoffleit by Pamela Mack, October 11, 1976. Interview with Peter Millman, March 29, 1983. Frances W. Wright, "Harlow Shapley—a Tribute to a Great Man," *Mercury* (March–April 1973):3–4.

14. Cecilia Payne to Henry Norris Russell, December 11, 1930, Henry Norris Russell Papers, Princeton University Library. I thank the Princeton Library for permission to cite this and other passages from the Russell Papers.

15. Frances Wright decribes "open nights" and the Bond Club in "Harlow Shapley." Talks Payne gave on Open Nights are mentioned in H. Shapley to J.E.D.

Seymour, September 23, 1925, Shapley Directorial Papers, HUA; *Harvard University Gazette* 31 (1935–36):19; and *Harvard University Gazette* 44 (1948–49):12. For more on the Bond Club, see Wright's article; "Bond Club," Shapley Directorial Papers, HUA; and C. H. Payne to V. M. Slipher, March 1926, Shapley Directorial Papers, HUA. Shapley discussed his informal seminars in *Through Rugged Ways to the Stars* (New York, 1969), 114–115. For an outsider's view, see G. Kuiper to O. Struve, March 31, 1944, Otto Struve Papers, Yerkes Observatory, Williams Bay, Wis. These and other lectures from the Struve Papers are cited by permission of the director of the Yerkes Observatory. Colloquia are listed in the *Harvard Gazette*.

16. For descriptions of Shapley's parties, see Wright, "Harlow Shapley"; Payne-Gaposchkin, *Autobiography*, 208; and David DeVorkin's interview with Helen Sawyer Hogg, August 17, 1979, Niels Bohr Library, American Institute of Physics. Payne-Gaposchkin mentions the Pinafore skit in her autobiography, and Henrietta Swope preserved the lyrics (Henrietta Hill Swope Papers, Schlesinger Library, Radcliffe College). Peter Millman recalled photographing the event in an interview of March 29, 1983.

17. The quotation is from H. Shapley to R. S. Dugan, September 14, 1934, Shapley Directorial Papers, HUA. See also Payne-Gaposchkin, *Autobiography*, 171–172, 183–185; and H. Shapley to Ada Comstock, September 14, 1928, Shapley Directorial Papers, HUA.

18. On early Radcliffe graduate work in astronomy, see P. Kidwell, "E. C. Pickering, Lydia Hinchman, Harlow Shapley and the Beginning of Graduate Work at the Harvard Observatory," *Astronomy Quarterly* 5 (1986):157–162. The quotation is from J. H. Moore to D. H. Menzel, March 23, 1934, D. H. Menzel Personal Papers, Harvard University Archives. Quoted by permission of the Harvard University Archives.

19. Deborah J. Warner, "Women Astronomers," *Natural History* 88, no. 5 (1979):22. M. Rossiter, *Women Scientists in America*, 140–141.

20. The careers of Hogg, Rieke and Vyssotsky can be traced from *American Men and Women of Science* (16th ed., 1986; 10th ed., 1960; and 6th ed., 1938, respectively), as well as from correspondence in the Shapley Papers.

21. On Payne's fondness for Eddington, see Payne-Gaposchkin, *Autobiography*, 156; and C. H. Payne to Margaret Harwood, August 16, 1924, Margaret Harwood Papers, Schlesinger Library, Radcliffe College. On her resentment of Shapley, see C. H. Payne to H. N. Russell, H. N. Russell Papers, December 11, 1930, H. N. Russell Papers, Princeton University Library.

22. Norbert Wiener to Bertha and Lee Wiener [June 1925], Norbert Wiener Papers, Institute Archives. I thank Margaret Rossiter for drawing my attention to this collection and the staff of the Institute Archives for looking up these particular letters. Wiener's letters are quoted by permission of the Institute Archives.

23. N. Wiener to Bertha Wiener, June 24, 1925, Norbert Wiener Papers, Institute Archives.

24. N. Wiener to Constance Wiener, July 5, 1925, Norbert Wiener Papers, Institute Archives.

25. N. Wiener to Constance Wiener, August 27, 1925, Norbert Wiener Papers, Institute Archives.

26. C. Payne to Margaret Harwood, September 1, 1925, Margaret Harwood Papers, Schlesinger Library, Radcliffe College; N. Wiener, *Ex-Prodigy: My Childhood and Youth* and *I Am a Mathematician* (Cambridge, Mass., 1964).

27. C. Payne to Margaret Harwood, September 1, 1925, Margaret Harwood Papers, Schlesinger Library, Radcliffe College.

28. Payne-Gaposchkin, *Autobiography*, 189–191.

29. For an outline of the proposed volume, see the fourteen-page typescript "Stellar Variability, by Boris P. Gerasimovic and Cecilia H. Payne," Shapley Directorial Papers, HUA. On Gerasimovic, see P. G. Kulikovsky, "Boris Petrovich Gerasimovich," *Dictionary of Scientific Biography* (New York, 1972) 5:363–364. On Russian astronomy of the era, see Robert A. McCutcheon, "The Purge of Soviet Astronomy: 1936–37 with a Discussion of Its Background and Aftermath" (master's thesis, Georgetown University, 1985).

30. C. Payne to H. Shapley, July 16, 1933, Shapley Directorial Papers, HUA.

31. C. Payne to H. Shapley, August 2, 1933, Shapley Directorial Papers, HUA. For a more extensive account of this journey, see Payne-Gaposchkin, *Autobiography*, 191–197.

32. This account is based largely on that given in S. Gaposchkin's thesis, as translated by Payne. See S. Gaposchkin to H. Shapley, August 14, 1933, Shapley Directorial Papers, HUA.

33. E. Hertzsprung to H. Shapley, June 10, 1933; and H. Shapley to E. Hertzsprung, June 28, 1933, Shapley Directorial Papers, HUA.

34. P. Guthnick to H. Shapley, July 19, 1933; and H. Shapley to C. Payne, August 2, 1933, Shapley Directorial Papers, HUA.

35. C. Payne to H. Shapley, August 12, 1933, Shapley Directorial Papers, HUA.

36. Sergei Gaposchkin to H. Shapley, August 14, 1933; H. Shapley to S. Gaposchkin, September 1, 1933; H. Shapley to S. Gaposchkin, October 10, 1933; S. Gaposchkin to H. Shapley, October 21, 1933; S. Gaposchkin to H. Shapley, November 7, 1933. All of these letters are from the Shapley Directorial Papers, HUA.

37. C. Payne to Svein Rosseland, January 29, 1934; Svein Rosseland to C. Payne, February 19, 1934, Svein Rosseland Papers, Institute for Theoretical Astrophysics, University of Oslo. H. Shapley to P. C. Galpin, February 26, 1934, Shapley Directorial Papers, HUA. I thank Karl Hufbauer for this reference.

38. Shapley said that he had no plans to continue Gaposchkin's temporary appointment past one year in H. Shapley to P. Guthnick, February 27, 1934, Shapley Directorial Papers, HUA.

39. H. Shapley to W.J.V. Osterhout, March 2, 1934; W.J.V. Osterhout to H. Shapley, March 2, 1934; W.J.V. Osterhout to H. Shapley, March 4, 1934; "Stars" to Cecilia Payne [March 5, 1934]; Mrs. W.J.V. Osterhout to H. Shapley, March 4, 1934; C. Payne-Gaposchkin to H. Shapley, March 6, 1934, Shapley Directorial Papers, HUA.

40. H. Shapley to Mrs. W.J.V. Osterhout, March 17, 1934, Shapley Directorial Papers, HUA.

41. Emma Payne to H. Shapley, March 10, 1934; H. Shapley to Emma Payne, April 11, 1934, Shapley Directorial Papers, HUA.

42. H. N. Russell to Joseph Boyce, March 15, 1934, H. N. Russell Papers, Princeton University Library.

43. C. Payne-Gaposchkin to R. S. Dugan, March 16, 1934, R. S. Dugan Papers, Princeton University Library. This letter is cited by permission of the Princeton University Library.

44. C. Payne-Gaposchkin to O. Struve, December 11, 1944, Struve Papers, Yerkes Observatory.

45. The quotation is from H. Shapley to H. H. Plaskett, October 16, 1935, Shapley Directorial Papers, HUA. On the Gaposchkin salaries, see Payroll of the Harvard College Observatory, HUA. On Payne-Gaposchkin's election to the American Philosophical Society, see H. Shapley to *Harvard Alumni Bulletin,* May 8, 1936, Shapley Directorial Papers, HUA. On her election to the American Academy of Arts and Sciences, see A. P. Usher to H. Shapley, June 3, 1943, Shapley Directorial Papers, HUA.

46. See B. P. Gerasimovich to H. Shapley, July 10, 1934; H. Shapley to B. P. Gerasimovich, January 4, 1935, Shapley Directorial Papers, HUA.

47. Payne-Gaposchkin, *Autobiography,* 199–202.

48. These comments about the interdependence of Cecilia and Sergei Gaposchkin are based on separate interviews with Sergei Gaposchkin. Katherine Haramundanis, Dorrit Hoffleit, Kenneth Jones, and Celia Eckhart, as well as Shapley and Struve correspondence.

49. O. Struve to W. W. Morgan, January 17, 1945, Struve Papers, Yerkes Observatory.

50. S. Chandrasekhar to O. Struve, November 21, 1941, Struve Papers, Yerkes Observatory.

51. S. Gaposchkin to Frank Schlesinger, December 11, 1936, Schlesinger Papers, Astronomy Department Records, Yale University Archives.

52. H. Shapley to C. Payne-Gaposchkin, March 15, 1940, Shapley Directorial Papers, HUA.

53. C. Payne-Gaposchkin to O. Struve, April 13, 1944, O. Struve Papers, Yerkes Observatory.

54. Katherine Haramundanis, Payne-Gaposchkin's daughter, presents her account of her family life in "A Personal Recollection," in Payne-Gaposchkin, *Autobiography,* 39–67.

55. H. Shapley to C. Payne-Gaposchkin, May 29, 1948, Shapley Directorial Papers, HUA.

56. C. Payne-Gaposchkin to Harvard Observatory Council [March, 1958], D. H. Menzel Directorial Papers, Harvard University Archives.

57. Haramundanis, "A Personal Recollection," 41.

58. An interview with Kenneth Janes on January 10, 1983, gave me some sense

of how the Gaposchkins worked together. For further evidence of Sergei Gaposchkin's attitudes toward women and his general personality, see his privately published autobiography, *Divine Scamper: Biography of Humanity,* 3 vols. (1974). I thank Owen Gingerich for loaning me these volumes. Payne-Gaposchkin's attempts to advance Sergei's career are especially evident in her correspondence with Otto Struve and Henry Norris Russell.

59. On Edward Gaposchkin, see *American Men and Women of Science,* as well as C. Payne-Gaposchkin to Martin Schwarzschild, November 29, 1958, Martin Schwarzschild Papers, Princeton University Library.

60. In a January 27, 1982, interview, Henry J. Smith reported that Payne-Gaposchkin's text was written to pay for Katherine's education. Katherine Gaposchkin Haramundanis described her subsequent career in an August 11, 1981, interview.

61. Personal communication, Stephen Kahler.

62. I thank Frances W. Wright for information about the Observatory Philharmonic Orchestra.

63. Minutes of the Harvard Observatory Council are preserved in papers of Bart J. Bok, HUA.

64. Meetings of the Harvard Observatory Forum on International Affairs were announced in the *Harvard Gazette.* Payne-Gaposchkin describes the organization briefly in *Autobiography,* 205–206.

65. O. Struve to H. N. Russell, November 4, 1938; H. N. Russell to O. Struve, November 7, 1938, O. Struve Papers, Yerkes Observatory.

66. O. Struve to P. Buck, December 24, 1944, O. Struve Papers, Yerkes Observatory.

67. H. N. Russell to B. B. Cronkhite, October 28, 1948, H. N. Russell Papers, Princeton University Archives.

68. I.R.S. Broughton to H. Shapley, March 7, 1941; and H. Shapley to I.R.S. Broughton, March 14, 1981, Shapley Directorial Papers, HUA.

69. Anne Peterson, "Politics: Would 'Liberate' Women Scholars," *New York Times,* October 31, 1937.

70. C. Payne-Gaposchkin, "Problems of the Woman Scholar," *Journal of the American Association of University Women* 33 (1940): 80–84. See also her "The Scholar and the World," *American Scientist* 31 (1943):329–337.

71. Emma Bugbee, "Dr. Gaposchkin Says Scholars Must Not Be Detoured by War," *New York Tribune,* June 17, 1941.

72. *Oakland Tribune,* June 17, 1941. A copy of the article is in the files of the Radcliffe Archives, Schlesinger Library.

73. "Harvard's Woman Astronomer Says We're Safe from Those Atomic Explosions on Bursting Stars: 293,000,000,000,000 Miles Insulate Us," *Boston Globe,* March 10, 1946.

74. Virginia Bohlin, "Find Housework Real Tough Job: Mrs. Gaposchkin Regards Mother's Task With Awe," *Boston Traveller,* November 3, 1949.

75. F. G. Strauss to Director of Publicity, Harvard College Observatory, June 4,

1954; C. Payne-Gaposchkin to V. Adams, June 18, 1954, D. H. Menzel Directorial Papers, Harvard University Archives.

76. "Harvard Astronomer Wins 1957 Achievement Award," *Journal of the American Association of University Women* 51 (1957): 15–17.

77. Haramundanis, "A Personal Recollection," 55–66. Personal communications, Stephen Kahler.

78. C. Payne-Gaposchkin, *The Garnett Letters* (privately printed, 1979).

79. C. Payne-Gaposchkin, "The Nashoba Plan for Relieving the Evils of Slavery," *Harvard Library Bulletin* 23 (1975):221–225, 429–461.

80. Payne-Gaposchkin, *Autobiography,* 238.

Chapter 12: Synergy or Clash

Acknowledgments: I am very grateful to the archivists of the following institutions for superb assistance with my research and permission to consult and refer to documents in their care: Sophia Smith Collection, Smith College, Northampton, Mass., and Shenkman Books, Inc., Cambridge, Mass.; Rockefeller Archive Center, North Tarrytown, N.Y.; Caltech Archives, Pasadena, Calif.; Center for the History of Physics, New York, N.Y.; Bertrand Russell Archives, McMaster University Library, Hamilton, Ontario; Cambridge University Library, Manuscript Room, Cambridge; Girton College, Cambridge; Lady Margaret Hall, St. Hilda's College, Somerville College, and Balliol College, all at Oxford; and the University Archives, Bodleian Library, Oxford.

Oral and written exchanges with the following scholars, among others, provided key insights into the life and work explored in this essay; I am very thankful to them all for their kindness and cooperation: Alison Duke, Registrar Emerita of the Roll, Girton College; Lady (Bertha) Jeffreys, Dame Mary Cartwright, FRS; Dorothy Moyle Needham, FRS; Dora Russell and Edith Shuttleworth, all members of Girton College; Sir Harold Jeffreys, FRS, of Cambridge, Dorothy Crowfoot Hodgkin, O.M., FRS, John and Flora Philpot, all of Oxford; G. J. Whitrow and E. Whyte of London; Margery Senechal and George Fleck of Smith College; Ruth Hubbard, George Wald, and John Edsall, all of Harvard University; and the late Alfred Shenkman of Cambridge, Mass.

My colleagues in the history of science, Ivor Grattan Guinness of the Middlesex Polytechnic and Karl Hufbauer of the University of California at Irvine, gave generous advice on source material.

Discussions with Margaret Rossiter and Joy Harvey over the last several years, and more recently with the contributors to this volume, were immensely helpful to my work on this essay. The author alone is responsible for the views expressed here.

1. The quotation is from M. C. Bradbrook, *That Infidel Place: A Short History of Girton College, 1869–1969* (Cambridge, 1984), 79; and "The Girton Girl: Social Images from Within and Without," 91–120. Other sources on attitudes toward ca-

reer and marriage among the first generations of college-educated women in Britain include A. Phillips, ed., *A Newnham Anthology* (Cambridge, 1979); M. Vicinus, "One Life to Stand Beside Me: Emotional Conflict in First-Generation College Women in England," *Feminist Studies* 8 (1982):603–628; idem, "Women's Colleges: An Independent Intellectual Life," in her *Independent Women* (London, 1985):121–162; see also its bibliography, 363–387; V. Brittain, *The Women at Oxford* (London, 1960).

2. For a comprehensive discussion of the struggle to win equality for women at Cambridge and Oxford, see R. McWilliams-Tullberg, *Women at Cambridge: A Men's University—Though of a Mixed Type* (London, 1975); A.M.A.H. Rogers, *Degrees by Degrees: The Story of the Admission of Oxford Women Students to Membership of the University* (London, 1938). See also T.E.B. Howarth, *Cambridge between Two Wars* (London, 1978).

3. For an overview of Wrinch's career, see *Girton College Register* (Cambridge, 1948; updated edition currently under preparation); M. Senechal, "A Prophet without Honor, Dorothy Wrinch, Scientist, 1894–1976," *Smith Alumnae Quarterly* 68 (1977):18–23; Senechal, ed., *Structures of Matter and Patterns in Science* (a symposium inspired by the work and life of Dorothy Wrinch, 1894–1976, held at Smith College, September 28–30, 1977; Cambridge, Mass., 1980). For specific episodes in Wrinch's career in the 1930s, especially her work on protein structure and her participation in the Theoretical Biology Club, see P. G. Abir-Am, "Brains and Beauty in the Early Days of Molecular Biology: The Career of Dorothy Wrinch, 1894–1976," paper read at the *Joint Atlantic Conference on the History of Biological Sciences* (Toronto, April 1979); idem, "Sciences's Reception of a Woman Theoretician in the 1930s," *Proceedings of the Sixteenth International Congress for the History of Science,* vol. B, Symposia (Bucharest, 1981), 221–227); idem, "The Biotheoretical Gathering in England, 1932–38, and the Origins of Molecular Biology" (Ph.D. diss., Université de Montréal, 1983), esp. chap. 4); idem, "The Biotheoretical Gathering, Transdisciplinary Authority and the Incipient Legitimation of Molecular Biology in the 1930s: New Perspectives on the Historical Sociology of Science," *History of Science* 25 (1987):1–70. See also J. T. Edsall and D. Bearman, "Historical Records of Scientific Activity: The Survey of Sources for the History of Biochemistry and Molecular Biology," *Proceedings of the American Philosophical Society* 123 (1979):279–292.

4. On Surbiton High School in Wrinch's time, see *The Girls' School Year Book* (London, 1911), 298–299; on girls' education in Edwardian England, see C. Dyhouse, *Girls Growing up in Late Victorian and Edwardian England* (London, 1981); F. Hunt, "Secondary Education for the Middle-Class Girl: A Study of Ideology and Educational Practice, 1870–1940" (Ph.D. diss., University of Cambridge, 1984); N. Michison, *Small Talk . . . an Edwardian Childhood* (London, 1973); B. Wootton, "Girls in Edwardian England," *London Review of Books* (March 19, 1982): 8–9; S. Wintle, "Prep-School Girl," *London Review of Books* (April 4, 1985):11–12.

5. *Girton College Register;* Dora Russell to the author, July 25, 1983.

6. D. M. Wrinch and H.E.H. Wrinch, 1923 and 1926 in n. 46, below. See also R. MacLeod and R. Moseley, "Fathers and Daughters: Reflections on Women, Science and Victorian Cambridge," *History of Education* 8 (1979):321–333.

7. *Cambridge University Reporter*, June 17, 1916 (results of tripos examination).

8. Edith Shuttleworth (Girton, 1915), personal communication to the author, June 18, 1983.

9. See for comparison the path of Wrinch's contemporary at Girton, Mary Slow, *Girton College Register;* Dame Mary Lucy Cartwright, FRS (1900–; first woman to have held, since 1935, a university lectureship in pure mathematics at Cambridge University, mistress of Girton College, 1949–69), personal communications to the author, June 1983 to June 1984.

10. Dorothy M. Wrinch to Mr. [Bertrand] Russell, September 9, 1914; May 12, 1916; June 16, 1916; July 22, 1916; August 26, 1916; Russell Archives, McMaster University Library, Hamilton, Ontario (hereafter Russell Archives). On Russell's preoccupations during Wrinch's association with him (1914–19), see I. Grattan-Guinness, "Russell's Logicism Versus Oxbridge Logics, 1890–1925: A Contribution to the Real History of Twentieth-Century English Philosophy," *Russell: Journal of the Bertrand Russell Archives* 2 (1986):101–131; idem, "Bertrand Russell's Logical Manuscripts: An Apprehensive Brief," *History and Philosophy of Logic* 6 (1985): 53–74; idem, "Notes on the Fate of Logicism from *Principia Mathematica* to Godel's Incompletability Theorem," *History of Philosophy and Logic* 5 (1984):67–78; P. A. Schilpp, ed., *The Philosophy of Bertrand Russell* (Chicago, 1944); B. Russell, *My Philosophical Development* (London, 1959). On Russell's life at that time, see *The Autobiography of Bertrand Russell*, vol. 2 (Boston, 1968); D. Russell, *The Tamarisk Tree, I. My Quest for Liberty and Love* (London, 1977); R. Clark, *The Life of Bertrand Russell* (London, 1975); K. Tait, *My Father Bertrand Russell* (New York, 1975); see also, in all these books, the index entries for Wrinch. On Russell's relationship with Jourdain, see I. Grattan-Guinness, *Dear Russell—Dear Jourdain* (London, 1977); idem, "Russell and Philip Jourdain," *Russell: Journal of the Bertrand Russell Archives* 8 (1973):7–13, and 9 (1973): 20–21; D. Russell, "Russell and Jourdain," *Russell: Journal of the Bertrand Russell Archives* 9 (1973):20. Wrinch's association with Jourdain, via Russell, led to the publication of her first articles, "Bernard Bolzano," in *The Monist,* and "Mr. Russell's Lowell Lectures," in *Mind,* both in 1917.

11. See B. Russell, D. Russell, K. Tait, and R. Clark (all in n. 10, above); Dora Russell to the author, July 25, 1983; Dora Russell to Bel Mooney in L. Caldecott, *Women of Our Century* (London, 1984), 56–83.

12. D. M. Wrinch to B. Russell, September 10, 1917; September 14, 1917; B. Russell to Miss Wrinch, September 12, 1917; September 17, 1917; all in Russell Archives. On Whitehead, see V. Lowe, *Alfred North Whitehead: The Man and His Work. I. 1861–1910* (Baltimore, Md., 1985). On Whitehead's collaboration with Russell in the *Principia Mathematica* (1910–1913), see Lowe, pp. 253–294; see also I. Grattan-Guinness, "The Royal Society's Financial Support of the Publication of Whitehead and Russell's *Principia Mathematica,*" *Notes and Records of the Royal Society of London* 30 (1975):89–104. On Nicod, see J. Nicod, "Mathematical Logic and the

Foundations of Mathematics," *Encyclopaedia Britannica,* 12th ed. (1922), 31, 874–876; J. van Heijenoort, ed., *From Frege to Gödel* (New York, 1973). On Wiener, see Norbert Wiener, *I Am a Mathematician* (Cambridge, Mass., 1964); S. J. Heims, *John von Neumann and Norbert Wiener: From Mathematics to the Technologies of Life and Death* (Cambridge, Mass., 1980).

13. Wrinch to B. Russell [August 1918a], May 10, 1918, Russell Archive. The reference to the "Sabine women" in Wrinch's letter of May 10, 1918, is unclear, but see a recent interpretation of the Sabine women's cultural representation as rape objects in N. Bryson, "Two Narratives of Rape in the Visual Arts: Lucretia and the Sabine Women," in *Rape in History,* ed. S. Tomaselli and R. Porter (Oxford, 1986), chap. 8.

14. Wrinch to B. Russell [August 1918a], Russell Archives.

15. Wrinch to B. Russell [August 1918b], Russell Archives.

16. Ibid. On G. H. Hardy, see his *A Mathematician's Apology* (London, 1940); E. C. Titchmarsh, "Godfrey Harold Hardy, 1877–1947," *Biographical Memoirs of Fellows of the Royal Society* 6 (1949):447–462.

17. Wrinch to B. Russell [August 1918b], Russell Archives.

18. Wrinch to B. Russell, August 20, 1918, Russell Archives.

19. Ibid. On Lady Ottoline Morrell (1873–1938), see R. Gathorne-Hardy, *Ottoline at Garsington: Memoirs of Lady Ottoline Morrell* (London, 1974); also Clark, *Bertrand Russell.* On the cultural hegemony of Bloomsbury at that time and its political resonance for circles such as Russell's, see R. Williams, "The Bloomsbury Fraction," in his *Problems in Materialism and Culture: Selected Essays* (London, 1980), 148–169; M. Holroyd, "Mrs. Webb and Mrs. Woolf," *London Review of Books,* November 7, 1985, pp. 17–19.

20. Wrinch to B. Russell, August 26, 1918, Russell Archives.

21. Ibid.

22. Wrinch to B. Russell, September 3, 1918; September 10, 1918a; 1918b; Russell Archive.

23. Wrinch to B. Russell, July 18, 1919, Russell Archives. On Wittgenstein, see B. McGuinness, ed., *Wittgenstein and His Times* (Oxford, 1982).

24. Wrinch to B. Russell, September 24, 1919, Russell Archives.

25. B. Russell, D. Russell, K. Tait, R. Clark (see n. 10, above); Caldecott, *Women of Our Century.* Russell's publications in the 1920s included *How to Be Free and Happy* (1924), *Education and the Good Life* (1926), *Why I Am Not a Christian* (1927), *Marriage and Morals* (1929), *The Conquest of Happiness* (1930), and *Education and the Modern World* (1932).

26. "Tender Emotions," Typescript, Folder 94, Dorothy Wrinch's Papers, Sophia Smith Collection, Smith College Library, Northampton, Massachusetts (hereafter DWP-SSC-SCNM). See also Wrinch's *The Retreat from Parenthood* (London, 1930; published under the pseudonym Jean Ayling), 57, where a character named Maud (Wrinch's middle name) loses a suitor to a friend belonging to their circle, an experience described as bitter.

27. Wrinch, *Retreat from Parenthood,* 61: "Being a woman given to clear think-

ing, she perceived that she wished to breed. She therefore married Mr. Grey, quickly and quietly, though not before the alternatives to Mr. Grey had also been carefully investigated."

28. On Nicholson, see *Balliol College Register,* 1916–67, entry for 1921; *Who Was Who* (London, 1960); *Biographical Memoirs of Fellows of the Royal Society* 2 (1956): 209–213. For information on the $\nabla^2 V$ Club see its minutes, *Project on the History of Quantum Physics,* for the History of Physics, American Institute of Physics, New York City. An excellent source on Cambridge physicists in the first two decades of this century is J. L. Heilbron and B. R. Wheaton, *Literature on the History of Physics in the Twentieth Century* (Berkeley, 1981).

29. See R. McCormack, "The Atomic Theory of John William Nicholson," *Archive for History of Exact Sciences* 3 (1966):160–184; M. J. Seaton, "Atoms, Astronomy and Aeronomy," *Quarterly Journal of the Royal Astronomical Society* 23 (1982):2–25.

30. Miss Grier (LMH Principal) to Professor Nicholson, June 18, 1923, Lady Margaret Hall Archive, Oxford, File Wrinch-Nicholson (hereafter LMHAO-WN). Previous correspondence of LMH principal with Wrinch herself dates from October 1922, when, in reply to an invitation of October 2, 1922, to take four new mathematical students, Wrinch replied that she would be happy to teach them "when it becomes possible for me to reside in Oxford," while further suggesting that her husband might be able to take those students (Wrinch to LMH principal, October 5, 1922, LMHAO-WN). The principal pursued that suggestion next day, Miss Grier to Professor Nicholson, October 6, 1922; Miss Grier to Mrs. Nicholson, October 6, 1922, LMHAO-WN.

31. J. W. Nicholson to Miss Grier, June 30, 1923, LMHAO-WN.

32. Ibid.

33. Miss Grier to Mrs. Nicholson, July 3, 1923, LMHAO-WN.

34. Mr. Newboult to Miss Grier, July 3, 1923; Miss Grier to Mrs. Nicholson, July 5, 1923, LMHAO-WN.

35. Miss Grier to Mrs. Nicholson, May 12, 1924, LMHAO-WN.

36. On the social scene and relationships between faculty wives and women dons, see S. Ardener, "Oxford Wives," in H. Callan and S. Ardener, eds., *The Incorporated Wife* (London, 1984), chap. 1.

37. See correspondence between Miss Grier and the principals of the other four women's colleges—Miss Penrose of Somerville, Miss Moberley of St. Hilda's, Miss Gwyer of St. Hugh's, and Miss Burrows of the Society for Home Students—in the period 1924–29 (LMHAO-WN). For the salary of Oxford University lecturers, see University of Oxford Stipend Ledger, UC/A/1/36, Keeper of the Archives, Bodleian Library.

38. D. Wrinch Nicholson to Miss Grier, March 1, 1927, LMHAO-WN.

39. *Oxford University Calendar,* 1927.

40. Vice-chancellor to Miss Grier, May 15, 1924, LMHAO-WN.

41. Secretary of the Finance Committee (Miss Anson) to Miss Grier, March 2, 1929, LMHAO-WN.

42. Miss Gwyer to Miss Grier, March 4, 1929; treasurer of Somerville College to Miss Grier, March 6, 1929, LMHAO-WN.

43. Miss Anson to Mrs. Nicholson, May 12, 1929; Mrs. Wrinch Nicholson to Miss Anson, May 13, 1929; Miss Grier to Mrs. Nicholson, May 16, 1929; secretary of LMH Council to Mrs. Nicholson, February 3, 1930, and November 24, 1930, LMHAO-WN.

44. Keeper of the Archives Office, University of Oxford, to the author, July 12, 1983, and August 19, 1986.

45. On Russell's social activism in the 1910s and 1920s, see B. Russell, D. Russell, and R. Clark (all in n. 10, above). See also J. M. Byles, "Women Experiences of World War I: Suffragists, Pacificists and Poets," *Women Studies International Forum* 8 (1985):473–487. H. Jeffreys, Wrinch's collaborator in the period 1919–23 has also recalled her socialist and feminist views in a personal communication to the author, June 1980 and July 1983. On a dozen socialist and feminist couples in British science in the interwar period, see G. Werskey, *The Visible College* (London, 1978), 210–211; yet in only one such couple (Enid Charles and Lancelot Hogben), who were friends and neighbors of Sylvia Pankhurst, did the wife retain her name. On S. Pankhurst see her *The Suffragette Movement* (London, 1931); C. Pankhurst, *Sylvia Pankhurst: Portrait of a Radical* (New Haven, 1986). On Dorothy Enid Charles (1894–1972), see her *The Practice of Birth Control* (London, 1932) and *The Twilight of Parenthood* (London, 1935). Charles was a contemporary of Wrinch at Cambridge (1913–16) who took the mathematics and economics tripos; see *Newnham College Register, 1871–1923* (vol. 1), 246.

46. These three papers were: "On Mediate Cardinals," *American Journal of Mathematics* 45 (1923):87–92; "On the Asymptotic Evaluation of Functions Defined by Contour Integrals," *American Journal of Mathematics* 50 (1928):269–302; and "The Hypergeometric Function with k Denominators," *Quarterly Journal of Pure and Applied Mathematics,* (1924):204–224. The other nine papers in pure mathematics, all published in *Philosophical Magazine,* a physicist's journal edited by Sir Oliver Lodge and Sir J. J. Thomson, are: "A Generalized Hypergeometric Function with n Parameters," 41 (1921); "An Asymptotic Formula for the Hypergeometric Function o 4(Z)," 41 (1921); "Some Approximations to Hypergeometric Functions," 45 (1923); "Table of the Bessel Function $I_n(x)$," with Hugh E. H. Wrinch, (1923); "Laplace's Equation and the Inversion of Co-ordinates," 50 (1925):1049–1058; "The Roots of Hypergeometric Functions with a Numerator and Four Denominators," with H.E.H. Wrinch, 51 (1926); "On the Spheroidal Harmonics as Hypergeometric Functions," 56 (1928):1117–1122; "On the Structure of Serial Relations," 58 (1929): 698–702; "On the Multiplication of Serial Relations," 58 (1929):1025–1042; "On Some Integrals Involving Legendre Polynomials," 60 (1930):1037–1043. The three papers in mathematical physics addressed to mathematical fora are: "Some Boundary Problems of Mathematical Physics," *Proceedings of the London Mathematical Society* 24 (1924):435–458; "On a Method for Constructing Harmonics for Surfaces of Revolution," *Proceedings of the International Congress of Mathematics* (Bologna, 1928); and "Harmonics Associated with Certain Inverted Spheroids," *Proceedings of the In-*

ternational Congress of Mathematics (Zurich, 1932). On the distinction between pure and applied mathematics, as viewed by Hardy, the most influential pure mathematician in interwar Britain, see n. 16, above. For an overview of mathematical fields, see G. Temple, *100 Years of Mathematics* (London, 1981); M. Kline, *Mathematics: The Loss of Certainty* (New York, 1980).

47. Lady Jeffreys (1903–), lecturer in mathematics and vice-mistress emerita of Girton College; personal communication to the author.

48. These papers were: "The Relation of Geometry to Einstein's Theory of Gravitation," with H. Jeffreys, *Nature* 106 (1921):806–809; "On the Lateral Vibrations of Bars of a Conical Type," *Proceedings of the Royal Society A* 101 (1922):493–508; "On the Orbits in the Field of a Doublet," *Philosophical Magazine* 43 (1922): 993–1014; "On the Rotations of Slightly Elastic Bodies," *Philosophical Magazine* 44 (1922); "On the Seismic Waves from the Oppau Explosion of 1921," with H. Jeffreys, *R.A.S. Monthly Notices,* Geophysics Supplement (January 1923):15–22; "The Theory of Mensuration," with H. Jeffreys, *Philosophical Magazine* 46 (1923):1–22; "On the Lateral Vibrations of Rods of Variable Cross-Section," *Philosophical Magazine* 46 (1923):273–290; "Some Problems of Two-Dimensional Electrostatics," *Philosophical Magazine* 48 (1924):692–703; "Some Boundary Problems of Mathematical Physics," *Proceedings of the London Mathematical Society* 24 (1924):435–458; "Some Problems of Two-Dimensional Hydrodynamics," *Philosophical Magazine* 48 (1924):1089–1104; "Fluid Circulation Round Cylindrical Obstacles," *Philosophical Magazine* 49 (1925):240–250; "On the Electric Capacity of Certain Solids of Revolution," *Philosophical Magazine* 50 (1925):60–70; "On the Pressure Distribution Round Certain Aerofoils of High Aspect Ratio," *Journal of Royal Aerodynamic Society* 30 (1926):129–142; "Electrostatic Problems Concerning Certain Inverted Spheroids," *Philosophical Magazine* 3 (1927):865–883; "On a Method for Constructing Harmonics for Surfaces of Revolution," *Proceedings of the International Congress of Mathematics* (Bologna, 1928). The two papers coauthored with Nicholson (with Wrinch as the first author) were: "Laplace's Equation and Surfaces of Revolution," *Proceedings of the Royal Society, A,* 108 (1925):93–104; and "A Class of Integral Equations Occurring in Physics," *Philosophical Magazine* 54 (1927):531–560. All but five of the seventeen papers in mathematical physics appeared in *Philosophical Magazine.*

49. See Nicholson's *The Theory of Optics: Papers in Mathematical Physics and Astronomy,* with Sir Arthur Schuster (London, 1924); and his *Scientific Papers of S. B. McLaren,* with Joseph Larmor et al. (London, 1925).

50. The twelve philosophical papers include four on scientific method written with H. Jeffreys: "On Some Aspects of the Theory of Probability," *Philosophical Magazine* 38 (1919):715–731; "On Certain Fundamental Principles of Scientific Inquiry," *Philosophical Magazine* 42 (1921):369–390; "On Certain Fundamental Principles of Scientific Inquiry," *Philosophical Magazine* 45 (1923):368–374. "The Theory of Mensuration," *Philosophical Magazine* 46 (1923):1–22. Other papers are: "On the Structure of Scientific Inquiry," *Proceedings of the Aristotelian Society* (1920–21): 181–210; "On Certain Methodological Aspects of the Theory of Relativity," *Mind*

31, n.s. (1922):200–204; "On Certain Aspects of Scientific Thought," *Proceedings of the Aristotelian Society* (1923–24):37–54; "Scientific Thought," *Mind* 33NS (1924):184–192; "Scientific Methodology with Special Reference to Electron Theory," *Proceedings of the Aristotelian Society* (1926–27):41–60; "The Relations of Science and Philosophy," *Journal of Philosophical Studies* 2 (1927): 153–166; "Aspects of Scientific Method," *Proceedings of the Aristotelian Society* (1928–29):94–122; "Scientific Method in Some Embryonic Sciences," *Proceedings of the Aristotelian Society* (1929–30):229–242.

51. The papers on relativity are: "The Relation of Geometry to Einstein's Theory of Gravitation," with H. Jeffreys, *Nature* 106 (1921):806–809; "Relativity," *International Congress of Philosophers* (Paris, 1921); "The Theory of Relativity in Relation to Scientific Method," *Nature* 107 (March 23, 1922); "On Certain Methodological Aspects of the Theory of Relativity," *Mind* 31NS (1922):200–204; "The Idealistic Interpretation of Einstein's Theory," *Proceedings of the Aristotelian Society,* a symposium with Professors Carr, Nunn, Whitehead, and Lord Haldane (1922–23).

52. For the six papers coauthored with H. Jeffreys, see nn. 48, 50, 51. For Jeffreys's views on scientific method in the 1920s, see his *Scientific Inference* (London, 1931; subsequent editions in 1957 and 1973). See also his *Collected Papers in Geophysics and Other Sciences* (London and New York, 1971–77). Sir Harold Jeffreys, FRS, Fellow of St. John's College, became Plumian Professor of Astronomy at Cambridge in 1946.

53. The quotation is from p. 163 of Wrinch, "The Relations of Science and Philosophy," *Journal of Philosophical Studies* 2 (1927):153–166.

54. See Wrinch, "Scientific Methodology with Special Reference to Electron Theory," *Proceedings of the Aristotelian Society* (1926–27):41–60; "The Relation of Science and Philosophy," *Journal of Philosophical Studies* 2 (1927):153–166; "Aspects of Scientific Method," *Proceedings of the Aristotelian Society* (1928–29):94–122; "Scientific Method in Some Embryonic Sciences," *Proceedings of the Aristotelian Society* (1929–30):229–242.

55. See, e.g., B. Russell's *The ABC of Relativity* (London, 1925); idem, *The Scientific Outlook* (London, 1931). On the appeal of the theory of (general) relativity to mathematicians and logicians in Britain (rather than to physicists), see G. J. Whitrow, "Theories of Relativity," *British Journal for the Philosophy of Science* 2 (1951): 61–68. Whitrow occasionally substituted for Wrinch's tutorials at Oxford in the late 1930s, personal communication to the author, April 1984. See also L. Graham, "The Reception of Einstein's Ideas in Britain and the Soviet Union: The Example of Two Contrasting Political Cultures," in *Albert Einstein, Historical and Cultural Perspectives,* ed. G. Holton and Y. Elkana (Princeton, 1982), 107–136.

56. Curators of the University Chest, minutes for 1931–33, UC/M/1/19, Bodleian Library, Oxford, entry for February 20, 1931, signed by F. H. Dudden, vice-chancellor: "Mr. J. W. Nicholson: In view of serious state of health preventing him from carrying out his statutory duties as University Lecturer, it was agreed that his office as Lecturer be regarded as terminated on 31 December 1931."

57. A Balliol student in the 1920s (B.A. 1929), John Philpot (1907–), recalled in personal communications to the author on April 13, June 6, and June 7, 1984, that his and other students' searches for Professor John William Nicholson for tutorials in mathematics and physics often ended in the porters' lodge with the porters' clarification "You must be looking for Billy Bottle, good luck."

58. Wrinch to B. Russell, July 14, 1930, Russell Archive.

59. D. M. Wrinch to Mr. Pyke (her solicitor), July 13, 1932; D. M. Wrinch to Balliol's home bursar, July 26, 1932; Balliol's home bursar to D. M. Wrinch, July 12 and 25, 1932; official solicitor to Balliol's home bursar, July 26, 1932, and December 1, 1932; Balliol's home bursar to official solicitor, November 28, 1932; Balliol's librarian to home bursar, December 2, 1932; "List of Books from the Library of J. W. Nicholson which have been taken over by Balliol College, Library," February 1933; all documents in Balliol College's Archive, Oxford.

60. Quotations are from Wrinch to Pyke, July 13, 1932 (see n. 59, above); see also "List of books selected by Dr. Wrinch in the Lecture Room, Balliol College," December 12, 1932, in Balliol College's Archive. She requested to be allowed to retrieve a number of her own books, two in mathematics and four in physics, which somehow remained in her husband's office. Among the other more general books she was allowed to pick, one finds many in natural history, since one of her husband's hobbies was entomology with a special interest in butterflies, as well as classical treatises in mathematics, physics, astronomy, and a smaller number of literary classics.

61. Wrinch, *Retreat from Parenthood;* see n. 26, above.

62. Ibid., xii–xiv, also chaps. 18–22.

63. Ibid., 109.

64. See n. 25, above.

65. B. Russell to Dot (D. Wrinch), May 29, 1930, Russell Archive.

66. Sybil Moholy-Nagy to D. Wrinch, February 22, 1937, DWP-SSC-SCNM. See also K. Passuth, *Laszlo Moholy-Nagy* (New York, 1985).

67. On Margery Fry, see E. H. Jones, *Margery Fry: The Essential Amateur* (London, 1966), esp. chap. 11 (principal of Somerville).

68. D. Wrinch Nicholson to LMH principal, February 24, 1930; secretary of the Rhodes Trust to Miss Grier, February 27, 1930; Miss Grier to Mrs. Wrinch Nicholson, April 23, 1930; Miss Grier to the warden of New College, May 1, 1930; warden of New College to Miss Grier, May 2, 1930, LMHAO-WN.

69. Copies of the application, including the letters of recommendation discussed here, are in Wrinch's Papers, DWP-SSC-SCNM.

70. Ibid.

71. Ibid.

72. Ibid.

73. Miss Grier to Mr. Hall, June 11, 1930; Miss Grier to Mrs. Wrinch Nicholson, July 17, 1933; Mrs. Wrinch Nicholson to LMH principal, July 18, 1930; D.

Wrinch to LMH principal, July 28, 1930; Miss Grier to Mrs. Wrinch Nicholson, July 29, 1930; Miss Grier to Mrs. Nicholson, July 30, 1930; Mr. N. F. Hall to Miss Grier, October 17, 1930.

74. Carr-Saunders to Miss Grier, August 3, 1930; Miss Grier to Carr-Saunders, August 5, 1930; D. Wrinch to LMH principal, November 28, 1931, LMHAO-WN.

75. Dr. Wrinch to LMH principal, May 11, 1931, LMHAO-WN; postcards from Pamela (Wrinch's four-year-old daughter) to her grandparents, Christmas 1931, in Wrinch's Papers, DWP-WHA-SCNM. On the major collective approach to philosophy of science in Vienna at that time, see *The Scientific Conception of the World: The Vienna Circle* (1929; reprint, Boston, 1973).

76. Wrinch to LMH principal, November 28, 1931; December 5, 1931; May 25, 1932; Miss Grier to Wrinch, December 7, 1931; February 10, 1932, LMHAO-WN.

77. On the Biotheoretical Gathering's activities in the 1930s, including documentation of its members' contributions to each meeting and interpretation of the avant-garde social structure of this group, see Abir-Am, "Biotheoretical Gathering in England" and "Biotheoretical Gathering in the 1930s" (see n. 3, above).

78. Wrinch to LMH principal, March 18, 1934 (interim report on the Suzette Taylor Fellowship); Miss Grier to Dr. Wrinch, March 24, 1934; Wrinch to LMH principal, February 1935, LMHAO-WN. The report also includes references to Wrinch's last mathematical publications in 1932 (see n. 46, above); to her first biological publication, "Chromosome Behavior in Terms of Protein Pattern" (*Nature* 134 [December 22, 1934]:978); and to biological work in press that appeared in 1935.

79. On the Rockefeller Foundation's contacts with this group, see P. Abir-Am, "The Discourse of Physical Power and Biological Knowledge in the 1930s: A Reappraisal of the Rockefeller Foundation's Policy in Molecular Biology," *Social Studies of Science* 12 (1982):341–382; idem, "Biotheoretical Gathering in England" and "Biotheoretical Gathering in the 1930s."

80. Tisdale to Wrinch, October 1, 1935, Rockefeller Archive Center, North Tarrytown, N.Y., Series 401D, Oxford, Theoretical Biology (hereafter RAC-OTB); Wrinch to LMH principal, July 16 and October 3, 1935, LMHAO-WN.

81. Miss Gwyer (St. Hugh's principal) to Miss Grier, November 14, 1935; Miss Grier to Miss Gwyer, November 16, 1935; Miss Grier to Dr. Wrinch, November 16, 1935, LMHAO-WN.

82. Miss Grier to Dr. Wrinch, November 19 and November 22, 1935, LMHAO-WN. It was the wise counsel of her benevolent principal that prevented her from circulating this document, while alerting Wrinch that she must decide whether she could fulfill her teaching duties in the new circumstances.

83. See Wrinch, "The Cyclol Hypothesis and the 'Globular Proteins,'" *Proceedings of the Royal Society of London, A,* 161 (1937):505–524; communicated by Sir William H. Bragg, O.M., P.R.S.

84. For an overview of the history of proteins, see P. R. Srinivasan, J. S. Fruton, and J. T. Edsall, eds., *The Origins of Biochemistry: A Retrospect on Proteins*

(New York, 1979), esp. pt. 1; see also Abir-Am's review in the *British Journal for the History of Science* 15 (1982):301–305; and P. Laszlo, "Proteins as Molecules," pt. 2 in his *Molecular Correlates of Biological Concepts* (Amsterdam, 1986), vol. 34A of *Comprehensive Biochemistry*. Wrinch's theory—described as "mistaken"—and reactions to it, especially by Pauling and Crowfoot Hodgkin, are discussed, out of context, in chap. 13.

85. See Wrinch, "The Pattern of Proteins," *Nature* 137 (March 7, 1936):411; idem, "The Cyclol Theory and the 'Globular Proteins,'" *Nature* 139 (June 5, 1937): 972–974.

86. See n. 83, above.

87. See Abir-Am, "Brains and Beauty," and "Science's Reception of a Woman Theoretician"; idem, "Disciplinary Passions in the Controversy Over Protein Structure in the Late 1930s: Champions of Mathematics, Physics and Chemistry Struggle to Monopolize the 'Secret of Life'" (forthcoming).

88. See the chapter with this title in M. Florkin, *A History of Biochemistry* (Amsterdam, 1972). For a more balanced view, see J. T. Edsall, "Proteins and Macromolecules: An Essay on the Development of the Macromolecule Concept and Some of Its Vicissitudes," *Archives of Biochemistry and Biophysics*, suppl. 1. (1962): 12–21; N. W. Pirie, "Patterns of Assumptions about Large Molecules," *Archives of Biochemistry and Biophysics*, suppl. 1 (1962):21–29.

89. See Wrinch's philosophical papers in the late 1920s (n. 54, above), especially her "The Relation of Science and Philosophy."

90. See J. D. Bernal, "The Structure of Proteins," *Nature* 142 (1939):631–636; idem, "W. T. Astbury, 1898–1961," *Biogr. Mem. FRS* 9 (1963):1–35.

91. See the correspondence of Wrinch and Bernal, 1936–39, in Bernal's Papers, Manuscript Room, Cambridge University Library, and in Wrinch's Papers, DWP-WHA-SCNM. The latter also includes correspondence between Wrinch and Crowfoot. On Bernal and Crowfoot's "ancestral roles" in protein X-ray crystallography, see Abir-Am, "Toward a Historical Ethnography of Scientific Rituals: Interpreting the 50th Anniversary of the First Protein X-ray Photo" (forthcoming). See also R. K. Merton, "Priorities in Scientific Discovery," in his *The Sociology of Science: Theoretical and Empirical Investigations* (Chicago, 1973), 286–324.

92. Correspondence of Bernal, Crowfoot, Fankuchen, Riley, Bragg, and Neville to Wrinch, Wrinch's Papers, DWP-WHA-SCNM.

93. Archivist of Somerville College, Oxford, to the author, July 15, 1983; also Wrinch's Papers, DWP-WHA-SCNM. On D'Arcy Thompson's crucial influence on members of the Biotheoretical Gathering and on Wrinch in particular, see Abir-Am, "Biotheoretical Gathering in England," chap. 4.

94. LMH principal to Dr. Wrinch, October 6, 1938; this was a reminder that Wrinch was given permission temporarily, during a time of stress (presumably the Munich crisis) to dine in Hall; since apparently she availed herself of this opportunity on a more permanent basis, she was reminded after having taught at LMH for six-

teen years, that permanent permission must be sought from the Senior Combination Room (SCR) with a proper application; see also LMH principal to Dr. Wrinch, February 13, 1939; Dr. Wrinch to LMH principal, June 2, 1939, LMHAO-WN.

95. The referee reports are in Wrinch's papers, DWP-SSC-SCNM; see also Wrinch to Neville, October 1940, DWP-SSC-SCNM, quoted below. Wrinch's tendency to cast her views in absolutist or conclusive terms, a legacy of her logicistic background, was counterproductive in the context of a controversy. For views, also questioning the universality of the peptide chain structure of proteins, yet presented more carefully as a tentative hypothesis, see K. Linderstrom-Lang et al., "Peptide Bonds in Globular Proteins," *Nature* 142 (1938):996.

96. Paper clippings from the *New York Times,* September 5, 1951, and other newspapers in Pauling's Papers, Caltech Archives, Pasadena, California.

97. Wrinch to Neville, October 1940, DWP-SSC-SCNM.

98. See Wrinch, *Chemical Aspects of the Structure of Small Peptides* (Copenhagen, 1960) and *Chemical Aspects of Polypeptide Chain Structure and the Cyclol Theory* (New York, 1965).

99. Wrinch to Neville, September 8, 1940, DWP-SSC-SCNM, File 97.

100. Ibid.

101. Ibid.

102. Ibid.

103. Ibid. On Langmuir, see L. S. Reich, "Irving Langmuir and the Pursuit of Science and Technology in the Corporate Environment," *Technology and Culture* 27 (1983):199–221; J. W. Servos, "Mathematics and the Physical Sciences in America, 1880–1930," *ISIS* 77 (1986):611–629.

104. Wrinch to Neville, December 10, 1940, DWP-SSC-SCNM, File 97.

105. Wrinch to Neville, October 1940, DWP-SSC-SCNM, File 97.

106. Marion P. Wrinch to Pauling, 1940, DWP-SSC-SCNM, File 16.

107. Wrinch to Neville, March 29, 1941, DWP-SSC-SCNM. On Veblen's and Einstein's efforts to assist refugee mathematicians in the United States in the 1930s, see N. Reingold, "Refugee Mathematicians in the United States of America, 1933–1941: Reception and Reaction," *Annals of Science* 38 (1981):313–338.

108. Glaser to Wrinch, December 14, 1940, DWP-SSC-SCNM, File 17.

109. Wrinch to Glaser, May 3, 1941, DWP-SSC-SCNM.

110. Wrinch to Neville, September 29, 1941, DWP-SSC-SCNM, File 97.

111. Ibid.

112. Newspaper clippings, DWP-SSC-SCNM, File 17.

113. Glaser to Wrinch, July 17 and August 17, 1942, DWP-SSC-SCNM.

114. Handwritten notes, DWP-SSC-SCNM, File 94.

115. See Bernal, "The Material Theory of Life," *Labour Monthly* 50 (July 1968):323–326; idem, "The Pattern of Linus Pauling's Work in Relation to Molecular Biology," in *Structural Chemistry and Molecular Biology,* ed. A. Rich and N. Davidson (San Francisco, 1968), 370–382; L. Pauling, "Structural Chemistry and

Molecular Biology," *Nature* 221 (1974):778–783. Ironically, Bernal and Pauling spent the intermission of the play *The Physicists* at Aldwych Theatre in London in 1963 in the company of the actress Leslie Caron. Their picture in the newspapers epitomized the cultural mold of associating brains with men and beauty with women, a mold their arch rival Wrinch tried so hard to challenge.

116. Wrinch to Neville, n.d., DWP-SSC-SCNM, File 97.

117. It is to be hoped that Wrinch's work and life in America, largely spent at Smith College, 1941–76, will be explored in the forthcoming dissertation by her former student, Sybilla Kennedy.

Notes on Contributors

The Editors:

PNINA G. ABIR-AM, a visiting scholar in the Department of the History of Science at Harvard University is a historian of molecular life sciences. She is principal investigator on a National Science Foundation–funded project on research schools of molecular biology in the United States, United Kingdom, and France and a founding member of the International Committee on Women in the History of Science, Technology, and Medicine. She is completing a book on collective creativity in a transdisciplinary group of avant-garde British scientists in the 1930s.

DORINDA OUTRAM, a lecturer in modern history at University College, Cork, Republic of Ireland, is a historian of eighteenth- and nineteenth-century French science. She is the author of *Georges Cuvier: Vocation, Science and Authority in Post-Revolutionary France* (1984) and is currently at work on a book to be titled "Liberty's Crimes: The Political Culture of Revolutionary Terror in France, 1789–1799."

MARIANNE GOSZTONYI AINLEY is an adjunct fellow of the Simone de Beauvoir Institute, Concordia University, Montreal, Quebec. She is at work on two projects on women scientists: a study of American women ornithologists and a study on Canadian women naturalists in the field-oriented natural sciences (geology, botany, and zoology).

JOY HARVEY, a visiting scholar in the Department of the History of Science at Harvard University, is a historian of nineteenth-century French biology and social science. She is on the editorial board of the *Journal of the History of Biology* and a founding member of the International Committee on Women in the History of Science, Technology, and Medicine. She is at work on a book on the sociopolitical context of evolutionary debates in nineteenth-century France.

PEGGY A. KIDWELL, a museum specialist in the Division of Mathematics of the National Museum of American History, Smithsonian Institution, Washington, D.C., is a historian of British and American astronomy and astrophysics. She is at work on projects related to Harvard astronomers in World War II and to the development of calculating machines in nineteenth-century America.

ANN HIBNER KOBLITZ, an assistant professor in the Department of History at Wellesley College, is a historian of Russian science. She is the author of *A Convergence of Lives, Sofia Kovalevskaia: Scientist, Writer, Revolutionary* (1983) and is at work on a book on the first generation of Russian women in science.

SALLY GREGORY KOHLSTEDT, teaches in the Department of History at Syracuse University, Syracuse, New York. She is the author of *The Formation of the American Scientific Community: The American Association for the Advancement of Science, 1848–1860* (1976) and coeditor, with Margaret W. Rossiter, of *Historical Writing on American Science* (1985).

REGINA M. MORANTZ-SANCHEZ is professor of history at the University of Kansas. She is the author of *In Her Own Words: Oral Histories of Women Physicians* (1982) and *Sympathy and Science: Women Physicians in American Medicine* (1985).

MARILYN BAILEY OGILVIE is an associate professor of natural science and chair of the Division of Natural Science and Mathematics at Oklahoma Baptist University, Shawnee, Oklahoma. She is the author of *Women in Science, Antiquity through the Nineteenth Century: A Biographical Dictionary with Annotated Bibliography* (1986).

HELENA M. PYCIOR is an associate professor in the Department of History of Science at the University of Wisconsin in Milwaukee and managing editor of *Historia Mathematica*. She is presently at work on a book on the history of British algebra and directing a National Science Foundation–funded project on using history in the mathematics classroom.

ANN B. SHTEIR is chairperson of the Department of Humanities at York University, North York, Ontario, and adviser to its president on the status of women. She is the editor of *Women on Women* (1978) and is at work on a book to be titled "Linnaeus's Daughters: Women and the History of British Botany."

NANCY G. SLACK is professor of biology at Russell Sage College, Troy, New York, and lecturer in an interdisciplinary program on science, technology, and human values. She is currently at work on a history of American botany in the late nineteenth and early twentieth centuries.

Index